ELECTRONIC POWER SUPPLIES

By
M.C. SHARMA

BPB PUBLICATIONS
B-14, CONNAUGHT PLACE, NEW DELHI-1

FIRST EDITION 2004, REPRINTED 2012

Distributors:

MICRO BOOK CENTRE
2, City Centre, CG Road,
Near Swastic Char Rasta,
AHMEDABAD-380009 Phone: 26421611

COMPUTER BOOK CENTRE
12, Shrungar Shopping Centre, M.G. Road,
BANGALORE-560001 Ph: 25587923, 25584641

MICRO BOOKS
Shanti Niketan Building, 8, Camac Street,
KOLKATTA-700017 Ph.: 22826518, 22826519

BUSINESS PROMOTION BUREAU
8/1, Ritchie Street, Mount Road,
CHENNAI-600002 Phone: 28410796, 28550491

DECCAN AGENCIES
4-3-329, Bank Street,
HYDERABAD-500195
Phone: 24756400, 24756967

MICRO MEDIA
Shop No. 5, Mahendra Chambers,
150 D.N. Road, Next to Capital Cinema
V.T. (C.S.T.) Station,
MUMBAI-400001 Ph.: 22078296, 22078297

BPB PUBLICATIONS
B-14, Connaught Place, **NEW DELHI-110001**
Phone: 23325760, 23723393, 23737742

INFO TECH
G-2, Sidhartha Building, 96 Nehru Place,
NEW DELHI-110019
Phone: 26438245, 26415092, 26234208

INFO TECH
Shop No. 2, F-38, South Extension Part-1
NEW DELHI-110049
Phone: 24691288, 24641941

BPB BOOK CENTRE
376, Old Lajpat Rai Market,
DELHI-110006 PHONE: 23861747

ISBN 81-7656-921-6

Published by Manish Jain for BPB Publications, B-14, Connaught Place, New Delhi-110 001 and Printed by him at Akash Press, Delhi.

Table of Contents

By the same Author:

41 Projects Using 741 Op Amp
51 Projects Using CD4011
555 Timer and its Applications
Batteries, Chargers and Emergency Lights
Build Your Own Disco Projects
Build Your Own Test Instruments
Care and Repair of Electronic Flash Guns
CMOS Integrated Circuits: Learning by Experiments
Desk Top Publishing on PC
Digital Integrated Circuits: Learning by Experiments
Easy to Build Electronic Alarms
Electronic Formulas, Tables and Symbols
Fast Servicing with Modern Multimeters
Fast Servicing with Oscilloscopes
Function Generators and Phase Lock Loops
Music & Sound ICs and Circuits
Op Amps Made Simple
Practical SCR and Triac Projects
Relays and Their Applications
Simple Audio Projects
The Little Book on Computers
Using Field Effect Transistors
Using Multimeters
Using Semiconductor Diodes
Voltage Stabilizers and Automatic Cutouts

कम्प्यूटर - एक पूर्ण परिचय *(अखिल भारतीय तकनीकी शिक्षा परिषद, भारत सरकार के तकनीकी पाठ्य पुस्तक योजना की प्रथम पुरस्कार विजेता)*

डैस्क टॉप पब्लिशिंग ऑन पीसी *(सूचना और प्रसारण मंत्रालय, भारत सरकार के भारतेन्दु हरिश्चन्द्र मान पुरस्कार से सम्मानित)*

वोल्टेज रेग्युलेटर्स एन्ड ऑटोमेटिक कटआउट्स

1

Introduction

All electrical and electronic equipment work on electric power. Many electrical gadgets like fans, heaters, refrigerators etc. use 230 V, 50 Hz a.c. supply directly. In contrast, most of the electronic equipment need different a.c. and d.c. voltages for their operation. For this reason, almost all electronic equipment have a special circuit built inside them that draws electric power from the mains supply, converts it into d.c. and a.c. at appropriate voltage levels and supplies needed currents to different circuits in the device. This special circuit (or section) is called a *power supply*. All electronic devices therefore have a power supply.

While many electronic circuits are tolerant to modest variations in voltages, there are others which need a strictly stable voltage within say ± 5%. Some circuits may work on voltages as low as 1.25 V (e.g., calculators and wrist watches), while those using cathode ray tubes may need several kilovolts for their operation. Portable devices need light-weight but efficient power supplies to economize on battery consumption, whereas high power audio amplifiers need power supplies capable of delivering several amperes of current. Therefore the design of a right type of power supply for an electronic device needs a lot of careful considerations.

Before we undertake a study of power supplies in details, let us have look at the basic principles of working and different elements that are involved in such designs.

DC Voltage

The simplest source of electric energy is a battery. A battery gives a smooth and clean d.c. voltage. This is represented graphically by a smooth horizontal line as shown in Figure 1.1. In this Figure, the horizontal axis represents time and the vertical axis represents voltage. The smooth horizontal line shows that there are no voltage variations or noise riding over the supply voltage.

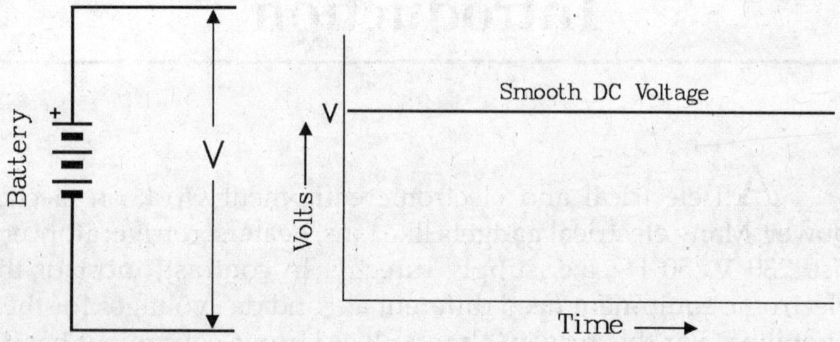

Figure 1.1: A battery supplies noise free smooth d.c. voltage.

Mechanical generators (dynamos), which use commutators and brush-gear provide a d.c. output with a lot of noise riding over the d.c. voltage as shown in Figure 1.2, because of the irregularities of brush contacts. This voltage requires considerable smoothing before it can be used to supply power to electronic circuits.

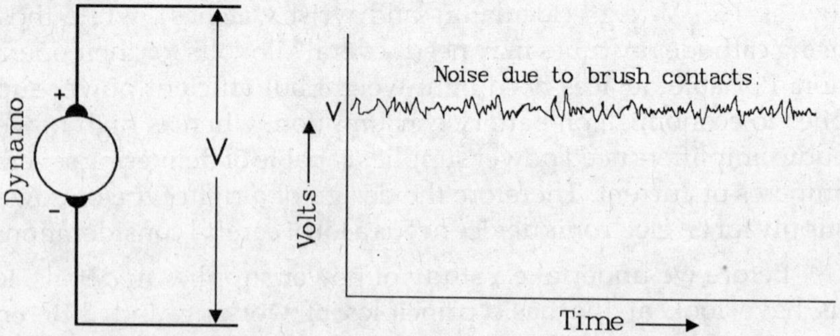

Figure 1.2: DC generated by a dynamo has a lot of noise riding on it.

The d.c. that is obtained by rectification of a.c. voltage is really only unidirectional. It has a large percentage of a.c. riding over d.c. as shown in Figure 1.3. This voltage is called *ripple*. This ripple has to be filtered out by using inductors, resistors and capacitors before the d.c. output becomes fit for supplying power to electronic circuits.

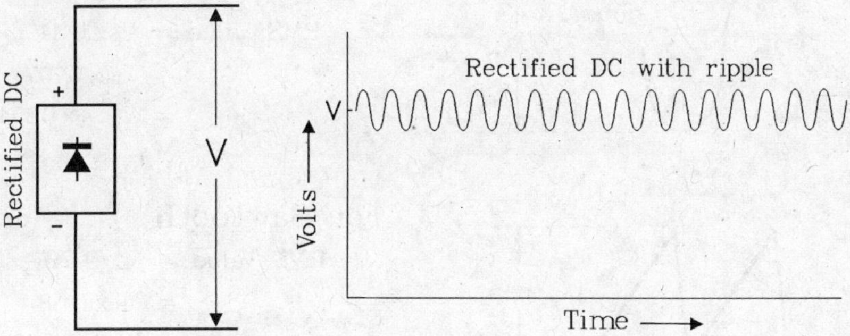

Figure 1.3: Rectified d.c. has a lot of a.c. content, called ripple.

If a polyphase a.c. supply is rectified to get d.c., the ripple content is lesser, but filtering of the ripple content is still necessary before it can be used for electronic circuits..

Ripple

A supply that is not perfectly smooth d.c. is said to have *ripple*. The ripple can be measured as current or as voltage. The ripple current or voltage can be quoted in terms of peak-to-peak (V_{p-p}) value, or it can be assumed to be a sinewave and its r.m.s. value quoted. For a sinewave, the r.m.s. value is 0.707 times the a.c. peak value (V_p) or 0.3535 times the a.c. peak-to-peak (V_{p-p}) value. as shown in Figure 1.4. The ripple content is sometimes also quoted as a percent of the rectified average d.c. voltage. The important thing is that the ripple content should be as small as practicable.

The ripple waveform is most unlikely to be a sinewave; a much more usual waveform is a sawtooth or triangular wave, but since the *comparative* ripple amplitude is the factor that is of interest, the r.m.s. value is more often quoted. The peak-to-peak value (V_{p-p}), which is more likely to be the measured value on an oscilloscope, is then found by multiplying the r.m.s. value by 2.8. The relationships between peak-to-peak and r.m.s. values for a sinewave and for a sawtooth are shown in Figure 1.4.

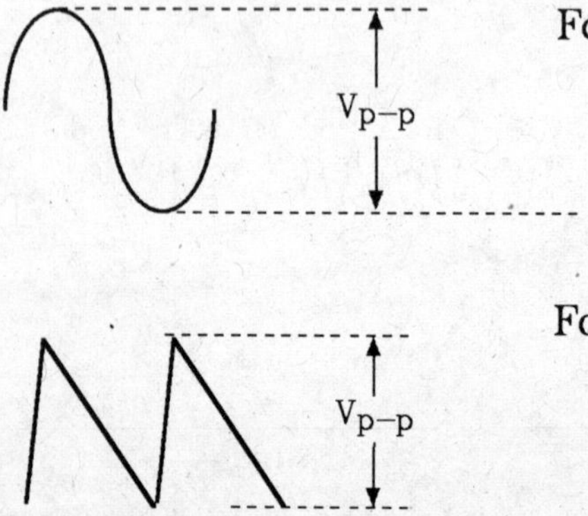

For Sinewave

$$\text{RMS Value} = V_p/1.414 = V_p \times 0.707 = V_{p-p} \times 0.354$$

For Sawtooth

$$\text{RMS Value} = V_p/1.73 = V_p \times 0.58 = V_{p-p} \times 0.29$$

Figure 1.4: Peak-to-peak and r.m.s. relationships.

Ripple is never constant, and it is misleading to quote ripple amplitude without also quoting the current load at which the ripple is measured. Many power supplies have their ripple amplitude quoted at low current outputs, though it is now becoming more common for manufacturers to show the amount of ripple when the supply is operating at full rated current. This is a much more useful and informative measurement.

The amplitude of ripple can be expected to increase considerably if the rated current output is exceeded, irrespective of whether the supply is electronically stabilized or not. Stabilization is not a cure-all for power supplies, and when the current rating of a stabilized supply is exceeded, the voltage drop in the output voltage can be excessive and the amount of ripple can also become very large.

In audio amplifiers, if the ripple is large, it can be heard as a low hum like tone in the background. It may not be noticeable at large volumes, but it becomes annoying when the audio signal is low or there is no audio signal input at all to the amplifier.

Regulation

An ideal power supply should maintain the output voltage constant irrespective of changes in the supply voltage and variations in load current. However, in practice, for any form of d.c. power supply, the output voltage will change as the input voltage changes and also as the load current changes. For example, the output voltage in the case of a battery supply falls as the battery runs down and its internal resistance increases. For d.c. supplies that draw power from a.c. mains, the output voltage changes as the line voltage changes, and also as the load current changes. The regulation of a supply measures how well it can cope with such changes.

Line Regulation

For line voltage changes, the *line regulation* can be quoted as the change in output voltage per change in a.c. input voltage. It is expressed mathematically by the equation:

$$\text{Line regulation} = \frac{\delta V_{out}}{\delta V_{in}}$$

where δ has the usual meaning of a small change in the quantity.

Thus, line regulation is given by dividing the change in output voltage caused by change in input voltage. This is a fraction, but it can be converted into percentage by multiplying the figure with 100. Ideally, this should be 0 percent.

Load Regulation

The effect of internal resistance of the supply on the output voltage of a power supply is illustrated in Figure 1.5. The source of supply in this case is a battery. The internal resistance of a battery, (or of a power supply) can be considered as a resistance (r) connected in series with the voltage source as shown in Figure 1.5. When there is no load connected to the supply, there is no voltage loss across the internal resistance r and the voltmeter shows a reading equal to the e.m.f. of the source.

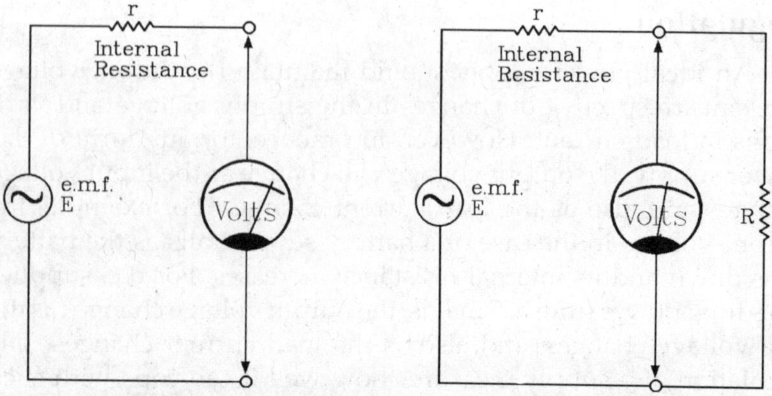

Figure 1.5: Effect of load current on output voltage.

When a load resistance R is connected, a current I equal to E/(R+r) flows in the circuit. This causes a voltage equal to I × r to be lost inside the battery. The output voltage therefore falls. The effect of load current is that the output voltage will drop as the load current increases. The lesser the drop in the output voltage, the better is the output regulation of the power supply. This will happen when the internal resistance of the battery (or power supply) is small.

A small change of output voltage caused by small change in output current is called the *internal resistance of the power supply* and this is expressed mathematically by the equation:

$$\text{Internal resistance} = \frac{\delta V}{\delta I}$$

The internal resistance of a supply should be as low as possible for good regulation. Ideally, it should be zero.

For example, if the open circuit output voltage of 9 V of a supply changes from 9 V to 8.85 V when 120 mA of current is drawn, then the internal resistance is:

$$\frac{(9-8.85)}{(0.12-0)} = \frac{0.15}{0.12} = 1.2\ \Omega$$

A stabilized supply can be designed so that, over a limited range of voltage and current, the output voltage will actually rise slightly when the output current increases. This corresponds to the internal resistance having a negative value. This is normally an unstable situation which will lead to oscillation, and therefore it is used only to compensate for other resistances that are present. Most stabilizer integrated circuits, however, feature very large internal gain, and are liable to oscillation unless care is taken in the design, such as use of suitable bypass capacitors.

Electronic Power Supplies

As stated earlier, all electronic equipment need a power supply. Electronic power supplies can broadly be classified into two categories: *unregulated* supplies and *regulated* supplies. The basic elements of an electronic unregulated power supply are the transformer to step the voltage up or down, rectifier to convert a.c. into unidirectional d.c., and filter to remove ripples and provide smooth d.c. These elements are shown in Figure 1.6.

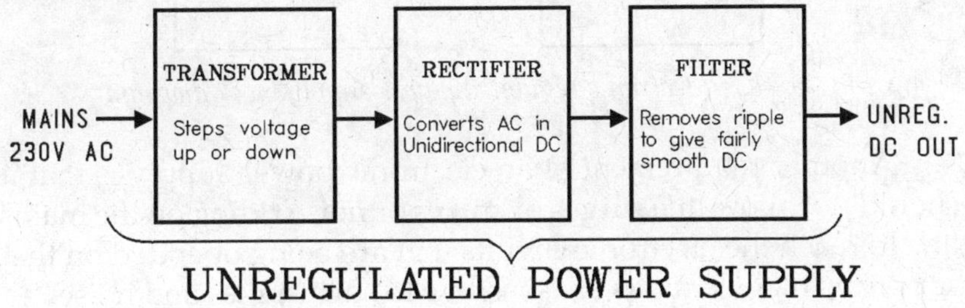

Figure 1.6: Electronic unregulated power supply block diagram.

The first element in a power supply is usually a transformer. A power transformer steps the incoming voltage voltage up or down, as needed. This a.c. output of the transformer is then rectified by rectifiers (mostly semiconductor diodes) and converted into unidirectional d.c. A filter circuit removes the ripples and makes the output voltage smooth enough so that it can be used for operating electronic circuits.

A large number of electronic devices are tolerant to variations in the supply voltage, i.e. they continue to operate satisfactorily in spite of of voltage variations in the supply. For such equipment an unregulated supply is good enough. Examples of this type of power supplies are battery eliminators for radios and cassette players and supplies for audio amplifiers.

There are many other applications where a variation in d.c. voltage is not acceptable. Digital circuits using ICs and sensitive test equipment are examples where regulated supplies are needed. A regulated power supply is basically an unregulated power supply followed by a regulating circuit as shown in Figure 1.7. The electronic regulating circuit tries to maintain the output voltage constant irrespective of line voltage and load current variations; of course, within limits.

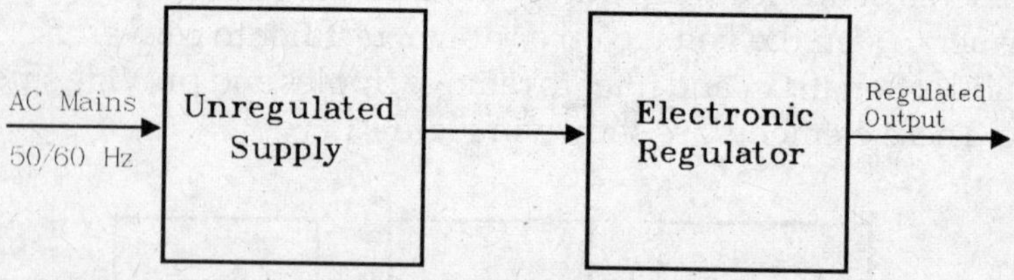

Figure 1.7: Electronic regulated power supply block diagram.

Another requirement of an electronic power supply is that it should not allow line surges or spikes or noise riding on the mains line to reach the electronic circuits that are being operated on that supply as many circuits are sensitive to such spikes and noise.

Protection

Power supplies need two kinds of protection. One is to protect the power supply itself in cases of very large variations in the supply voltage, voltage transients and overloads. The second requirement is that if the power supply fails, no damage should be caused to the equipment being supplied power by it. This is often referred to as *fail-safe* operation. Protective devices include fuses, circuit breakers and electronic crow bar protection.

2

Cells and Batteries

Electronic equipment in yester years used vacuum tubes. They were bulky, heavy and consumed a lot of electrical power. They dissipated a lot of heat too. Invention of transistor in 1948 reduced the size and power requirements of electronic circuits drastically because the job of a vacuum tube then could be done by a tiny transistor. Many types of electronic equipment, which had been hitherto operating on a.c. mains supply were redesigned to operate on batteries to provide convenience as well as portability. Later developments in semiconductor technology led to the development of MOS devices and integrated circuits which further reduced the size and power requirements of electronic circuits and devices. Today, a large number of electrical and electronic equipment like toys, calculators, watches, remote controls for TV and VCRs, mobile phones etc. operate on dry cells and batteries.

In recent years, there has been a great increase in the variety and size of appliances operating on batteries. They use primary as well as secondary batteries of different kinds. At one end of the scale are the minute hearing aids, tiny medical devices, and some little-larger-than-tablet devices to be implanted into the body to signal conditions such as the acidity of the blood and heart beat rate. At the other end are devices like battery backups for computers, uninterrupted power supplies (UPS), high power emergency lights and traction using batteries.

Manufacturers are now developing battery systems to meet the operational criterion of modern devices. For things like pocket calculators and watches, batteries are designed for low drain with long runtime. Batteries for devices like flash guns and UPS for computers should be able to provide heavy load current for short durations. To meet this varied requirement, practical advances are taking place, involving design of appropriate cells and batteries to suit a specific application. As a result, the field of battery design is exploding in many directions simultaneously. Figure 2.1 shows the design of a flat type of battery for use in smart cards. Appendix B gives detailed information about sizes and shapes of primary cells.

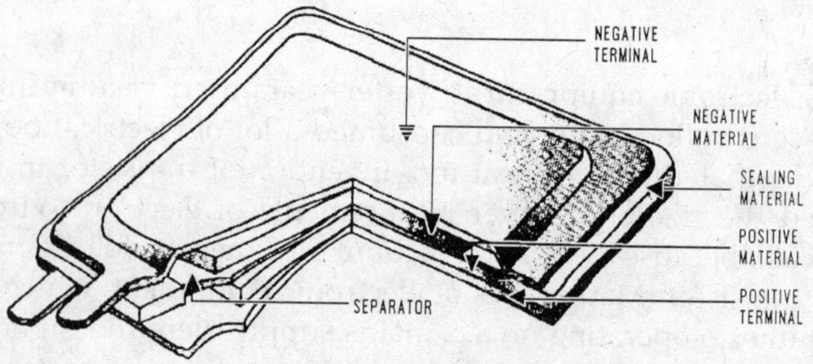

Figure 2.1: A thin flat battery for a Smart Card.

Cells and Batteries

A cell is an electro-chemical system that converts chemical energy into usable electrical energy. When chemical reactions are irreversible, it is called a *primary* cell; and when the reactions are reversible, it is known as a *secondary* cell or a *rechargeable* cell.

A cell has two electrodes (one positive and one negative), chemicals for the reaction, and a container to house these elements. The chemicals inside the cell are in the form of liquid and the solution is called *electrolyte*. Such cells have to be used in upright position. The so called 'dry cells' have electrolyte in the form of paste in a sealed container so that it does not spill out. This kind of cells can be used in any position.

Containers for cells are usually designed in standard sizes and shapes so that cells from different manufacturers can be fitted easily in the space provided for them in a device. Figure 2.2 shows some of the shapes and sizes of dry cells. However, special applications need special designs. For example, button cells are made in different sizes for use in calculators, wrist watches etc.

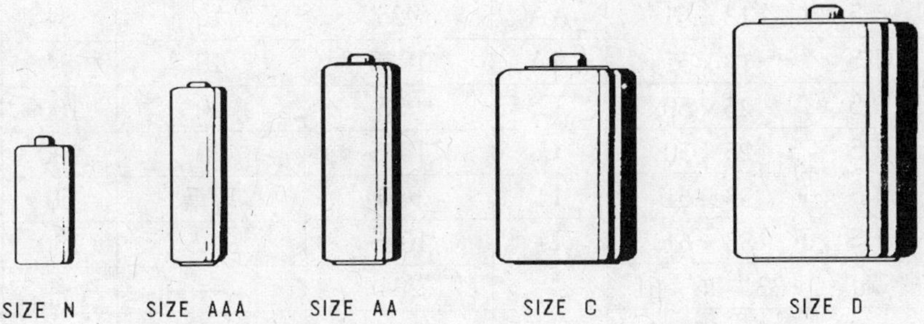

Figure 2.2: Comparison between different sizes of dry cells.

A number of cells can be connected in series or in parallel to make a *battery*. Cells are connected in series to get higher voltages. Parallel connections are used to get higher current capacity or longer life. Only cells of same voltage output can be connected in parallel. For convenience of use, ready-made battery packs are available in different voltages and sizes. Some of these are shown in Figure 2.3.

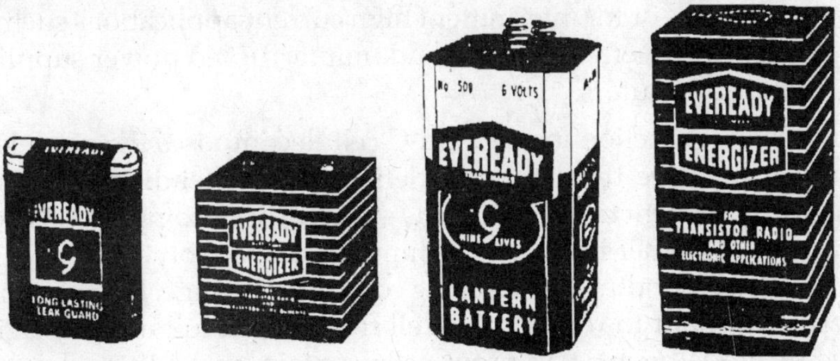

Figure 2.3: Battery packs are available in different shapes and sizes.

Table 2.1 gives information about numbers and dimensions of a few types of cells and batteries.

Table 2.1: Parameters of different types of primary cells.

EMF (Volts)	**Dia. × H or (L × W × Dia)**	**Size Design**	**Eveready Number**	**Current (mA)**	**Time (Hrs.)**
1.5	13 × 51	AA	925	10	84
1.5	13 × 51	AA	1025	10	84
1.5	25 × 50	C	935	20	100
1.5	25 × 50	C	1035	20	100
1.5	33 × 61	D	950	37.5	50
1.5	33 × 61	D	1055	37.5	150
6	33 × 26 × 61	—	234	—	—
9	26 × 17 × 49	F22	216	10	25
9	66 × 54 × 90	—	276-P	20	230
9	56 × 54 × 154	—	286	20	500
22.5	26 × 16 × 50	—	412	—	—

The materials used for electrodes and the type of chemicals used in the cell determine its terminal voltage which can be from 1.25 V to 3 V. The proportion of chemicals within the container determines the type of service the battery is meant for, i.e., low current longtime application for use in devices like calculators and wrist watches, or for intermittent high current applications such as flash lights, camera flash guns and uninterrupted power supplies (UPS) for computers.

Chemical action inside a Zn-C cell decomposes the zinc electrode. Therefore the cell has a definite life depending upon the discharge current rate at which the electrode is decomposed. When the zinc container has been consumed, the electrolyte often leaks into the surrounding area causing corrosion damage. It is therefore very important to remove this cell from the circuit as soon as its useful life is over. Leak-proof cells contain an additional metal cover to overcome this problem.

Cells of similar sizes and shapes can have totally different operating characteristics. Say one can be for low current long time application and the other for high-current for short durations application. This is illustrated in Figure 2.4. The characteristics are controlled by proportion of the different chemicals inside the cell. You should therefore be careful to choose the right kind of cell for an application.

Figure 2.4: Cells of same size but with different characteristics

Primary or Secondary?

It is becoming increasingly harder to choose a battery for a particular application due to the availability of various options. Each application demands different performance characteristics. Choice is dictated by design, operating environment, and power needs.

A primary battery is useful if the power requirement is low or in applications where the battery will be used infrequently over a long period of time. For frequent applications, the choice depends on the user's preference, assuming that primary and secondary types of batteries are available that can meet the requirements.

One-time-use primary batteries are cheaper, available in a wide range, have better shelf life, good charge retention capacity, and are useful for low to moderate power applications. They are convenient light-weight choice for portable devices. However, frequent replacements are likely to prove expensive.

Secondary batteries cost more initially and need frequent charging, but prove to be economical in the long run. Hence, for long time operation with high current drain, these are a better choice.

Selecting Primaries

If you decide to choose primaries, the choice of the batteries is narrowed to C-Zn, alkaline, Mn, Hg, Ag-O, Li, etc. Capacity of a cell (discharge current and maximum discharge time), operating temperature range, self-discharge, cell size, need for surplus capacity, etc., form the relevant factors to be considered for each application. Go through the battery selection brochure provided by the manufacturers depicting discharge current vs. operating time for various brands.

Special applications require Mg, Li, Zn-air, or solid-state batteries. C-Zn type batteries are inexpensive. Available in a variety of shapes and sizes, these batteries are widely used and are useful for general applications having a light or intermittent duty cycle. Disadvantages include low shelf life and variable output voltage.

Alkaline-Mn dry cells offer a better option for heavy-duty applications requiring continuous operation at high current. These have better energy-to-weight ratio with long shelf and service life. Other special features include low temperature characteristics and low voltage dropping. But these batteries are more expensive and not good at lower current drains.

For a high energy-to-volume ratio and flat discharge characteristics, Hg battery is a better choice for reference voltage applications and miniature devices. Ag-oxide cells, though expensive, have better discharge features and higher voltage than Hg cells.

The merits of Li batteries include long life, reliability, good mechanical and electrical properties. Ideal for CMOS circuits, these have better life expectancy, open-circuit and load voltages, capacity, temperature range, flat discharge profile, and low discharge. Li batteries have the highest energy density, followed by Ag-O, alkaline, Zn-Cl etc.

Selecting Secondaries

Among secondaries, Nickel-Cadmium batteries, or in short, the Ni-Cd (or nicad) batteries have become very popular because they are both rechargeable and sealed. They can deliver high current and can be cycled many times for recharging. During its charging and discharging no gases are evolved and no water need to be added Also, they can be stored for a long time, even when discharged, without any damage. The advantages offered by Ni-Cd cells can be summarized as below:

- ☐ Maintenance free
- ☐ Hermetically sealed and can be used in any position
- ☐ Very low internal resistance (high load currents)
- ☐ Withstands high current charge and discharge
- ☐ Can be stored in charged or discharged condition for long periods
- ☐ Wide range of temperatures (-20°C to +55°C).

Ni-Cd cells offer such features as extended operation, low energy density, ruggedness, and suitability for portable applications. Costing high initially, these prove to be economical in the long run. Table 2.2 shows specifications of a few types of Ni-Cd cells.

Table 2.2: Standard specifications of some Ni-Cd cells.

Type	Nominal Voltage (V)	Nominal Capacity (5 hr rate) (mAh)	Dimensions		Charging Current for 15 hrs (ma)	Approx. Weight (gms)
			Dia. (mm)	Height (mm)		
P-500AA	1.2	500	14.0	50.0	50	25
P-2000C	1.2	2000	26.0	50.0	200	80
P-2500D	1.2	2500	34.0	44.0	250	115
P-4000D	1.2	4000	34.0	61.0	400	160

Alkaline-Mn batteries offer features similar to their primary version, with limited charge/discharge cycles and good charge retention. Initial cost is lower than Ni-Cd, but higher than C-Zn, bridging the gap between low-cost disposable cells and expensive recyclable Ni-Cd batteries.

Ag-Zn batteries offer higher currents, level voltages, and higher Wh capacity than Ni-Cd/Pb-acid versions. These are sensitive to overcharge and have shorter life cycle. This more expensive battery is used where low weight and high performance are mandatory.

In comparison to Ni-Cd, Ni-MH offers longer service life, higher energy density, more environment-friendly features, and faster charging capabilities.

Lithium-ion rechargeable versions have the highest energy density with a very low self-discharge. These are stable and safe. However, cost can be a major consideration.

Compact sealed lead acid (SLA) batteries are resistant to vibration and shock. These have exceptional leak resistance and provide full and reliable power. Small size SLA batteries find applications in communication systems, security systems, UPS, smaller power tools, toys etc. Mid-size versions are used in large power tools, UPS, communication equipment, security systems, etc. Large versions are used as alternatives to moderate power generators. Characteristics of primary portable batteries are summarized in Appendix B.

Battery Design

Ever since its inception by Luigi Galvani and Allessandro Volta two centuries ago, battery technology has been in a dynamic state, with sustained improvements in the related fields of their electrochemistry, materials science, electronics and packaging techniques. Traditional batteries, having reached their performance limits, are being replaced by newer ones with better outputs and functions. There has been an increase in the variety, size, and applications of primary and secondary batteries.

New combinations followed the traditional carbon-zinc cells. Magnesium cells use a magnesium alloy anode instead, for their magnesium-manganese dioxide system. Another example of cell modification is magnesium plus magnesium perchlorate or bromide plus metadinitrobenzine. The organic depolarizer provides the same number of watt-hours per unit price as electrolytic manganese dioxide. This factor becomes important as supply sources of battery-grade natural ores diminish. Modifications like aluminium plus aluminium, chromic chlorides, and ammonium chromate plus manganese dioxide are attractive proposals on account of lower density and electro-chemical equivalent value of aluminium, in comparison to zinc.

Manganese alkaline cells, using alkaline potassium-hydroxide electrolyte, are superior and less corrosive, and permit the use of a wider range of metals. The cell comprises zinc anode with a large surface area and high density manganese dioxide. It delivers electrical energy at a high rate. This kind of cell has the ability to withstand high, short-term load current. Its versions are useful in portable devices requiring high current drain. Examples of such applications are small motors, razors, model boats, electronic flashes, etc.

Mercury oxide cell, another alkaline version, has high capacity per unit volume, good storage characteristics and a relative constant output voltage. An important feature of this cell is the flat voltage-time discharge characteristic. Its relatively high energy-to-weight ratio and ease of miniaturization has enabled its use in portable aerospace and communication systems. Silver oxide or silver-zinc cell, an alkaline version, has become significant. An example is the tiny button cell used in electronic watches. It has high energy-to-weight ratio and is used for missile and torpedo applications where light weight and high performance parameters are critical. The cells have flat voltage and steady current-discharge capabilities, and are amenable to miniaturization in the form of button and coin packages. Because of these features and qualities these are suitable for use in hearing aids, photoelectric exposure device, etc.

Zinc-air cells, initially used to power telephones, were further developed by researchers in NASA. These cells are useful where high current for long duration is needed, such as in railway applications. These have higher capacity than corresponding mercury and silver oxide cells.

Lead-Acid

Lead-acid secondary battery, as we know well, has widespread applications in the automotive field. These versions have been developed in new forms different from the traditional versions and are being trimmed down from their large versions for automobiles to small sealed units for electronic applications. These batteries offer high cell-voltage, increased capacity (due to pulse-discharge technique), high discharge efficiency, long life, good reliability and wide temperature range.

In applications such as engine starting, batteries must deliver high output intermittently. In such cases, the cells rest between the pulses. The acid diffuses through the separator back to the working plates, replenishing the electrolyte at the plate and delivering higher capacity output. In addition to automobiles, these cells are also used as standby source for telecom, IT installations, and solar power systems.

Industrial Lead-Acid Batteries

'Plante' type batteries use pure lead with fine-grooved surface, designed for maximum performance and durability, as the positive plate. Negative plate is made of lead alloy grid filled with active material. Microporous PVC separators prevent internal short-circuiting of the plates. These are useful for continuous float operations, such as in telephone exchanges and standby applications, requiring a reliable d.c. supply.

Tubular cells are suited to applications requiring frequent charge/discharge cycles, such as emergency lighting applications and telecommunications. The positive is encased in a one-piece multitude woven gauntlet of high-tensile, acid-resistant terylene.

The negative contains additives and expanders. Sintered or micro-porous PVC is used as the separator. Sealed maintenance-free valve-regulated lead-acid (SMF VRLA) batteries are used in many applications. These include process industry, telecommunications, solar photo-voltaic applications, alarm systems, emergency lighting, safety and security services, power plants, distribution networks, control equipment, railways, UPS , etc.

In the conventional lead-acid battery, electrolysis of water produces hydrogen at cathode and oxygen at anode. This results in water loss, needing regular topping up of the battery. Oxygen is evolved from anode before hydrogen is evolved from the cathode. In the VRLA version, the oxygen generated at the anode travels through the open pores in the separator and reacts with spongy lead at the cathode to form lead oxide. Lead oxide is reduced to lead by nascent hydrogen at cathode, preventing evolution of free gas. No venting is needed. Hydrogen and oxygen generated from the electrolyte combine to form water, eliminating water loss and obviating the need of topping up with water in the conventional type.

SLA (Sealed Lead Acid) batteries are characterised by leak-proof design, long service life, maintenance free operation, exceptional deep discharge recovery, low self-discharge, high discharge-rate, and also non-generation of corrosive gas. These lead-calcium SLA batteries withstand tough operating conditions such as overcharge and deep discharge.

Cyclic applications include portable camcorders, TV, record players, vacuum cleaners, cash registers, cordless portable telephones, grass and hedge trimmers, cordless drills, electronic instruments, electronic cameras strobes, movie lights, toys, etc. Trickle or float-charge applications include fire/burglar alarms, communications systems, UPS systems for computers, sequencers, electronics, medical equipment, etc. Batteries are charged using the constant voltage, constant current or a combination of the constant voltage and constant current methods.

Lithium Cells

This miniature power source packs about ten times the energy density of carbon zinc cell with twice the cell voltage (namely 3V). It has improved shell life and temperature parameters. Cells employ lithium metal as the anode. Various electrode combinations can produce voltage ranging from 1.5 to 4V. Non-aqueous organic electrolytes, such as propylene carbonate and inorganic venyl chloride are used. Cathode's active materials include sulphur dioxide, carbon monofluoride, vanadium, pentaoxide, copper sulphide, etc.

The long shelf life is due to the fact that lithium anode is passive during long storage periods. It can operate over a wide temperature range. The flat discharge characteristic is beneficial for circuit operations. High reliability is guaranteed by the lithium-iodine and lithium-silver chromate versions used in the medical industry. The cell is an ideal power source for CMOS based circuits with integrated power supply built-in on a permanent basis. Lithium batteries can be mounted on the PCB. These are ideal for aerospace, portables, medical devices, CMOS-based communication devices, industrial controls, sensing data processing and portable instrumentation. Lithium-vanadium-disulphide battery, developed by bell Laboratories, is a rechargeable version and has the ability to be charged several times without chemical deterioration.

Rapid growth in demand for high specification, portable electrochemical power sources in the last decade, driven by the advent of micro electronics-based products, has proved to be a boon for lithium battery manufacturers.

Ni-MH

This 12V alkaline battery features energy density about 50 percent higher than the Ni-Cd battery. It uses more environment-friendly materials, and is capable of high-current charging and discharging. It resists overcharging/discharging, and has a long life. The power generation elements of this battery are the positive nickel electrode with nickel hydroxide as an active substance. The negative electrode uses hydrogen absorbed by the hydrogen-absorption alloy. The electrolytic is an alkaline solution which is made mainly from potassium hydroxide. A separator electrically isolates the positive and negative electrodes. The high energy of this battery stems from the use of high-capacity, high-density metal alloy for the negative electrode. High capacity is contributed by an improved positive electrode, which is made up of a highly porous nickel substrate filled at high density with nickel oxide as the active material.

Manufacturers are steadily increasing the volume energy density using improved materials and battery composition techniques. Battery capacity is rising 10 percent every year. The present volume energy-density of 350 Wh/unit is nearly equal to that of the Li-ion rechargeable batteries. Using the same components, energy density has been improved in several ways. Engineers have introduced a dense filling method for positive and negative materials, trimmed the thickness of the electrodes' conductive core material and separator, cut the dead space inside the battery materials, devised a thin sealing plate, and improved the scattering of active material filling.

The hydrogen-absorption negative electrodes are made of AB5 alloys based on rare-earth nickel. Elution (chemical removal by washing with solvent) of compound elements and susceptibility to surface oxidation are the problems related to this alloy. Efforts to extend the life cycle and improve the discharging performance of the negative electrode are on. Improved alloy manufacturing, surface treatment, formation of nickel layer as electrochemical catalyst for stimulating the elution of the metal in alkaline solution, removal of non-conductive metal-oxide layer, reduced cobalt content in alloy, improved alloys with a higher capacity for absorbing hydrogen, etc. are being tried out to boost the performance. In anodes, the surface of the active electrode (nickel hydroxide) covered with an oxide of cobalt, resists oxidation, extends battery life, and improves reliability. Ni-MH batteries are future candidates for high-mobile-power equipment such as hybrid electric vehicles.

Mobile Phone Batteries

Gradual size reductions of mobile phones have been accompanied by an increase in the number of functions that the phone is capable of performing. As a result, the demands placed on portable power supplies have increased. The selection of the battery technology for the mobile is of critical importance in terms of size, weight, and capability to perform numerous functions of the phone. Three primary battery technologies are available to the mobile phone designer, namely, the nickel-cadmium (Ni-Cd), nickel-metal-hydride (Ni-MH), and lithium (Li) version.

The higher voltage of lithium-ion battery offers advantages over the lower voltage of the nickel-based versions, contributing to a higher energy-density and fewer cell interconnections in a battery pack. Lithium-ion cell can deliver more than 1200 cycles and 100 percent depth of discharge, and still operate at more than 80 percent of its initial capacity. The discharge-rate performance of lithium batteries is inferior to those of Ni-based ones, due to their lower electrolytic conductivity.

Lithium-ion is more suitable for applications operating at high temperatures. Below 10°C, decline in performance of lithium-ion cell is not as appreciable as the Ni-Cd version. Presence of the heavy cadmium in Ni-Cd poses environmental problems. Toxicity of cadmium can pose problems during disposal or recycling. Lithium batteries contain organic solvents and need careful handling when cells are disposed of, or recycled. To counter those factors, battery manufacturers incorporate sealed construction modes, followed by extensive safety testing. This ensures that the mobile phones incorporating these cells are safe to use.

Batteries for UPS

The type of service demanded from UPS systems is different compared to stationary applications such as telephone exchanges, engine starting, etc. In cases of UPS systems, the current demand rate is high, but for a shorter time at a constant power. This is because the critical load connected to the UPS is constant with regulated output voltage. The average discharge time for the UPS is about 15 minutes and not 8 to 10 hours (typical of other applications). Batteries supplying UPS systems are required to supply additional current during discharge, since the terminal voltage drops while the power demand is constant.

High-voltage batteries, ranging from 12 to 400V, are used for various UPS systems. Some of the typical voltages in use for UPS applications are 12V, 24V, 48V, 60V, 106V, 210V, etc. Lead-acid, sealed maintenance-free, and Ni-Cd batteries are extensively used in UPS systems because of their proven reliability. At the same time, those are costly and occupy more floor space. Minute quantities of calcium, alloyed with lead in lead-calcium cells render the cell robust, compact, and suitable for use as vertically constructed cells.

Advantages of Ni-Cd batteries lie in their long life, minimal temperature effect, superior deep-discharge characteristics, sealed maintenance-free construction, and capability to withstand increased deep-discharge cycles. Drawbacks include cost, hazardous nature of Cd, etc. Ambient temperature, battery chemistry, cycling, and capacity affect the UPS battery life. Any variation in the operating temperature of battery from 25°C alters the performance and rated output. High temperatures reduce battery life.

Batteries for Emergency Power

When power breakdown occurs, only one thing matters: will the emergency system work? Given the cost of potential failure in many applications—airports, hospitals, security, and computer circuits, for instance—it is surprising how little emphasis is placed on the choice of batteries for starting standby systems, be it generators or UPS. The prime cause of this unreliability is the tendency to use automobile batteries for starting. These batteries lead a list of causes with a failure rate of 14 every 100,000 hours.

Automotive batteries, good as they may be in their own applications, were not designed for emergency power situations. They are, however, cheaper than industrial secondary batteries, and are often used to start standby generator systems. The choice of stationary batteries is usually offered as an optional extra. The onus is on the consumer to specify the correct battery. Emergency lighting is mandatory in public organisations. This requirement covers conference halls, hotels, theaters, schools, colleges and hospitals. Batteries used in these situations should be robust, reliable, capable of withstanding variations in ambient temperature during float charge, and withstand a high degree of ripple content.

The inherent long life of the nickel-cadmium battery is one of the reasons why it has been regarded as economical for emergency purposes despite its initial cost. When compared with other types, the nickel-cadmium cells give an extremely good high-rate discharge performance. Thus, a small capacity battery is adequate for high-performance applications such as engine starting. The ripple content to which batteries may be subjected in high-technology applications does not affect the nickel-cadmium cell. Maintenance of nickel-cadmium batteries is simple and infrequent, since the electrolytic plays no active part in the reaction. Finally, installation costs are low. Automobile batteries, which are permanently connected to a trickle charger, are the least suitable choice for anything except automotive starting, lighting, and ignition functions.

Precautions

Precautions in handling batteries result in long service life. Quality chargers should be used with set parameters. The settings include float voltage at 13.65V per 12V battery, maximum charging current at 10 percent of Ah capacity of the battery, and charge duration of a minimum of 16 hours.

Battery should not be discharged below 10.5V. Discharged battery should be charged as soon as possible and kept away from heat generating sources. Ambient temperature should be maintained below 25°C. For series and parallel connections, batteries of similar Ah ratings and same batch codes should be used. Vent plugs should not be blocked.

The lead-acid battery is capable of supplying a very heavy current. Hence, avoid shorting the battery terminals to prevent damage to internal cells. The battery should be put on supplementary charge every two to three months, if it is to be stored for longer periods, in a cool and dry place.

Testing of Batteries

DO NOT try to estimate the current capacity of a battery by shorting its terminals. A fair indication of the condition of a battery is obtained by measuring its terminal voltage on load conditions. Modern multimeters have a special range of volts to test 1.5V cells. This range has a resistance which gets connected in parallel to the battery to serve as a load. The voltage indicated shows the status of the battery condition.

Lead-acid batteries tested by a special meter (often called as 'tongs tester') which puts a heavy load on the battery and measures its voltage in that condition.

Another method used is to measure the specific gravity of the electrolyte. The specific gravity of a charged battery is higher than that of discharged battery. In a fully charged automotive cell, the mixture of sulphuric acid and water results in a specific gravity of 1.280 at room temperatures of 70°F to 80°F. As the cell discharges, more water is formed lowering the specific gravity. When it is down to 1.150, the cell is completely discharged.

Solar Cells

Solar cells work in a completely different way. There are no chemicals to react, or to be consumed, and there are no gases evolved. These cells convert the sun's light energy directly in electricity, which can be used as a source of electrical power.

The solar cell is a semiconductor device that converts solar energy directly into electrical energy. It is also called a *photovoltaic cell.* A single cell can deliver only a small amount of power, but many cells can be inter-connected in series-parallel combinations to supply more power at a higher voltage.

The reliability, convenience and present cost of photovoltaic systems have brought solutions to power supply problems in fields such as telecommunications, navigation, pipeline protection and railway signalling. Smaller systems are being rapidly adopted in many countries where remote and holiday homes have no other source of power to charge 12V battery supplies. Modern satellites generate all the needed power by solar panels installed on then as illustrated in Figure 2.5.

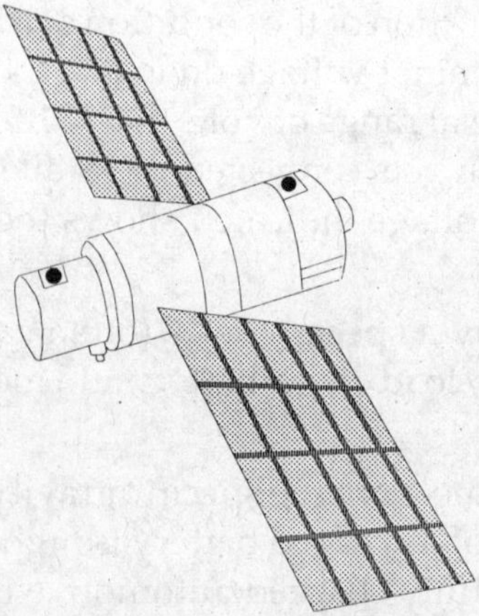

Figure 2.5: Modern satellites generate all the needed electrical power by solar panels.

At the opposite end of the scale, large quantities of solar modules are being used in calculators, watches, and earphone radios in place of primary batteries. Here, the bulk and cost of the power supply, combined with a low power equipment, means that a compact and inexpensive solar module can enhance the product.

Basically, a solar cell is a photovoltaic cell made out of semiconductors. Silicon photovoltaic cell consists of a thin wafer of silicon with its back surface completely silvered. Light waves entering the wafer force the electrons to move towards the rear surface. accordingly, a photovoltaic cell is capable of supplying electrical energy when exposed to light. Since light must pass through the front surface to create cell reaction, the front contacts must not completely cover the surface and so a grid pattern is used. The rear contact can cover the whole of the cell's black surface. Figure 2.6 shows the electrode grid patterns of solar cells.

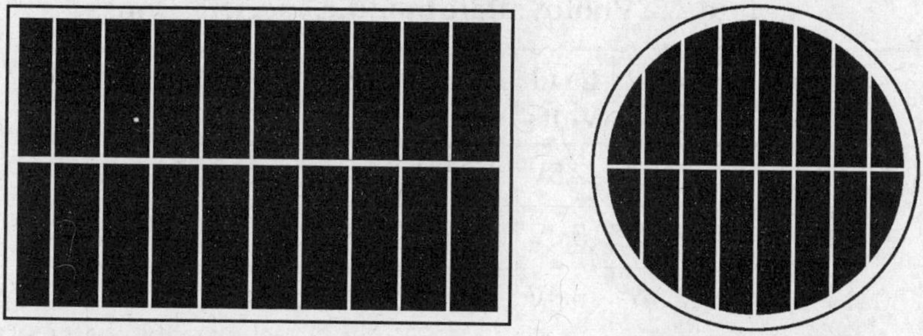

Figure 2.6: Electrode grid pattern on surface of solar cells.

Individual solar cell output is approximately 0.8 volts, (open circuit), corresponding to 0.55 volts on load for a cell width of 1 cm. The on-load current under bright sunlight is up to 10 mA per cm of cell's length, giving efficiencies in the range of 4 to 4.5 percent.

Two types of materials are used in construction cells, the amorphous and the crystalline. The major advantage of amorphous silicon modules is the technique used for interconnection of individual solar cells to produce a useful voltage and current. The p-i-n structure cells are fabricated by deposition onto an electrically conductive, tin-oxide coated glass sheet which is subsequently coated with a metallic back contact. The final module voltage is achieved by a set of stepped laser-scribed divisions which allow rear to front series connections between adjacent cells. This kind of monolithic interconnections allows a wide variety of configurations without the need for the laborious soldering required with conventional discrete solar cells. Table 2.3 shows some solar modules specifications.

For more information different kinds of batteries, see the book 'Batteries, Chargers and Emergency Lights' by the author, which is published by BPB Publications.

Table 2.3: Photovoltaic module specifications.

Size	Area	Load Volts	Load Amps	Application
10×50mm	5cm^2	2.5V	10mA	Credit card, calculator
20×100mm	20cm^2	5.0V	20mA	Desktop calculator
50×130mm	65cm^2	6.0V	50mA	Solar torch (4W bulb by AA NiCd rechargeable battery)
100×100mm	100cm^2	5.5V	100mA	Battery charger (2 AA Ni-Cd rechargeable batteries)
150×150mm	225cm^2	7.5V	150mA	Garden light (0.5W bulb with 1.2Ah 6V sealed battery)
100×300mm	300cm^2	15.0V	100mA	12 Volt trickle charger for battery up to 30Ah capacity
300×300mm	900cm^2	15.0V	300mA	12 Volt trickle charger for battery up to 90Ah capacity
300×900mm	2700cm^2	15.0V	750mA	Garden light (16W bulb with 2Ah 12V NiCd battery pack)

The Future

Ever since its invention by Allessandro Volta, a quiet but real revolution has been taking place in the battery industry. Batteries today are smaller, more efficient and powerful. A battery system backed by photovoltaic chargers is an obvious and desirable application in desert/remote conditions where power may be needed for a communication link and conventional electricity sources are impracticable. Research, aimed at optimising the power at minimal cost, goes on.

3

Transformers

By far, most d.c. power supplies used in electronic equipment draw power from a.c. mains supply. Mains voltages have been standardized at 115 V and 230 V. The power supply for the equipment requires the a.c. to be transformed to the correct voltage level, rectified to get the unidirectional current, and then smoothed so as to supply pure d.c. This arrangement is shown in Figure 3.1.

The task of drawing a.c. power from the a.c. mains and supply it at appropriate voltage levels is performed by a transformer. The transformer can step-up or step-down the input voltage. A single transformer can have different independent outputs to serve as a step-up transformer as well as a step-down transformer. Figure 3.2 shows typical transformers used in electronic power supplies.

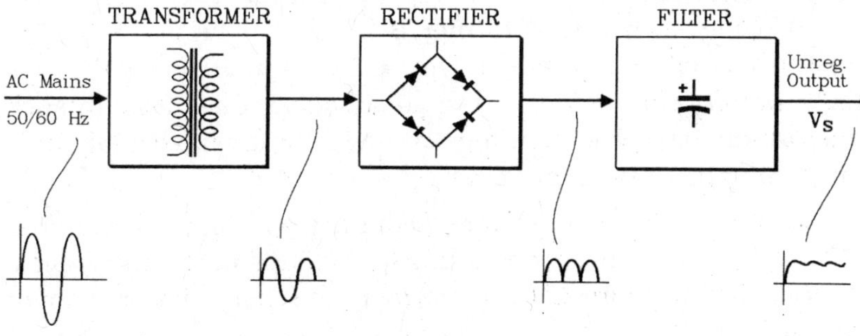

Figure 3.1: Components of an unregulated power supply.

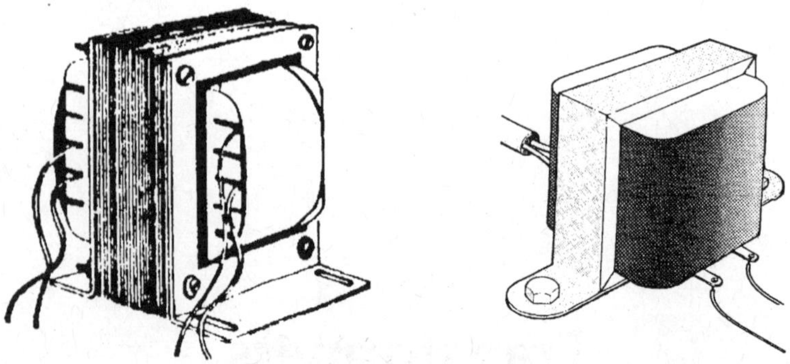

Figure 3.2: Typical transformers used in power supplies.

A transformer has a number of coil windings on a common iron core. The coil which receives power from the source is called the *primary* winding. All other windings to whom power is supplied are called *secondary* windings. The electrical energy drawn from the source by the primary winding is transferred to the secondary windings without any electrical connection between the two. The transfer of energy takes place by magnetic induction. A transformer with independent primary and secondary windings therefore, also serves another important purpose. It isolates the device to be operated on the power supply from the mains and thus reduce the chances of electric shock.

The transformer shown on the left side in Figure 3.2 is typical of the type used in power supplies of equipment using vacuum tubes. It supplied high voltages for plates of vacuum tubes and low voltages for heating the filaments of vacuum tubes. A power supply using such a transformer is shown in Figure 14.3. The transformer on the right side in Figure 3.1 is a small one of the type used in battery eliminators and small power supplies. Usually such transformers have one winding only which may have a tap at its center point or a number of taps.

A transformer can have one secondary winding, or a number of secondary windings or windings with taps on them. Typical symbols used to represent transformers in schematics are shown in Figure 3.3.

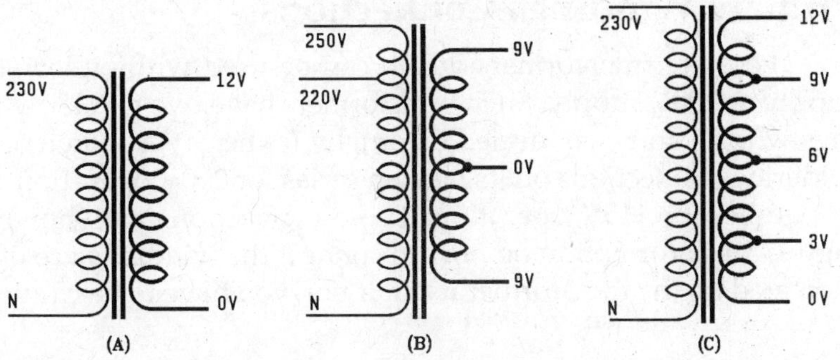

Figure 3.3: Typical symbols used for representing transformers.

A transformer is a wonderful device in the sense that it is self regulating. If there is no load on the secondary winding, the primary winding presents a large inductive reactance and hence, draws only a very small current from the mains. This current is often referred to as *idle current*. Hence, a transformer, if left connected to the mains without load for long durations, remains cool.

When a load is connected to a secondary winding, the voltage induced in the secondary causes a current to flow in the load. This current affects the magnetic field in the core in such a way that the inductance of the primary winding is reduced and the primary starts drawing more current from the mains and transfer that power to the secondary.

Transformer Efficiency

The capacity of a transformer to transfer power is indicated by its VA (product of voltage and maximum load current) rating. A transformer does not transfer all the electrical energy drawn by primary winding from the source to secondaries. In other words, a transformer does not have 100 percent efficiency. A part of the electrical energy drawn gets wasted within the transformer and that results in heating up of the transformer. Small transformers used in electronic equipment have efficiencies of about 80 to 90 percent. This fact should be taken into consideration while choosing or designing a transformer for a power supply.

Primary Windings Connections

There are transformers which can be used with either 115V supply or 230V supply. Such transformers have two similar windings which can be connected externally (either by a switch or by soldering connections on its tabs) in series for 230V operation and in parallel for 115V operation. These connections are shown in Figure 3.4. Information on how to connect the windings are often indicated on the transformer itself. If not, you have to be cautious.

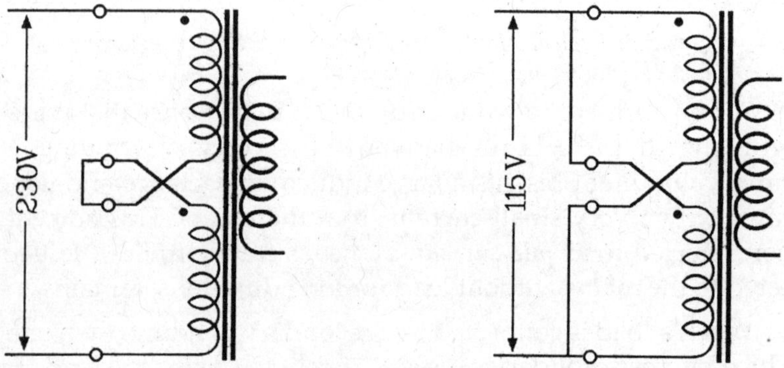

Figure 3.4: Connecting primary for 230 V and for 115 V operation.

If you happen to connect the coils in a wrong way, the inductance of one winding gets cancelled by the other. In that case, if the transformer is connected to the mains supply, it will behave like a short-circuit, draw a heavy current and blow up the fuse.

A safe method for testing such connections for their correctness is to use a 25 W or a 40 W lamp in series with the primary when you insert the plug for the first time after making connections. If the lamp glows dimly, the connections are correct. You can now remove the series lamp and connect the transformer directly to the mains.

On the other hand, if the lamp in series glows bright, it indicates that the connections are incorrect. In that case, simply reverse the connections of one of the windings and try again. It should be correct this time.

Turns Ratio

A transformer is used to step up or step down a.c. voltages. The voltages in the primary and secondary windings are directly proportional to the turns ratio of the two windings. If V1 is the voltage applied to primary, V2 is voltage induced in the secondary, N1 is the number of turns in the primary and N2 is the number of turns in the secondary (see Figure 3.5), then the relationship between them is given by the expression:

$$\frac{V1}{V2} = \frac{N1}{N2} \quad \text{-------------------------} \quad (3\text{-}1)$$

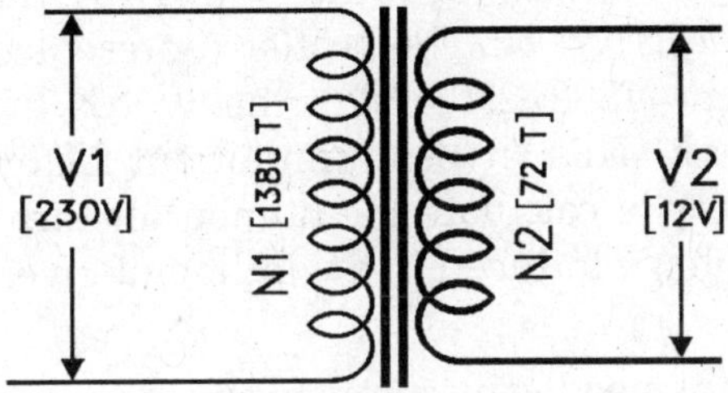

Figure 3.5: Relationship between primary and secondary number of turns and voltages in them.

If any three things out of V1, V2, N1 and N2 are known, equation 3-1 can be used to calculate the fourth one. For illustration, in Figure 3.5, the primary has 1380 turns and the secondary has 72 turns. If 230 V a.c. is applied to the primary winding, the secondary voltage is given by:

$$\frac{230V}{V2} = \frac{1380}{72}$$

or

$$V2 = \frac{230 \times 72}{1380} = 12V$$

The formula 3-1 can also be used to calculate voltages (or number of turns) of secondary windings when a transformer has more than one secondary winding. Calculations are repeated with primary and one of the secondaries.

Since power is V×A, if the transformer steps down the voltage, the current available is more, and vice versa. This relationship is given by the equation:

$$\frac{N1}{N2} = \frac{I2}{I1}$$

In the above illustration, if the primary takes 0.1 amperes, the secondary can deliver 1380 × 0.1 / 72 = 1.9A.

Secondary windings in a step down transformer are therefore wound with thick wire; and in a step up transformer, with a thin wire. The size of the wire is chosen according to the current to be handled.

If you divide the number of turns of primary (N1) by the voltage applied (V1), you *get turns per volt*. This is often represented by "n". In the above example, n = 1380/230 = 6. In other words, the windings have 6 turns per volt. This "n" is an important factor. Once you know n, you can easily calculate the number of turns needed for any voltage, or voltage induced in a winding of given number of turns.

The number of turns per volt needed by a transformer depends on the size of the core, frequency of the a.c. supply and the permeability of the magnetic material used for core, as shown by the following equation:

$$n = \frac{10^8}{4.44 \times f \times B_{max} \times A} \text{ - - - - - - - - - - - - - - - - - (3-2)}$$

where,

N = number of turns per volt
f = frequency of a.c. supply, and
B_{max} = maximum number of lines of magnetic flux per square inch of the core

This equation is used in designing transformers. For more information you may refer to the book "Designing Coils and Transformers" by the author and published by BPB Publications.

4

Rectifiers

As most of the electronic equipment need a d.c. supply, the stepped-up or stepped down a.c. available from the transformer has to be converted first into a unidirectional voltage by rectification and then the ripples removed by adding a filter circuit to get a smooth d.c. voltage as shown in Figure 3.1.

Rectifiers

A rectifier is a device which allows current to flow in one direction only. It has two electrodes called *anode* and *cathode*, and that is why it is called a *diode*. When the anode is positive with respect to cathode, the diode conducts. If a voltage is reversed, i.e. the anode is made negative with respect to cathode, the diode offers very high resistance and very little current flows.

Vacuum Tube Rectifiers

In yester years, electronic equipment were built around vacuum tubes and vacuum tube diodes were used for rectification. In most cases, the rectifier vacuum tube had two diodes built inside one envelope for full-wave rectification using a center-tapped transformer. Figure 4.1 shows a vacuum tube type double diode and the symbol used to represent it in electronic circuits.

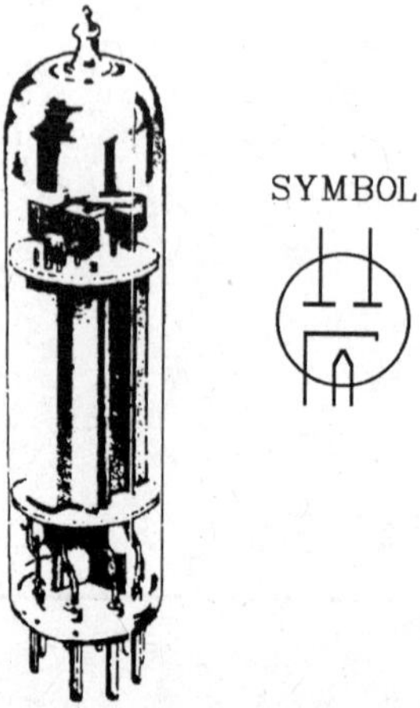

Figure 4.1: A vacuum tube double diode and the symbol for the same.

Vacuum tubes use a heated cathode. To use these rectifier tubes the transformer had a high voltage winding with a center tap. In addition to this high voltage winding, the transformer also had low voltage windings for supplying 5 V for the heaters of the rectifier tubes and 6 V for heaters of other vacuum tubes. Because of high voltages involved, the heater windings had to be insulated very well from the high voltage secondary winding. Figure 14.3 shows the circuit diagram of a power supply using vacuum tube rectifier diode.

Vacuum tube rectifiers are characterized by high forward voltage drops and inherently poor regulation, but they were immune to a.c. line transients that can destroy other type of rectifiers.

Mercury Vapour Rectifiers

The mercury vapor rectifier was an improvement over the vacuum tube rectifier in that the electron stream from cathode to anode could ionize the vaporized mercury in the tube and greatly reduce the forward voltage drop within the rectifier. Since ionized mercury is a much better conductor of current than a vacuum, these tubes can carry relatively large currents. As a result, they were popular in transmitters and r.f. power amplifiers.

Mercury vapor rectifiers had to be treated with special care. When power was initially applied, the tube filament had to be turned on first to vaporize condensed mercury before high-voltage a.c. could be applied to the plate. This could take from one to two minutes. Also, if the tube was handled or the equipment transported, filament power would have to be applied for about a half hour to vaporize any mercury droplets that might have been shaken onto tube's insulating surfaces. Vacuum tube and mercury vapour rectifiers have mostly been replaced by silicon diodes.

Selenium Rectifiers

The selenium rectifier was the first of the solid-state rectifiers to find its way into commercial electronic equipment. Selenium rectifiers were used for many applications because they did not need a heater for their filaments. They were fast to act because no heating delay was there. Offering a relatively low forward voltage drop, selenium rectifiers found their way into the plate supplies of test equipment and accessories, which needed only a few tens of milliamperes of current at about a hundred volts, such as vacuum tube voltmeters (VTVM).

Selenium rectifiers had a relatively low reverse resistance and were therefore inefficient. Voltage breakdown per rectifying junction was only about 20 V. The selenium rectifier has been discarded because of its bulk and weight. However, experience has proved that selenium rectifiers cannot be dispensed with in many applications. It has a number of desirable characteristics which makes it useful for many practical applications. Figure 4.2 shows typical selenium rectifier stacks.

There are differences in the characteristics of each individual silicon diode. Hence, if they are connected in series, problems like unequal voltage sharing due to unbalance of voltage-current characteristics and earth capacitance, and unequal voltage sharing due to difference of reverse recovery times, result in failure of silicon diodes. This only increases the overall size, and the claim of compactness of silicon diodes is destroyed.

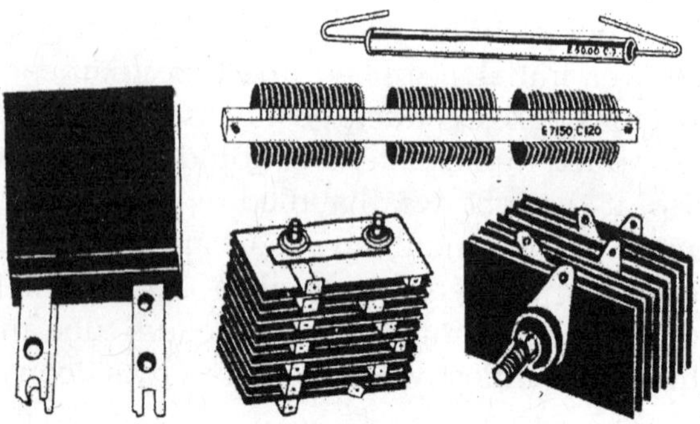

Figure 4.2: Typical selenium rectifier stacks.

Selenium rectifier, on the other hand, does not require any protection against surge voltages. This has a polycrystalline structure and has a large capacitance inherent in itself. This inherently large self-capacitance becomes self compensating for the steep transient wave front. Furthermore, selenium rectifiers have a large thermal mass compared to silicon rectifiers and therefore, are very reliable where overloads also occur frequently. These rectifiers, specially designed for high voltage applications have made the design and manufacture of high voltage supplies simpler and compact. Since these rectifiers do not require any heating operation as in vacuum tubes, there is saving in power and the warm period is avoided. These rectifiers can easily replace H.V. tubes with the added advantages of compactness, ruggedness, increased life and reliability. Typical applications of these rectifiers include X-ray machines, electronic dust collectors, photo-multiplier tubes, capacitor breakdown testers etc.

Germanium Diodes

Germanium diodes were the first of the solid-state semiconductor rectifiers. They have an extremely low forward voltage drop. Germanium diodes are relatively temperature sensitive, however. They can be easily destroyed by overheating during soldering, for instance. Also, they have some degree of back resistance, which varies with temperature.

Germanium diodes are used for special applications where the very low forward drop is needed, such as signal diodes used for detectors and ring modulators.

Silicon Rectifiers

Silicon diodes are the main choice today for virtually all rectifier applications. They are characterized by extremely high reverse resistance, forward drops of usually a volt or less and operations at high temperatures. Typical shape, symbol and its characteristics are shown in Figure 4.3. Note that the the scales for forward and reverse characteristics are different.

The threshold voltage, below which no current flows in the forward direction, is about 0.7 V and is a quantity that is determined by the nature of silicon and the use of a PN junction. Germanium diodes have a lower threshold of about 0.2 V. Once, conducting, the shape of V/I characteristic is exponential. That means a small increase in applied voltage results in a large current. In other words, a considerable increase in current is required to increase the voltage drop across the diode. Manufacturers usually specify the value of forward voltage at various values of forward current, or quote an 'average' value of slope resistance.

In the reverse direction, the silicon diode is non-conductive with negligible leakage current until the reverse breakdown voltage is reached, when the diode junction breaks down. The amount of current that can flow after breakdown is limited only by the total resistance in the circuit. The reverse breakdown is usually destructive, causing enough current to flow to overheat the diode and burn it out. Other components in the circuit may also get damaged in the process by this excessive current.

Because of the importance of avoiding reverse breakdown, diodes are usually rated for V_{RRM}. It is the maximum reverse voltage that can be applied repeatedly to a diode without breakdown. V_{RRM} is not the same as the r.m.s. voltage of the a.c. that is being rectified, and its value depends on the type of rectifying and filtering circuit that is used.

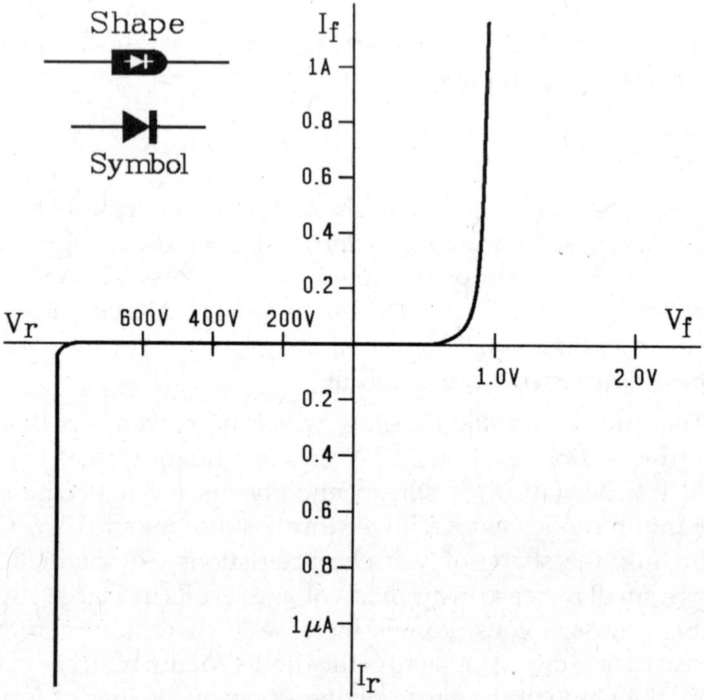

Figure 4.3: Silicon diode characteristic curve.

Because of the forward voltage drop V_f in the diode, power gets dissipated inside the diode when it conducts. The amount of power dissipated in watts is given by:

$$P = V_f \times I_f$$

where V_f is in volts (usually around 0.8 V) and I_f is in amperes.

For example, if the forward voltage drop across a diode at 5A is 0.8 V, and the average forward d.c. current is 5 A, the power dissipated in the diode is $0.8 \times 5 = 4.0$ W. The diode must be capable of dissipating this much amount of power; otherwise overheating will occur. Mounting a rectifier diode on a heat sink improves cooling and is often used to prevent overheating and consequent breakdown of the diode.

Small silicon rectifier diodes like 1N4006 can deliver up to 1A of average rectified d.c. For higher rectified currents stud mounted type rectifiers shown in Figure 4.4 are available which can be conveniently mounted on chassis or on heat sinks.

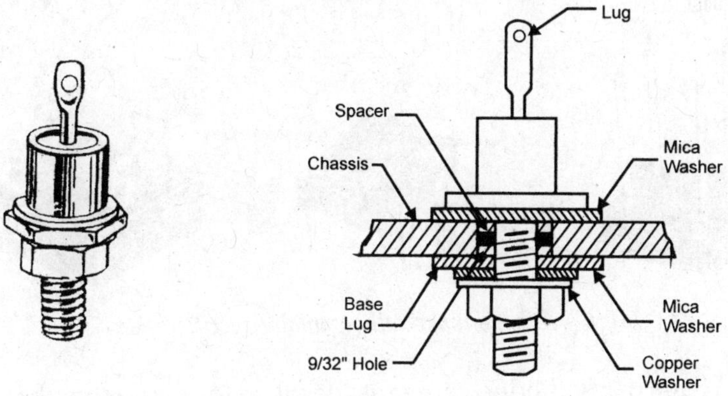

Figure 4.4: High power rectifier diodes and their mounting.

These diodes are available in normal polarity as well as in reverse polarity which makes it possible to mount a group of them on a common heat sink without any insulating washers when used in a bridge configuration. Ready made packages for full-wave and bridge rectification are also available commercially. Some of these are shown in Figure 4.5.

In the forward direction, the use of a filter capacitor following the rectifier places high demands on the current capacity of the diode. This capacitor is initially fully discharged. When the a.c. voltage is applied for the first time, a heavy current flows through the diode to charge the capacitor. This is a one time requirement and is denoted as I_{FSM}. Modern silicon diodes like 1N4006, which have an average forward current rating of 1 A, have a rating of I_{FSM} of 40 A. Once the capacitor is charged to its full voltage, only the lost charge has to be replenished between the next successive a.c. cycles. Diodes like 1N4006 have a rating of 10 A for this repetitive demand for every half-cycle of a.c. Manufacturers quote these values in the data sheets of their diodes.

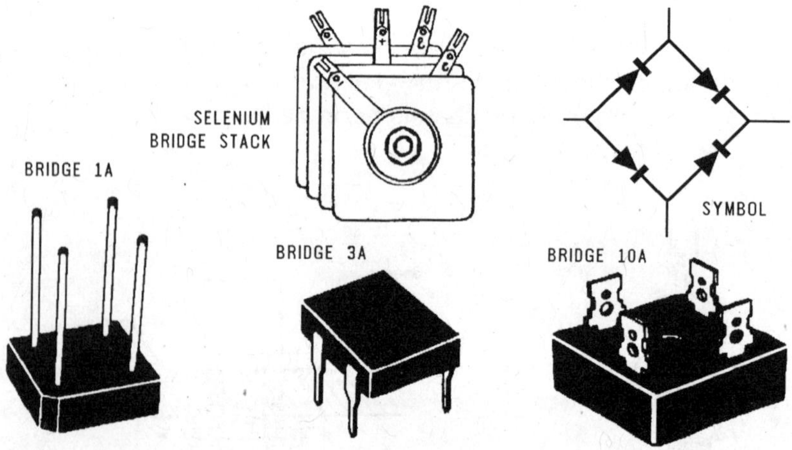

Figure 4.5: Commercially available rectifier packs.

While using diodes, care must be taken to ensure that the peak ratings are not exceeded. A diode may be used at a current which, on average, is within the rated value. Most silicon diodes can withstand very large current peaks provided that these are brief, but where very low transformer winding resistance values exist and the diode is connected directly to a low-resistance electrolytic capacitors (see Chapter 5), then it is possible to exceed the peak current rating of a diode, causing breakdown and failure. A current limiting choke or resistor is necessary then.

Schottky Diodes

The normal junction silicon diode is often replaced for high current rectification applications by Schottky diodes because it has lower forward voltage drop. The conventional silicon diode uses a junction between P- type and N- type silicon, but the Schottky diode uses N-type silicon in a junction with metallic aluminium. Unlike the PN junction, a Schottky junction is homopolar, meaning that only one polarity of carrier (electrons in this case) is used

For mains-frequency rectification, a more important effect is that the forward voltage is significantly lower even at high currents. Typical examples range include diodes with forward voltage levels of 0.475 V at 3 A, up to 0.78 V at 300 A for a matched pair of high-current diodes. These values should be compared with voltage levels of 1.4-1.6 V for silicon junction diodes of the same currents capabilities.

Rectifier Circuits

For single-phase rectifier circuits, silicon diodes can be used in half- wave, bi-phase half-wave (full-wave with center-tapped transformer), and full-wave bridge circuits as shown in Figures 4.6, 4.7 and 4.8 respectively. The rectified output is a set of half-waves of roughly sine shape as shown in these figures. This rectified output is unidirectional, it can be used as it is for applications like battery charging and driving some kinds of motors. This makes it important to know the value of the average output voltage in d.c. terms as this is the voltage read by a moving coil meter connected to measure the rectified output.

Half-Wave Rectifier

A half-wave rectifier (see Figure 4.6) requires one secondary winding and only one diode. If we neglect the voltage drop across the diode, the average level of output is $0.45 \times V_{ac}$, where V_{ac} is the r.m.s. value of the a.c. output from the transformer. For a 10 V r.m.s. a.c. input, for example, the equivalent d.c. is 4.5 V. Since this is unsmoothed, it is unsuitable for most electronics applications though it can be used for some battery charging applications.

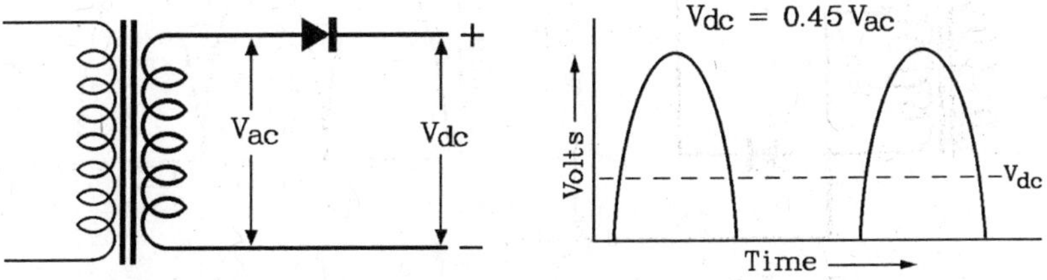

Figure 4.6: Single-phase half-wave rectifier and its waveform.

The important thing about this value is that it is the equivalent amount of d.c. that is still present when all smoothing fails, and is the value to which the voltage will fall when excessive current is taken from a half-wave smoothed supply. When the diode drop is considered, it makes the level smaller by about that amount. And if a further voltage drop is caused by the resistance of transformer windings, connections, fuses etc. it also has to be subtracted to get the available output voltage.

The average output current is about $0.45 \times I_{ac}$ where I_{ac} is the r.m.s. value of current supplied by the transformer.

Full-Wave Rectifier

The bi-phase half-wave type of rectifier shown in Figure 4.7 gives full-wave rectification. It requires a center-tapped secondary winding in the transformer. A transformer with two similar secondary windings can be connected in series to act as a tapped winding. Care should be taken not to connect such windings out of phase so that their voltages oppose each other — always check the open-circuit output voltage with an a.c. meter before connecting such a transformer to the rectifiers. It should indicate a voltage equal to the sum of the voltages in the individual windings.

The output voltage is equivalent to 0.9 of the r.m.s. a.c. voltage. The improvement is due to full-wave rectification. The average value of d.c. output current is 0.9 of the r.m.s. a.c. current value.

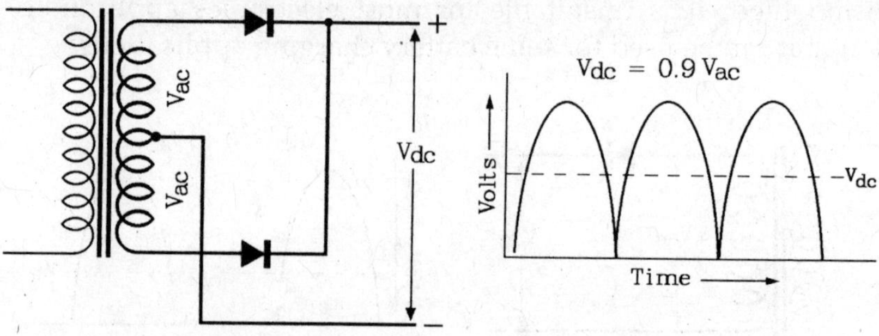

Figure 4.7: A full-wave rectifier using a center-tapped winding and its output waveform.

The Bridge Rectifier

The bridge rectifier circuit uses four diodes connected in a bridge form as shown in Figure 4.8. The advantage here is that it does not need a tapped winding on a transformer. Such circuits therefore, can also be used directly across the mains supply line to get d.c. directly from 230 V mains. Bridge rectifiers as a single pack are available commercially (see Figure 4.5) and their ratings range from 50 V to 1200 V peak reverse, and from under 1 A to 60 A forward current, covering the vast majority of electronics equipment requirements.

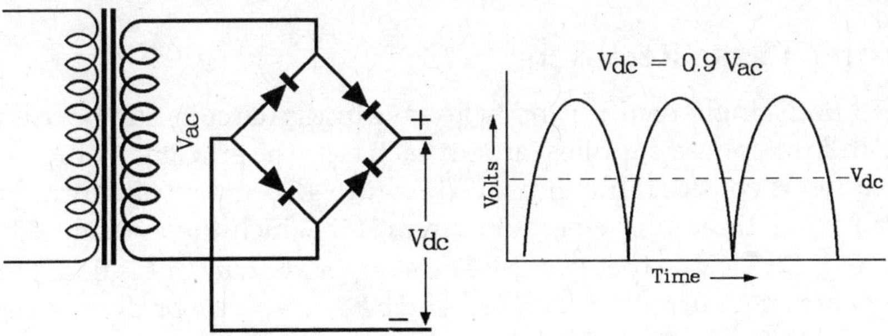

Figure 4.8: A full-wave rectifier using a bridge rectifier and its output waveform.

Many circuit diagrams depict bridge rectifiers in different ways. Two examples are shown in Figure 4.9. Figure 4.9A shows one method of drawing a bridge rectifier. Figure 4.9B shows a method of representing a bridge rectifier where the space is limited. Electrically, they all are same.

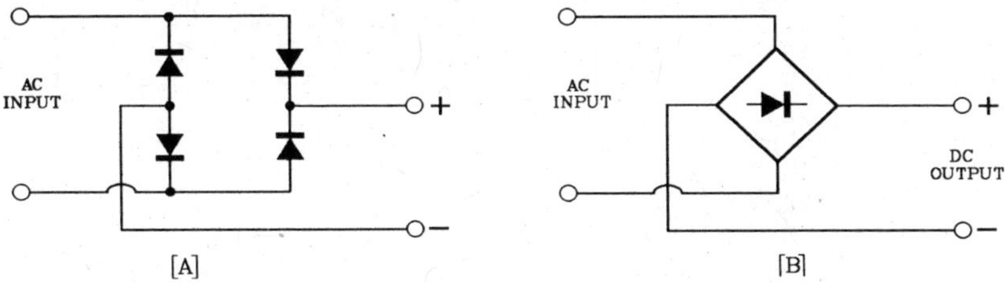

Figure 4.9: Alternative symbols for representing a bridge rectifier.

If you trace the path of current through the diodes on each half-cycle of the a.c. input voltage, you will find that the output current always flows through two diodes. The voltage drop due to the diodes in this case is twice as great as that for other two circuits.

As is evident from the output waveforms of the above type of rectifiers, the ripple content in the case of full-wave rectifiers is much less than that in half-wave circuits. The ripple content goes very much down in cases of rectifiers using poly-phase a.c. supplies. Circuits for three-phase rectification are shown in Figures 4.10 and 4.11. The output waveforms are also shown and it is easy to see that the output is almost like d.c. with very small ripple.

Three Phase Rectifiers

Both single-rectifier and full-wave bridge circuits can be used with three-phase supplies, and in each case the efficiency of rectification is considerably improved. Figure 4.10 shows one form of half-wave three-phase rectifier circuit for which the average d.c. level is 1.17 × V_{ac}, (per phase). The average d.c. current is equal to 1.22 of r.m.s. current in each phase. The three-phase bridge circuit shown in Figure 4.11 gives an average d.c. voltage of 1.35 × V_{ac} per phase, and d.c. current of 1.72. times a.c. r.m.s. in each phase.

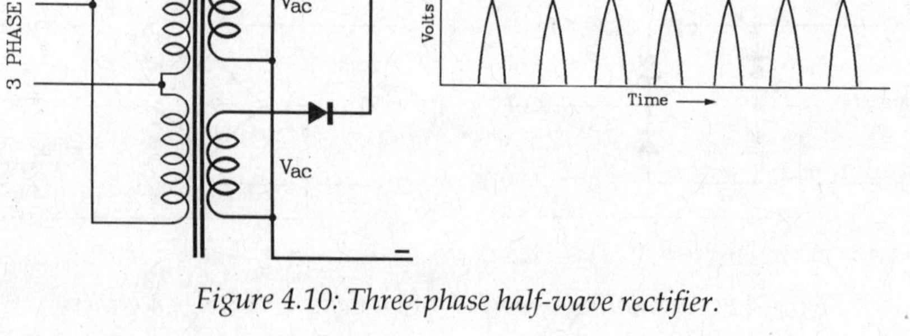

Figure 4.10: Three-phase half-wave rectifier.

The output from a three-phase bridge circuit is notably smooth, with ripple content of only 0.042 of peak d.c. output as opposed to the 0.48 of output for a single-phase bridge circuit when each is loaded by an inductor. Three-phase power supplies, however, are not common for electronics use. But this type of supplies are good for battery charging applications because the ratio between peak current and average current is very close to unity.

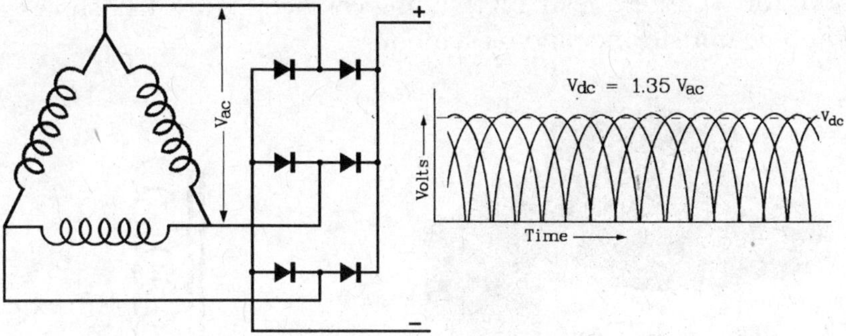

Figure 4.11: Three-phase full-wave rectifier and output waveforms.

Fast Recovery

When an a.c. voltage is applied to a rectifier diode, it should stop conduction the moment input polarity is reversed. Unfortunately, solid-state diodes do not cease to conduct instantaneously. The current continues to flow in the reverse biased diode for a short time. This is of no importance at power line frequencies of 50/60 Hz. But dc-to-dc converters and switch mode regulators operate at 25 kHz and higher frequencies where it becomes important.

In a bridge rectifier or a full-wave rectifier, conducting diodes in reverse direction can short circuit the converter for several microseconds. This puts excessive strain on the switching transistors and creates high current spikes leading to electromagnetic interference. Semiconductor manufacturers have recognized this problem and their *fast recovery diodes* can recover in about 50 ns or less. This information is included in their data sheets. Diodes used for flywheeling (in SMPS) should also be fast recovery diodes.

Power Control

The transformer we have used in the above circuits have a fixed voltage step-down ratio. If a variable or adjustable a.c. output is needed, it can be done by taps on the primary or secondary windings as shown in the symbols in Figure 3.2. However, the simplest method of getting variable output it is by using a *variac.* A variac is an auto-transformer in which the tap can be moved up and down continuously by a sliding contact. Figure 4.12 shows a variable transformer and its symbol.

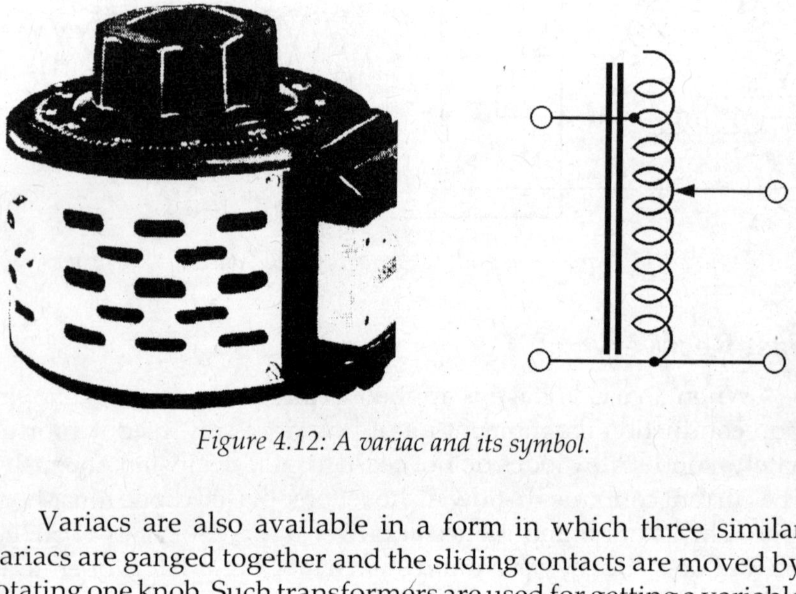

Figure 4.12: A variac and its symbol.

Variacs are also available in a form in which three similar variacs are ganged together and the sliding contacts are moved by rotating one knob. Such transformers are used for getting a variable output voltage in three-phase circuits.

Silicon Controlled Rectifiers (SCR's)

Another method of getting a variable output is by controlling the firing angle of Silicon Controlled Rectifiers (SCRs). However, Before we study the characteristics of SCRs, let us have another look at the working of the half-wave rectifier shown in Figure 4.6.

The diode can be considered as a switch that is automatically turned on as soon as the mains voltage exceeds 0.6 volts in one direction. The switch then remains on for the remaining half-cycle (180°) till the applied voltage drops down to nearly zero i.e. for of the input a.c. sinewave. Then, as soon as the voltage polarity is reversed in the next half-cycle, the diode stops conducting. In other words, the switch is turned off automatically. For the next half cycle the switch remains open. Figure 4.7 uses two such switches which turn on and off alternately, giving a conduction angle of 360°, that doubles the amount of power going into the load.

If we can delay the instant of turning on of the switch, i.e. of the diode. from 0° to anything up to 180°, we can control the power going into the load from full power to zero power. The silicon controlled rectifier is such a switch and is used to rectify as well as control power going into load.

Silicon controlled rectifiers are three-terminal devices. Like a diode, they have the *anode* and the *cathode*, but they also have a third terminal called *gate*. Figure 4.13 shows some of the types of SCR encapsulations and the symbol used to represent them.

The diode part of an SCR normally remains off irrespective of the polarity of voltage applied to its anode and cathode terminals, but it can be triggered to full conduction by applying a trigger signal between gate and cathode at a time when anode is positive with respective to cathode. Once switched into the conducting state, the SCR will remain conducting (i.e. it is latched in the on state) until the forward current is removed from the device.

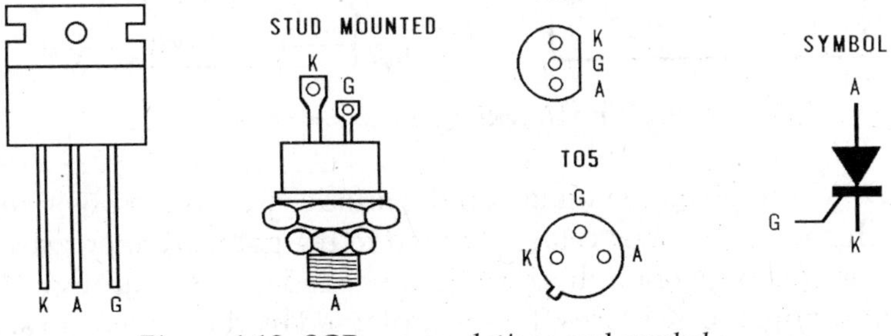

Figure 4.13: SCR encapsulations and symbol.

In d.c. applications where the anode voltage is always positive with respect to its cathode, this necessitates the interruption (or disconnection) of the supply before the device can be reset into its non-conducting state. Where the device is used with an a.c. supply, the SCR will automatically become reset (non-conducting) whenever the a.c. passes through its zero-crossing state and the voltage reverses in the next half-cycle. This characteristic of an SCR can be used for switching and controlling a.c. power.

SCRs can switch very rapidly from a non-conducting to a conducting state. In the off state, the SCR exhibits negligible leakage current whilst, in the on state the device exhibits very low resistance. This results in very little power loss within the SCR even when appreciable power levels are involved.

The SCR can be triggered into conduction by a low power pulse having correct polarity. The time after the zero-crossing point of the a.c. cycle at which this trigger pulse is applied controls the conducting angle of the SCR. In circuit shown in Figure 4.14, the gate resistor R_g controls the conduction angle of the SCR.

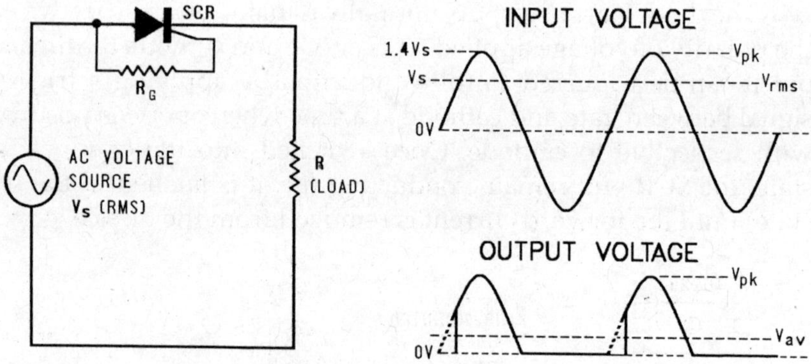

Figure 4.14: Controlling the firing point of SCR by a resistor.

As the positive voltage on the anode rises in the positive half of the a.c. cycle, the voltage applied to the gate also increases, and when it is high enough to pass the needed current through R_g for triggering, the SCR fires. This is explained by the waveform shown in the figure. The firing point can be delayed by increasing the value of Rg, but it cannot be made more than 90 degrees, i.e. beyond the positive peak on the positive half of the a.c. cycle.

Figure 4.15 shows how, with the addition of a capacitor, the firing point can be delayed further to almost 180 degrees. In other words, the power can be reduced from full power to almost zero power by delaying the firing point of the SCR.

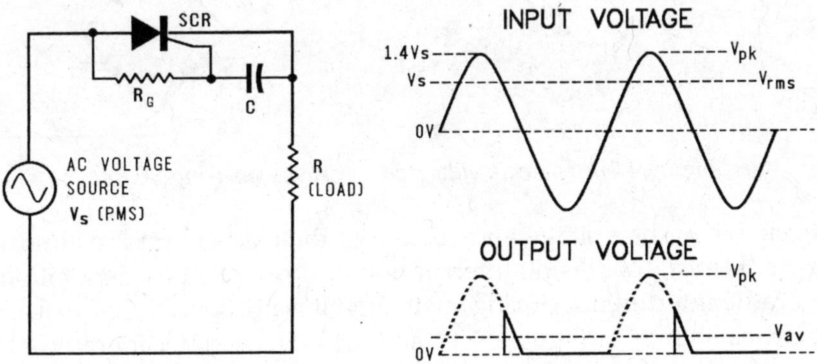

Figure 4.15: Controlling the delay of SCR firing by a resistor and capacitor.

As the positive voltage of the positive half cycle of a.c. increases, capacitor C starts charging through R_g. But the charging of a capacitor through a resistance takes time and the triggering of the SCR is therefore delayed. The delayed triggering is shown in the waveform in Figure 4.15. The larger the capacitor or R_g, the greater is the delay.

As the name implies, SCRs (silicon controlled rectifiers) are rectifiers first. They can be used as rectifiers in half-wave, in full-wave and in bridge circuits. Ready made packages are available for use in such applications. Some of them are shown in Figure 4.16. The diagrams show the internal connections of the device to help the user to connect

It should be noted that in Figures 4.6 and 4.7, the diode switch turns on at a point very near to the zero-crossing point of the a.c.

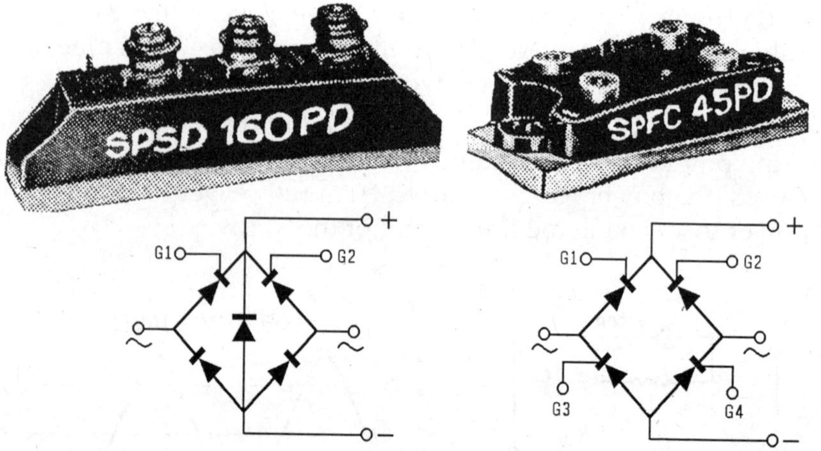

Figure 4.16: Typical ready-made packages for using SCRs.

cycle when the voltage and current in the load circuit are almost zero. Hence, there is no interruption of power and no generation of radio interference during switching. It is not so in the case of an SCR as shown in Figures 4.14 and 4.15. Hence, considerable r.f. interference is generated in these circuits which must be suppressed.

One method of reducing this interference is to add suppressing filters by putting inductors in series with the line and connecting a bypass filter capacitor as shown in Figure 4.17.

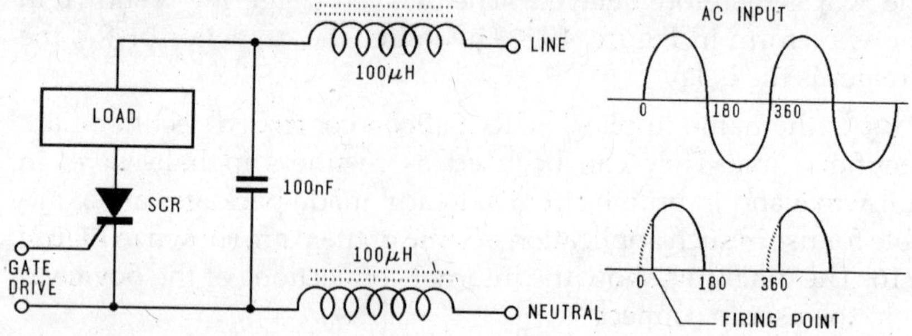

Figure 4.17: Reducing r.f. interference generated by SCR switching.

Triacs

Triacs are refinements of the SCR for a.c. applications. A triac can be triggered into conduction in the positive half of the a.c. cycle, as well as in the negative-half of the a.c. cycle of the applied a.c. voltage by an appropriate triggering signal on its gate. In other words, a triac is a controlled rectifier for a.c. applications. It can be used to control power in a.c. circuits by controlling the delay of the trigger pulses.

Like an SCR, a triac also has three terminals, but these are known as *main terminal one (MT1)*, *main terminal two (MT2)* and *gate (G)*. Figure 4.18 shows triac encapsulations and the symbol for representing a triac.

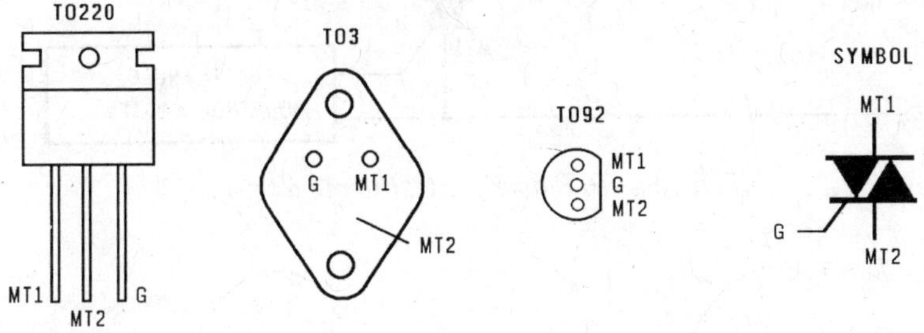

Figure 4.18: Triac encapsulations and symbol.

Triacs can be triggered by both positive and negative voltages present at the gate. Triacs thus provide full-wave control and offer superior performance in a.c. power control applications.

In order to simplify the design of triggering circuits, triacs are often used in conjunction with *diacs* (a bi-directional triggering device). The symbol for a diac is two diodes connected back-to-back. A typical diac conducts heavily when the applied voltage exceeds approximately ± 32V. Once in the conducting state, the resistance of the diac falls to a very low value and thus a large value of current will flow and trigger the triac.

Figure 4.19 shows a circuit for controlling power in a load by using a triac and varying its phase angle of firing. The firing can be controlled manually to get a variable output, or it can be controlled automatically by some electronic circuit to regulate the power in load automatically. Special integrated circuits have been developed for varying the phase angle (see Chapter 19).

Figure 4.20 shows the waveform to explain the manner in which power is controlled by a triac. As in the case of SCRs, this circuit also generates considerable r.f. interference and a suppressor is needed if the interference is considered undesirable.

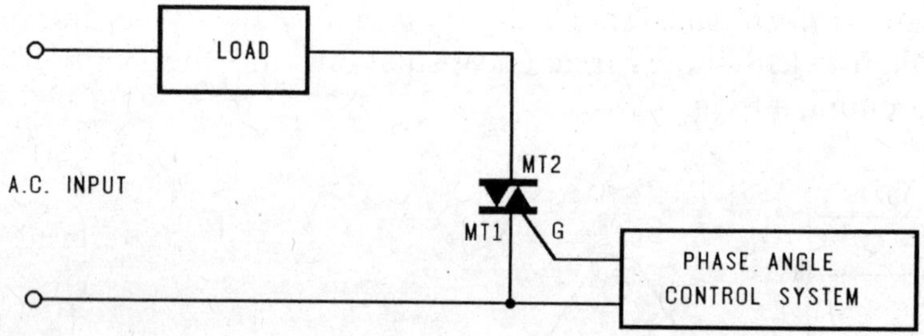

Figure 4.19: Power control using a triac.

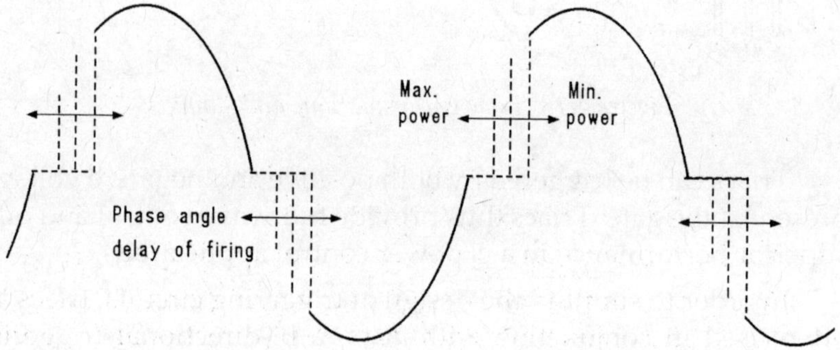

Figure 4.20: Waveforms to explain control of power by a triac.

See also the book "Practical SCR/Triac Projects" by the author, and published by BPB Publications.

5

Filters

Rectifiers described in the previous chapter allow the current to flow in one direction only. Hence, after rectification, a.c. becomes unidirectional voltage or current. Waveforms after rectification were shown in Figures 4.6, 4.7, 4.8, 4.10 and 4.11. Mathematical analysis shows that these waveforms contain a certain amount of d.c. mixed with some ac. This a.c. content is called *ripple*. The ripple has to be removed before the d.c. available can be used.

Ripple content is maximum in half-wave rectified waveform shown in Figure 4.6, and is least in three-phase full-wave rectifier. shown in Figure 4.11. If the frequency of supply is f, the ripple frequency is $1f$ for half-wave rectification, $2f$ for full-wave rectification, $3f$ for three-phase half-wave rectification, and $6f$ for three-phase full-wave rectification.

A waveform with ripple is not acceptable for most electronics applications. This ripple content must therefore be minimized to an acceptable level before the d.c. available can be used for electronic circuits; though, for some thyristor circuits, a raw rectified supply is necessary to allow the thyristor to attain a non-conducting state at zero crossings. The process of removing or reducing th ripple and making the d.c. output smooth like a battery voltag (see Figure 1.1) is called *filtering*. Filtering is done by using resistors, inductors and capacitors. We shall therefore study the characteristics of these components before using them for filtering.

Types of Capacitors

In general terms, a capacitor is an arrangement of conductors that are insulated from each other so that electric charge can be stored. The simplest example of a capacitor is two metal plates arranged parallel to each other and separated by air as shown in Figure 5.1. Figure 5.1 shows two metal plates separated by some distance and connected to a battery through a switch

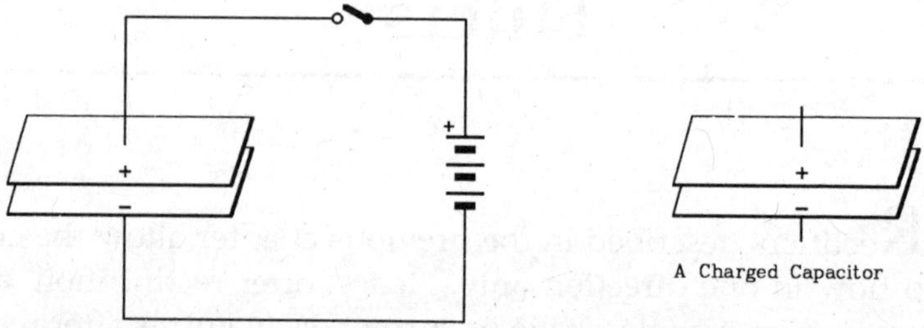

Figure 5.1: A parallel plate capacitor.

At the instant the switch is closed, electrons will be attracted from the upper plate to the positive pole of the battery, and the same number of electrons will be repelled into the lower plate from the negative battery terminal. This flow of electrons is the charging current. Enough electrons move into one plate and out of another plate to make the emf between them the same as the emf of the battery. When that happens, the charging of the capacitor stops. The plates hold the charge even after the switch is opened.

Thus, a capacitor possesses the property of storing electricity in the electric field between the two plates. During the time the electrons are moving — a capacitor is being charged or discharged – a current is flowing in the circuit even though the circuit is "broken" by the gap in between the capacitor plates. However, the current flow is only during the charge and discharge, and this time is generally very short. There can be no continuous current flow or direct current "through" a capacitor, but an alternating current can easily pass through easily if the frequency is high enough because of rapid charge and discharge of the capacitor plates.

The charge or the quantity of electricity that can be placed or stored on a capacitor is proportional to the applied voltage and to the capacitance of the capacitor. The larger the plate area and the smaller the spacing between the plates, the greater is the capacitance. The capacitance of a capacitor also depends upon the kind of insulating material between the plates. It is smallest with air insulation between the plates, but substitution of other materials such as mica, polyethylene, polystyrene etc. for air may increase the capacitance many times. The ratio of the capacitance with some material to the capacitance of the same capacitor with air insulation, is called the *dielectric constant* of that particular insulating material. The material itself is called a *dielectric*.

Units

The charge Q stored in a capacitor and the voltage V across the capacitor are related. The relationship is:

$$Q = C \times V \quad \text{------------------------} \quad (5\text{-}1)$$

or $$C = Q/V$$

where C is the capacitance of the capacitor.

The unit for measuring electric charge is *coulomb*, the unit of measuring voltage is *volt*, and the unit of measuring capacitance is a *farad*, which is represented by F. One coulomb divided by one volt gives 1 farad. However, a farad is too large a unit for most of our practical purposes. Smaller units used in practice are submultiples of a farad. They are *millifarads*, *microfarads*, *nanofarads*, and *picofarads* also called *micro-micro farads*. Their relationships with basic unit farad are:

millifarad (mF)	=	10^{-3} farad
microfarad (μF)	=	10^{-6} farad
nanofarad (nF)	=	10^{-9} farad
picofarad (pF)	=	10^{-12} farad
micromicrofarad (μμF)	=	10^{-12} farad

Capacitor Connections

Capacitors can be connected in series or in parallel, much in the same way as we connect resistors. When a number of capacitors are connected in parallel as shown in Figure 5.2, the total or effective capacitance of the group is equal to the sum of the individual capacitances. The equation for calculating the total capacitance C obtained by capacitances C1, C2, C3 etc. connected in parallel is:

$$C = C1 + C2 + C3 + C4 \ldots \qquad (5\text{-}2)$$

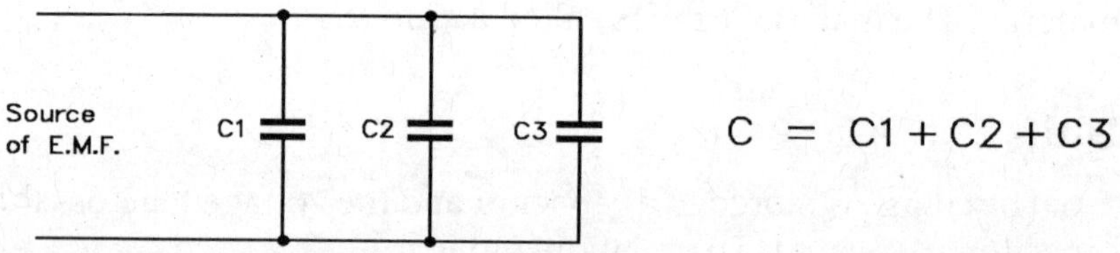

Figure 5.2: Capacitors connected in parallel.

This type of connection is used to get higher capacitances. The voltage rating of each capacitor should be sufficient to withstand the voltage that will be applied to them in practical circuits.

If two or more capacitors are connected in series as shown in Figure 5.3, the total capacitance is less than that of the smallest capacitor in the group. The value of the total capacitance C can be calculated by the equation:

$$C = \frac{1}{\frac{1}{C1} + \frac{1}{C2} + \frac{1}{C3}} \qquad (5\text{-}3)$$

Capacitors are connected in series to increase the total voltage withstanding capability. The total voltage applied (E) gets divided across the capacitors as shown in Figure 5.3. The voltage across any capacitor is in inverse proportion of the capacitance, i.e. the smallest capacitor has the highest voltage share across it, and the largest capacitor has the least voltage across it.

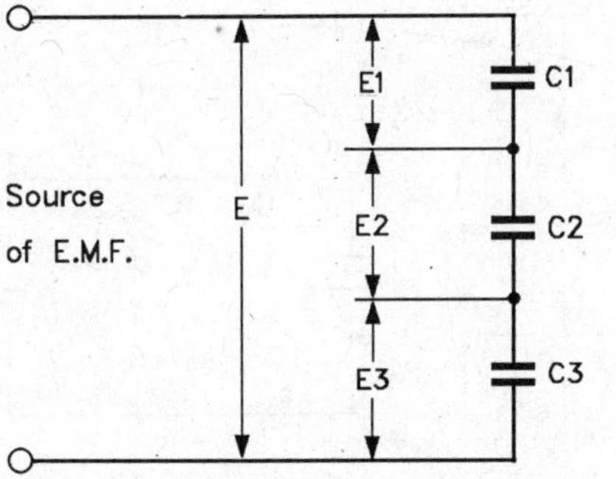

$$C = \frac{1}{\frac{1}{C1} + \frac{1}{C2} + \frac{1}{C3}}$$

$$E = E1 + E2 + E3$$

Figure 5.3: Capacitors connected in series.

The leakage resistance of capacitors can also affect the voltage distribution. The unequal distribution can result in application of excessive voltage across one of the capacitors and break it down. Once that happens, the remaining ones also break down one by one. In many circuits, therefore, high value resistors are connected in parallel with all capacitors to ensure that the voltage is distributed equally. These resistors should have enough wattage rating to dissipate heat generated by the flow of current through them.

Charging of a Capacitor

Figure 5.4 shows a capacitor which can be connected to a battery of E volts through a switch Sw. The capacitor is completely discharged at the beginning. As soon the switch Sw is put ON, a current V/R starts flowing through R and starts charging the capacitor. As the capacitor gets charged, the build up of voltage across its terminals opposes the charging voltage E. The current therefore goes on becoming lesser and lesser. Theoretically, the capacitor is fully charged to a voltage E only after an infinite long period. The voltage V across the capacitor after a time interval t is given by the equation:

$$V = E \times (1 - e^{-\frac{t}{RC}}) \quad \text{-------------------- (5-4)}$$

where e is the natural number (2.718).

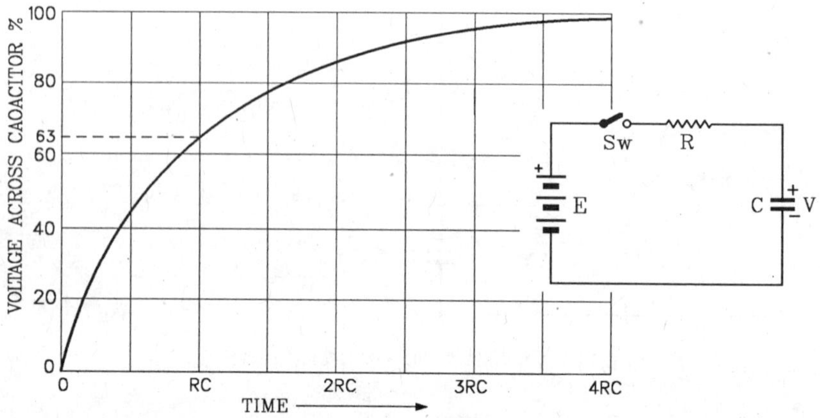

Figure 5.4: Charging a capacitor through a resistor.

At the beginning, t is equal to zero and the value of $1\text{-}e^{0}$ is zero. If t equals RC, the value of $1\text{-}e^{-1}$ is 0.63. Hence, when a capacitor charges through a resistance, it reaches a voltage equal to 63% of the charging voltage after charging for a time period equal to RC as shown in Figure 5.4. The product of R and C is called *time constant*. If R is in megohms and C is in microfarads, the product RC will be in seconds.

Discharging of a Capacitor

A capacitor charged to a voltage E will retain that charge for infinite time, if there are no losses. However, in practice, all capacitor have some leakage resistance, and the charge slowly leaks through it. In general, this leakage is too small and can be neglected in most of the calculations.

If a resistor R is connected across a charged capacitor, it starts discharging through it. Initially, the discharge current is E/R; but as the capacitor discharges, the voltage across the capacitor falls, and so does the discharging current. Theoretically, a capacitor discharges fully after an infinite time as shown in Figure 5.5.

The equation of voltage V across a capacitor after discharging for a time period t is given by the equation:

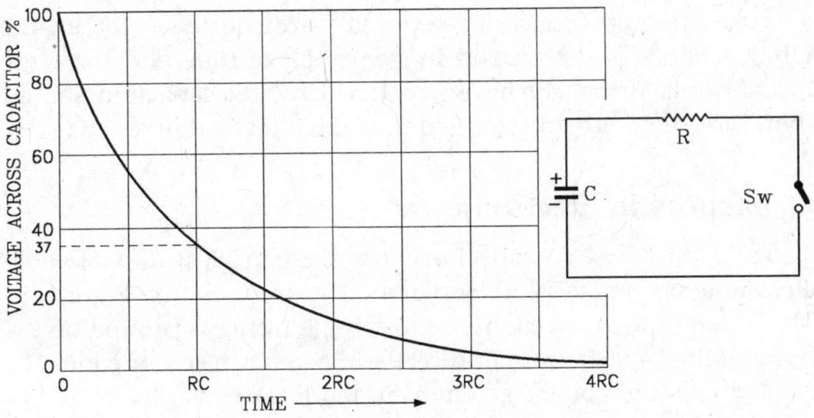

Figure 5.5: Discharging a capacitor through a resistor.

$$V = E \times e^{\frac{-t}{RC}} \quad \text{-------------------------------} \quad (5\text{-}5)$$

As before, if the value of t is equal to RC (time constant), then t/RC equals 1 and the value of e^{-1} is 0.37. Hence, after lapse of a time period equal to RC the voltage across a charged capacitor falls to 37% of its initial value.

Energy Stored in a Capacitor

The amount of charge-energy stored in a capacitor is given by the equation:

$$\text{stored energy} = 0.5CV^2 \quad \text{-------------} \quad (5\text{-}6)$$

where, if the capacitance is in farads and voltage is in volts, the stored energy is in joules.

For example, a 250 μF capacitor charged to 400V will carry a stored energy of $0.5 \times 250 \times 10^{-6} \times 400 \times 400 = 20$ joules, and this is the amount of power of 20W acting for 1 second.

Note: A capacitor can have enough charge to be harmful even after it has been disconnected from the supply. It should be noted that larger capacitors can store a lot of charge at a voltage level that can cause severe shock or burning, or may even be lethal.

If a charged capacitor's terminals are shorted together, the whole energy is discharged in a very short time and the large discharging current can be destructive. Charged capacitors should therefore, always be discharged through a low value resistor.

Capacitors in Electronics

Several types of capacitors are used in electronics and in electronic power supplies. Some of these are shown in Figure 5.6. The value of the capacitance and voltage rating is printed on the body of the capacitor either directly in text, or in colour code. The capacitor colour code is given in Appendix A.

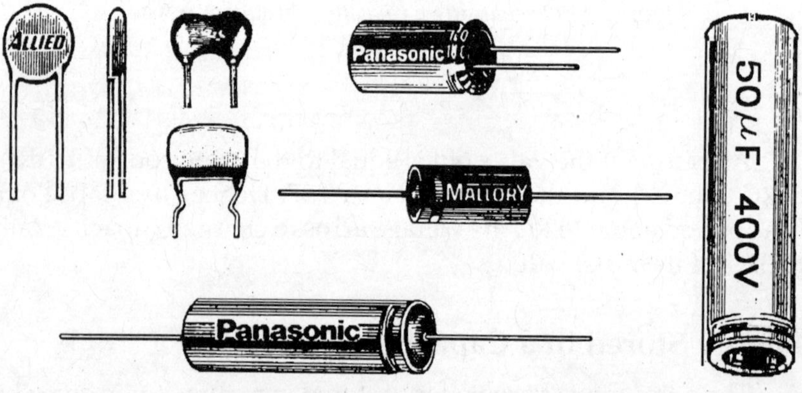

Figure 5.6: Capacitors used in electronic power supplies.

Variable value capacitors are usually made with air dielectric and are mostly used as tuning capacitors in radio receivers. Fixed value capacitors are made with thin foils separated by a dielectric. They are available in many sizes, shapes and voltage ratings.

Electrolytic Capacitors

Large capacitors are used in power supply circuits for filtering ripples. We know that the capacity of a capacitor increases as the distance between the plates is decreased. Electrolytic capacitors have a large capacitance because this separation is very small.

The principle of working of electrolytic capacitors is that several metals, notably aluminium and tantalum, can have very thin films of their respective oxides formed on the surface when a voltage is applied in the correct polarity (metal positive) between the metal and a slightly acidic or alkaline liquid. These very thin films then insulate the metal from the conducting liquid, the electrolyte, forming a capacitor, an *electrolytic* capacitor. Because the insulating films are very thin and of very large surface area, the capacitance value can be very large for a component that is physically small.

The most common type of electrolytic capacitor makes use of aluminium foil, which can be corrugated to increase the effective area, rolled and enclosed in an aluminium can which is filled with a slightly alkaline solution of ammonium perborate in jelly form. The capacitor is *formed* by applying a slowly-rising d.c. voltage to the capacitor, with the foil positive and the case negative until the voltage reaches its rated level and the d.c. current falls to a minimum, indicating that the insulation is as good as it is ever likely to be. This forming of electrolytic capacitors is done by the manufacturer of electrolytic capacitors.

When an electrolytic capacitor is used in a circuit, it must have a d.c. (polarizing voltage applied in the same polarity so as to maintain the insulating film. If the capacitor is used with the voltage reversed, the film will be dissolved, removing any insulation and allowing large currents to pass through the liquid, which will vaporize, destroying the can. The electrolyte is usually in jelly form, but the devastation that can be caused by an exploding electrolyte (not to mention noise) ensures that no one who has achieved this is willing to try again.

Tantalum Capacitors

Tantalum electrolytics invariably use solid electrolytes along with tantalum metal, and have much lower leakage than the aluminium types. This makes them eminently suitable for purposes such as signal coupling, filters, timing circuits, decoupling as well as for use in power supplies. The usual forms of these electrolytics are miniature beads or tubular axial types. The voltage range is 6.3 to 35V, with values of 1.0 μF to 100 μF. The temperature range is -55°C to +85°C. Tantalum electrolytics can be used without any d.c. bias and can also accept a small reverse voltage, typically less than 1.0 V. A minimum leakage current of 1μA and power factors of 0.02 to 0.2 can be expected.

The use of tantalum as the metal of an electrolyte allows for a very different form of construction, in which the oxide film is stable and is able to withstand moderate reversals of voltage. Tantalum capacitors can be used without a steady polarizing voltage, can be run with the electrolyte virtually dry, and have generally better characteristics than the traditional aluminium type of electrolytic capacitors. They are, however, more expensive.

Leakage in Electrolytics

Electrolytic capacitors suffer from a large amount of leakage. For that reason, leakage current at rated voltage is quoted by the manufacturer. You can approximately calculate the leakage current in microamperes as one percent of the product of capacitance and voltage, with a minimum of 3 μA. For example, a 500 μF capacitor of 10 V can have a leakage current of $0.01 \times 500 \times 10 = 50\mu A$.

When used for smoothing a d.c. supply, the leakage current is not of much consideration because the power supply can supply this much of current without any problem. However, if the power supply is not used for a long time, the film may get dissolved resulting in a large leakage current; but the film usually builds up rapidly and the leakage current falls to its previous value.

Using Electrolytic Capacitors

Electrolytic capacitors are used predominantly as reservoir and smoothing capacitors for mains-frequency power supplies, so that their most important parameters, other than capacitance and voltage rating, concern the amount of ripple current that they can pass. For each capacitor the manufacturer will quote a maximum ripple current (typically at 100 or 120 Hz), effective series resistance

(ESR) and impedance. The ESR is typically 50 milliohms, for low-frequency currents, and this value may set a limit to the ripple current that can pass; also to the effectiveness of the capacitor for smoothing. The other parameter is the effective impedance in milliohms, caused by self-inductance, measured at 10 kHz and at 20°C, which is used to measure how effectively the capacitor will bypass currents at higher frequencies. If an electrolytic capacitor is used in a decoupling circuit which is likely to handle a large frequency range, other capacitor types should be used in parallel to deal with frequencies higher than 10 kHz,

Inductors

Basically, inductors are coils that have large inductance. They oppose the change of current flowing through them, and are used in power supplies for filtering out ripples. Their principle of working is explained below.

We know that the flow of current through a conductor is accompanied by magnetic effects. The transfer of electrical energy to the magnetic field represents work done by the source of emf. Power is required for doing work, and since power is equal to current multiplied by voltage, there must be a voltage drop in the circuit during the time in which energy is being stored in the field. This voltage "drop" (which has nothing to do with the voltage drop in any resistance in the circuit) is the result of an opposing voltage "induced" in the circuit while the field is building up to its final value. When the field becomes constant the induced emf or *back emf* disappears, since no further energy is being stored.

Since the induced emf opposes the emf of the source, it tends to prevent the current from rising rapidly when the circuit is closed. The amplitude of the induced emf is proportional to the rate at which the current is changing and to a constant associated with the circuit itself, called the inductance of the circuit.

Inductance depends on the physical characteristics of the conductor. If the conductor is formed into a coil, for example, the inductance is increased. A coil of many turns will have more inductance than one of a few turns. Also, if a coil is placed on an iron core its inductance will be greater than it was without the magnetic core.

The polarity of induced emf is always such as to oppose any change in the current in the circuit. This means that when the current in the circuit is increasing, work is being done against the induced emf by storing energy in the magnetic field. If the current in the circuit tends to decrease, the stored energy of the field returns to the circuit, and thus adds to the energy being supplied by the source of emf. This tends to keep the current flowing even though the applied emf may be decreasing or be removed entirely.

The unit of measuring inductance is a *henry*. Inductance of several henrys are used in power supply circuits and to obtain such values of inductance it is necessary to use coils of many turns wound on iron cores. They are often referred to as *inductors*. When used in a filter circuit, an inductor is often referred to as a *choke*. Chokes in power supply circuits working at 50 Hz appear like small transformers shown in Figure 3.2. Inverters and switching mode power supplies work at higher frequencies and need smaller inductances. Mostly, in those cases, toroidal windings on ferrite cores are used.

Inductors and Resistors

A comparable situation exists between the charge and discharge of capacitors through resistors and when a resistor and an inductor are in series with a source of emf. Figure 5.7 shows an inductor L connected in series with a resistance R, a switch Sw and a battery of emf E.

First consider L to have no resistance and also assume that R is zero. Then closing switch Sw would tend to send a current through the coil. However, the instantaneous transition from no current to a finite value, however small, represents a very rapid change of current, and a back emf is developed by the self inductance of L, that is practically equal and opposite to the applied emf. The result is that initial current is very small.

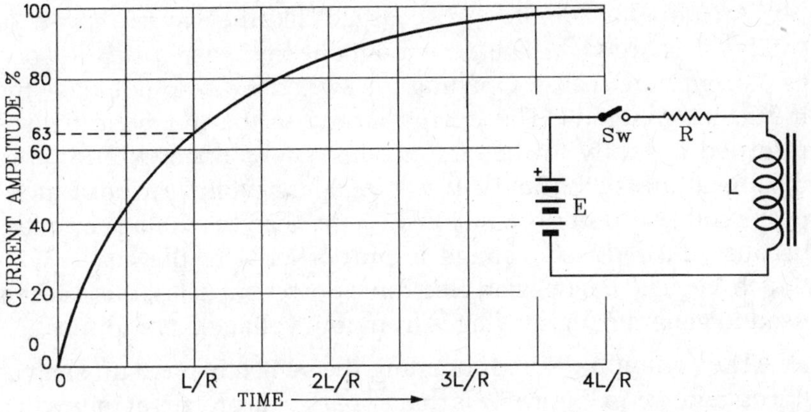

Figure 5.7: Growth of current in an LR circuit.

The back emf depends upon change in current and would cease to offer opposition if the current did not continue to increase. With no resistance in the circuit (which would lead to infinitely large current, by Ohm's law) the current would increase forever, always growing just fast enough to keep the emf of self-induction equal to the applied emf.

When resistance is in series, Ohm's law sets a limit to the final value that the current can reach. The back emf generated in L has only to equal the difference between E and the drop across R, because that difference is the voltage actually applied to L. This difference becomes smaller as the current approaches the final Ohm's law value. Theoretically, the back emf never quite disappears and so the current never quite reaches the Ohm's law value as shown in Figure 5.7.

The time constant of an inductive circuit is the time in seconds required for the current to reach 63 percent of its final value. The formula is:

$$\text{time constant} = \frac{L}{R} \qquad \text{(5-7)}$$

where L is the inductance of coil in henrys and R is the resistance in ohms.

An inductor cannot be "discharged" like the way as a capacitor can be discharged, because the magnetic field disappears as soon as current flow ceases. Opening the switch Sw does not leave the inductor "charged". The energy stored in the magnetic field is returned instantly to the circuit when Sw is opened. The rapid disappearance of the field causes a very large voltage to be induced in the coil — ordinarily *many times larger* than the voltage applied, because the induced voltage is proportional to the speed with which the field changes. In electronic power supplies, this effect is used to generate high voltages from low voltage d.c. supplies.

The common result of opening the switch in a circuit such as the one shown in Figure 5.7 is that a spark or arc forms at the switch contacts at the instant of opening. If the inductance is large and the current in the circuit is high, a great deal of energy is released in a very short period of time. It is not at all unusual for switch contacts to burn out or melt under such circumstances. The spark or arc at the opened switch can be reduced or suppressed by connecting a suitable capacitor and resistor in series across the contacts.

In Figure 5.7, if the battery is replaced by an a.c. voltage source, the inductor offers a high impedance to the flow of a.c. The value of this *inductive reactance* X_L is given by the expression:

$$X_L = 2\pi f L \quad \text{(5-8)}$$

where

X_L is the inductive reactance in ohms,
f is the frequency of the applied a.c. voltage
and L is inductance in henries.

Filtering

It was stated earlier that the rectified a.c. consists of d.c. mixed with a.c. This a.c. component is called *ripple*. In order to make this kind of voltage suitable for operating electronic circuits, the a.c. component (the ripple) should be removed. The process of removing a.c. component is called *filtering*, and the group of components interconnected to do this is collectively called a *filter*.

The technique of minimizing the a.c. component is by inserting an inductor in series to provide an easy path to d.c., but to offer a high impedance to a.c.; and a capacitor in shunt to provide a low impedance path to a.c. without affecting d.c.

Effect of Adding a Capacitor

If a d.c. meter is connected at the rectified output in Figure 4.6, it will read the average value which is 0.45 times the r.m.s. value of the applied voltage. For example, if Vac is 100 V r.m.s., the meter will read about 45 V. But, if a capacitor is connected at the output as shown in Figure 5.8, the situation becomes very different.

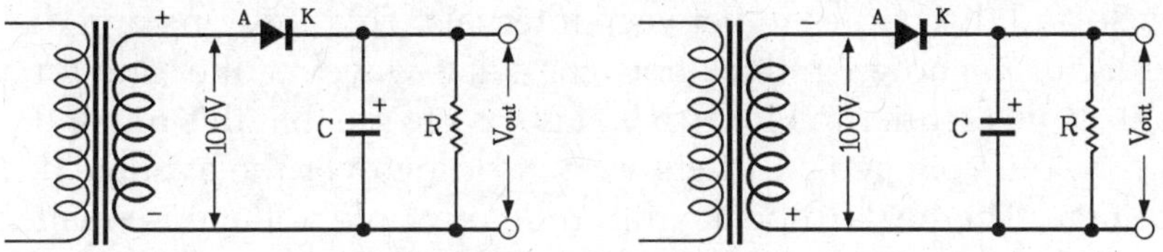

Figure 5.8: Effect of connecting a capacitor in a rectifier circuit.

Let us presume that the load resistor R is not connected. The peak value (V_{pk}) of 100 V r.m.s of a.c. voltage is 141 V, and its peak-to-peak value ($V_{pk\text{-}pk}$) is 282 V. On the positive peak of the a.c. voltage, the upper end of the secondary winding of the transformer goes positive to as high as 141 V as shown on the left side of Figure 5.8, and this charges the capacitor to its peak value of 141 volts. A d.c. meter connected to the output terminals will therefore, read V_{out} as 141 V.

On the next half cycle, (the negative-going half-cycle) the upper end of the secondary goes negative to -141 V as shown on the right in Figure 5.8. The diode is not conducting. The anode is now at -141 V and the cathode is already at +141 V because the capacitor is already charged. This builds up 282 volts across the diode and the diode has to withstand the stress of 282 V repeatedly. If it can't, it will breakdown. Rectifier diode manufacturers therefore always quote the value of the repetitive peak reverse voltage

(V_{PRV} or V_{RM}) for their diodes in their data sheets. For example, the peak reverse voltage for 1N4001 diode is 50 V, while that for 1N4007 is 1000 V. V_{RM} is an important parameter, and should be taken into account for choosing the right type of diode for a particular application. This characteristics makes 1N4001 diode suitable for use in low voltage circuits such as battery eliminators, while 1N4007 can be used to rectify the 230V mains voltage as well. Sometimes a number of diodes are connected in series to increasing their V_{RM}. Resistors in parallel equalize the voltage distribution.

When a load R is connected at the output, the charged capacitor starts discharging through R when the a.c. voltage starts going down from its peak value, and tries to maintain the voltage. The capacitor recharges to the peak voltage on the next positive half-cycle and discharges in between intervals. This charging and discharging is shown in thick lines for a half-wave rectifier and for a full-wave rectifier in Figure 5.9. It is easy to see that this forms the a.c. ripple voltage (V_R in Figure 5.9) riding over the average d.c. voltage. The peak-to-peak amplitude of ripple voltage is smaller for full-wave rectification than it is for half-wave rectification. In other words, half-wave rectifiers need more filtering.

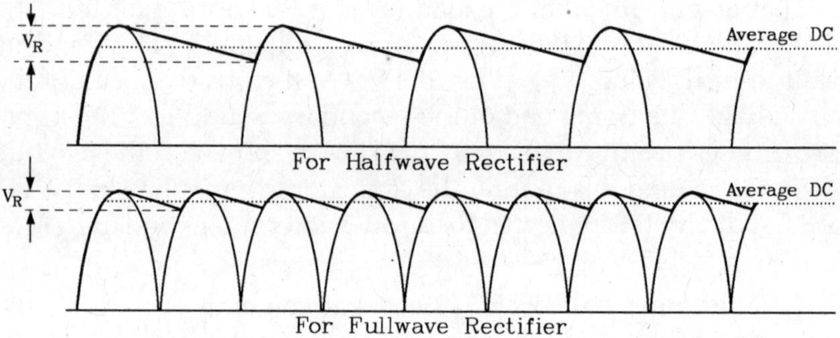

Figure 5.9: Ripple in rectified a.c. when a capacitor is added.

Also note that the frequency of the ripple in the case of full-wave rectifier is two times that of the half-wave rectifier. This means that a smaller filter capacitor can be used when full-wave rectification is being used.

Types of Filters

Power supply filters fall into two categories, capacitor-input filters and choke input filters. If a rectifier circuit directly feeds its output into a capacitor as shown in Figure 5.8 is called a capacitor-input filter. The capacitor acts as a reservoir to store charge and supply load current when the a.c. voltage goes low. The larger the capacitor, the better is the filtering. In choke-input filters, the rectifier circuit feeds the reservoir capacitor through a choke.

Choke Input Filters

If the reservoir capacitor is connected to the rectifier through a choke as shown in Figure 5.10, it forms a choke-input filter. A choke coil, because of its characteristics, does not allow the current to grow fast, and the reservoir capacitor therefore, does not get charged to the peak value of the rectified voltage. For this reason, the output voltage is lower, but the voltage regulation is better. The output voltage does not fall much when the load is increased. This characteristic was desirable to feed power to class-B type of audio amplifiers where the load current varies with the level of audio output of music or speech.

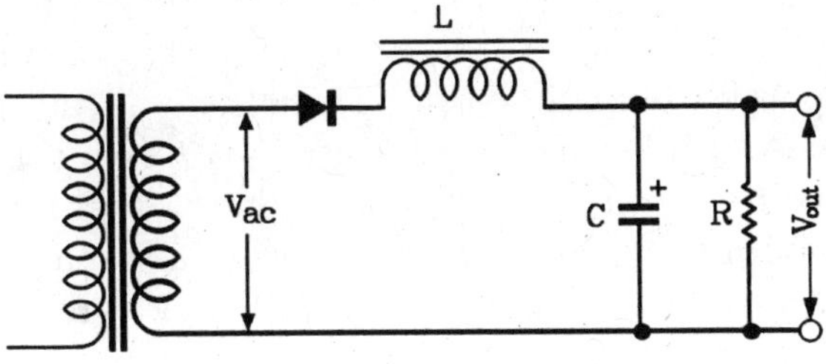

Figure 5.10:Choke-input filter

Choke input filters were used with vacuum tube rectifiers because vacuum tube diodes were not able to deliver high peak currents. In a choke input filter, higher load current and currents for charging the reservoir capacitor usually can be drawn without exceeding the peak current ratings of the rectifier.

A choke input filter will tend to act as a capacitor input filter unless the input choke has at least a certain value of minimum inductance called the critical value. The critical value is given by:

$$L_{crit}\ \text{(henrys)} = \frac{E\ \text{(volts)}}{I\ \text{(mA)}} \qquad \text{(5-9)}$$

where E is the output voltage of the supply, and I is the output current being drawn through the filter.

If the choke has at least the critical value, the output voltage will be limited to the average value of the rectified wave at the input to the choke (see Figures 4.6 and 4.7) when the current drawn from the supply is small. This is in contrast to the capacitor-input filter where the output voltage tends to soar toward the peak value of the rectified wave at light loads.

From the formula shown above, it is obvious that if no current is drawn from the supply, the critical inductance will be infinite. So that a practical value of inductance may be used, some current must be drawn from the supply at all times the supply is in use. From the formula we find that this minimum current is:

$$I\ \text{(mA)} = \frac{E\ \text{(volts)}}{L_{crit}}$$

A minimum permanent load is provided by connecting a load resistor in parallel with the supply. This resistor is called the *bleeder* or *bleeder resistance*. In a majority of cases it will be most convenient to adjust the bleeder resistance so that the bleeder will draw the required minimum current. Form the formula, it may be seen that the value of critical inductance becomes smaller as the load current increases.

The bleeder resistor also serves to discharge the high capacity reservoir capacitor(s) when the power supply is turned off. This prevents chances of electric shock because filter capacitors can keep the charge stored for quite some time even after switching off.

In a power supply using a choke input filter and properly designed choke and bleeder resistor, the no-load voltage across the filter capacitors will be about nine-tenths of the a.c. r.m.s. voltage.

Nevertheless, it is advisable to use capacitors rated for the peak transformer voltage. This large safety factor is suggested because the voltage across the capacitors can reach this peak value if the bleeder should burn out and there is no load on the supply.

Capacitor Input Filters

Capacitor input filters described earlier (see Figure 5.8) are characterized by a relatively high output voltage in respect to the transformer voltage. However, as the diodes conduct only for a short duration, they draw high current in pulses. The rectifier diode should therefore, be capable of withstanding a large current requirement for the first time when the power supply is turned on to charge the reservoir capacitor from zero to high d.c. voltage, and then repetitively supply high current in pulses to give the needed d.c. current. The ratio of peak current to average current is high. However, modern silicon diodes can meet all these requirements and capacitor input filters are used for most of the requirements.

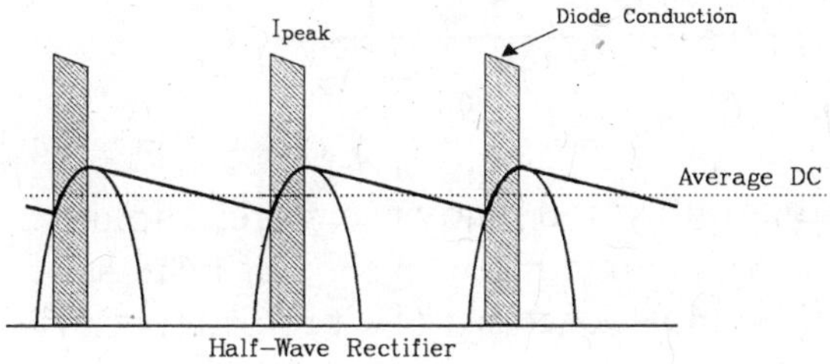

Figure 5.11: Diode conducts in pulses.

Voltage regulation in capacitor input types of power supplies is poor. The voltage drops as the load is increased, but the arrangement is satisfactory for most of the applications. Where a fixed voltage is a requirement, voltage regulator circuits are used.

Filter capacitors can be connected in series to make them withstand higher voltages, but at the same time, the effective capacitance is reduced (see equation 5-3). Moreover, the leakage resistances of two electrolytic capacitors are not likely to be equal, and therefore equalizing resistors have to be connected in parallel to make sure that the high voltage gets divided across them evenly.

Two Stage Filtering

A single capacitor, or a choke-capacitor filter may not reduce the ripple content to the required value. For example, audio power amplifiers can tolerate a certain amount of ripple, but the preamplifiers must get a very pure kind of d.c. Filter circuits can be connected in series to provide better filtering. In a multi-stage filter, outputs can be taken from a number of points to provide highly filtered and less-filtered types of d.c.

Figure 5.12 shows a capacitor input filter followed by a choke input filter. Less sensitive stages can receive power from V_{out1} and more sensitive stages can use d.c. output from V_{out2}.

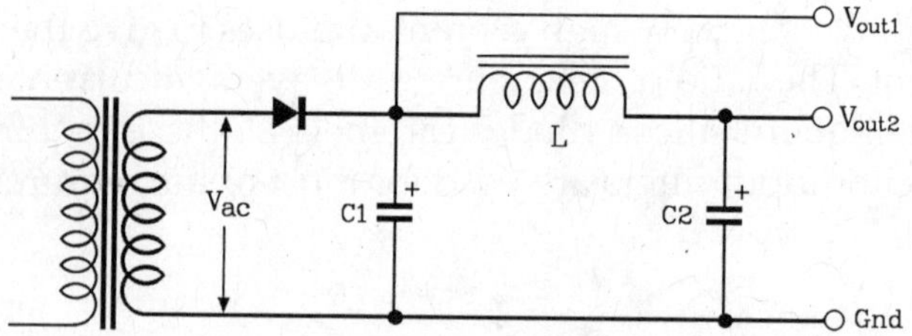

Figure 5.12: A two-stage filter.

Filter chokes are rather bulky and many users prefer to use a resistor in place of choke as shown in Figure 5.13. This results in a voltage loss in resistor R. That is not much of a problem because the circuits fed from V_{out2} usually have low current drain and can work at lower voltages too. The use of a resistor in place of a bulky choke makes the power supply compact.

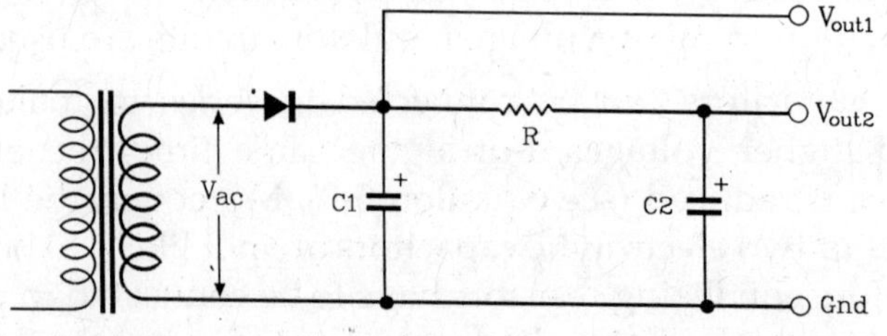

Figure 5.13: A two-stage filter using a resistor in place of choke.

In case of high current through R, the wattage rating of the resistor must be adequate enough to dissipate heat generated. Wire-wound resistors are often used in such applications.

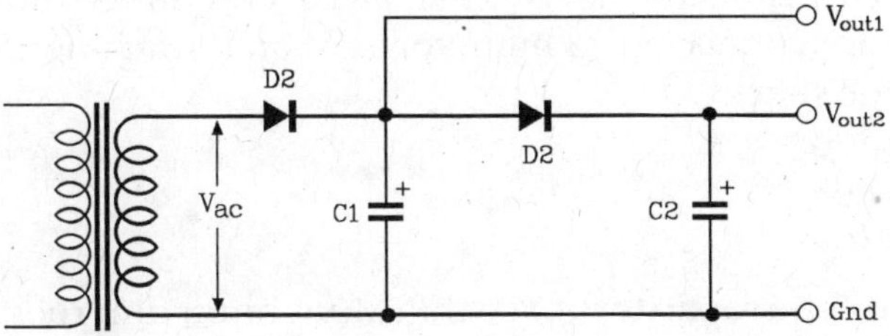

Figure 5.14: A two-stage filter using a diode for isolation of outputs.

Many designers prefer to use a diode in place of resistor as shown in Figure 5.14. The low voltage drop in a diode overcomes the voltage drop and heat generation problems, and also provides some isolation between the two supplies V_{out1} and V_{out2}.

Negative Lead Filtering

It is almost a universal practice to place filter choke, resistors, diodes etc. in the positive line as shown in above Figures. But there is no reason why they cannot be placed in the negative line. Figure 5.15 shows a negative voltage supply with a full-wave rectifier.

In such an arrangement there may be some difficulty in mounting of can type electrolytic capacitors. Axial lead type of capacitors have both terminals isolated and the can is covered by an insulating material. Can type capacitors have their positive

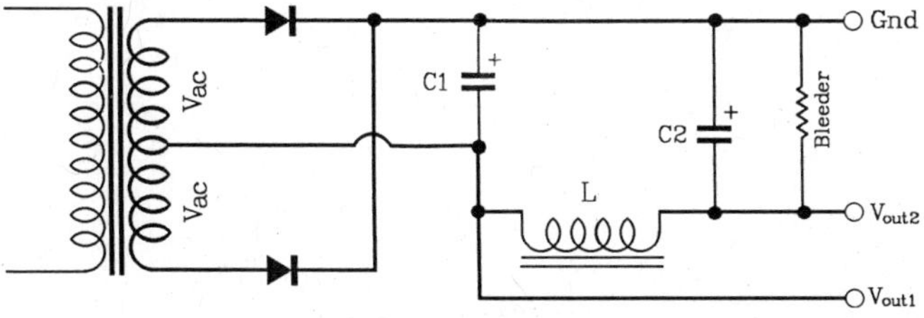

Figure 5.15: A two-stage filter with choke in the negative lead.

terminal at the center and the can is negative. This type of capacitor has therefore to be mounted on the chassis insulated from the chassis. If the voltage involved are high, the body of the capacitor should be covered by an insulating material. Commercial units having such arrangements have the can covers painted red to show that they are connected to high voltages and touching them can give electric shock.

RF Bypass

Electrolytic capacitors have thin aluminium film rolled inside the can. This type of construction makes it inductive like a coil. The inductive reactance is negligible at power line frequencies, but it does become significant at radio frequencies. Such capacitors therefore, cannot bypass high frequencies to ground because of their inductive reactance. The solution is to connect another polyester capacitor of smaller value in parallel with the electrolytic, or across the d.c. output line.

Filter chokes or inductances are wound on iron cores, with a small gap in the core to prevent magnetic saturation of the core at high currents. When iron becomes saturated, its permeability decreases, and consequently the inductance also decreases. Despite the air gap, the inductance of a choke usually varies to some extent with the direct current flowing in the winding; hence it is necessary to specify the inductance at the current which the choke is intended to carry. Its inductance with little or no direct current flowing in the winding will usually be considerably higher than the value when full load current is flowing.

6

Power Supply Protection

Power supplies, as implied by the name, deliver power for use by electronic circuits in devices. It is therefore of utmost importance to ensure that the power supply continues to works reliably. In case of overload or failure of some component, the power supply should keep itself as well as the components in the load circuit fully protected. This is called *fail-safe*.

If no protective measures are taken, a failure of power supply can cause excessive damage to itself; but what is worse is that on failure it can pass on high voltages to the circuits being powered by it and cause extensive damage there as well. For these reasons, the components used in the power supply should be reliable and extra components or circuits should be incorporated to ensure full protection. We shall first consider the components used.

Switches

Power supplies are turned on and off by switches. Therefore they become part of a power supply circuit. Switches are required to make a low-resistance connection in the ON setting, and a very high-resistance insulation in the OFF setting. A wide variety of switches are used in power supplies, but the toggle type switch is most common. Figure 6.1 shows some of the types of switches used in power supplies.

Figure 6.1: Power on/off switches used in power supplies.

When the switch is in its ON state, the resistance of the switch is determined by the switch contacts, the moving metal parts in each part of the circuit which will touch when the switch is turned on. This amount of contact resistance depends on the area of contact, the contact material, the amount of force that presses and holds the contacts together, and also in the way that this force has been applied. For a power supply, the main consideration in switch contacts is the surge of current as contacts are made (due to charging of capacitors or other non-linear devices present in the load) and any reverse voltage surge that occurs when the switch is opened (because of inductances present in the load e.g., primary winding of a transformer) which often causes arcing at the contacts.

Large contact areas are used only for high-current operation and the contact areas for low-current switches like those used for electronics circuits will be small. The actual area of contact of electrical connection will not be the same as the physical area of contacts, because it is not possible to construct contacts that are precisely flat. When the switch is turned ON, the contacts can push or slide over one another. In the later case, when the contacts scrape against each other in a wiping action, the contact resistance can often be much lower than that can be achieved when the same force is used simply to push the contacts straight together.

The bulk material used for making switch contacts is chosen for its mechanical characteristics. The contact surfaces are often electroplated with a more suitable contact material to provide low contact resistance. This technique makes it possible to use materials such as gold and platinum for contacts and still keep the cost of switches down. The usual choice of material for mains switches for power supplies is silver or silver-plated nickel. It should be noted that the ratings for such switches always assume a resistive load, and derating should be applied, i.e., over-rated switches are used for inductive loads.

Switch housings are made from a material of good physical strength and high insulation properties. Use of polished surfaces reduce the possibility of leakage due to accumulation of dirt and dust on the exposed surfaces.

Electrical connections to a switch can be made by soldering, welding, crimping or by using various connectors or other plug-in fittings. By far, the most common connection for panel switches, as distinct from PCB mounted switches, is crimping, because this is very much better adopted for production. Welded connections are used where robot welders are employed for other connection work, or where military assembly standards insist on the greater reliability of welding.

Switch Ratings

Switch ratings are always quoted separately for a.c. and for d.c. The a.c. rating often allows higher current and voltage limits, particularly for inductive circuits. If switches are used on d.c. lines of a power supply, some care over specifying the switch is needed. When d.c. through an inductor is interrupted, a high voltage of reverse polarity is produced and the peak of this pulse can be very large, so that arcing is almost certain, unless some form of arc suppression is used.

Arcing in Switches

Arcing tends to reduce the life of a switch. During the time of an arc very high temperatures can be reached causing the metal of the contacts to vaporize, and be carried from one contact to the other. This effect is very much more serious when the contacts carry d.c. because the metal vapor will also be ionized, and the charged particles will always be carried in one direction only.

If the circuits that are being switched are low voltage circuits with no inductors, arcing will not be a problem. However, with inductance, arcing may be present even in low-voltage circuits, and these include relays, electric motors, and chokes. Circuits in which voltages above 50 V are switched, and particularly if inductive components are present, are most susceptible to arcing problems. An arc suppression is desirable in such cases.

Contact Protection

Arcing can occur across the contacts in free air whenever the voltage exceeds 300V. Several methods are in use to reduce the generated voltage below a value at which arcing will not occur. These have used components such as fixed resistors, semiconductors, capacitors, permanent magnets, and arc blowout coils. Semiconductor devices have proved to be most effective and have the additional advantage of small size, long life, low cost and ease of maintenance.

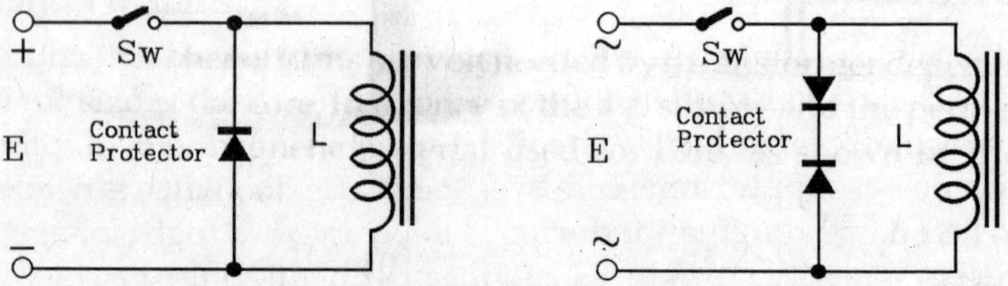

Figure 6.2: Contact protection with semiconductor protectors; polarized on left and non-polarized on right.

Figure 6.2 shows two methods of contact protection using semiconductor contact protectors (see Appendix D) manufactured by Usha Rectifier Corporation. The circuit on the left is for polarized protection, and that on the right is for non-polarized protection. Several models are available to meet different requirements. The choice depends on three factors: supply voltage, maximum permissible induced voltage, and maximum allowable time delay in de-energizing the coil.

Connectors

Connectors often use the plug and socket arrangement and present the same problems of contact resistance as switches. However, it can usually be assumed that the connectors will never be connected or disconnected while the voltages are present, and hence the problems of wear out or arcing is not a consideration. The important point is the safety standards for mains connectors. For safety reasons the receiver of power is always a male plug and the supplier of power is always a female plug (socket).

Fuses

A fuse is a protective device whose performance affords high returns for its cost. This cost is insignificant compared to the cost of the equipment which it is protecting.

Around the turn of the century, electrical circuit "protection" consisted of a piece of undersized wire which was designed to be the first component to break down in case of overcurrent. With higher currents, this piece of wire would sometimes explode and damage the very equipment it was designed to protect. It wasn't long before the enclosed fuse was developed. It was designed to contain the products of volatilization caused by the overcurrent and interrupt the current safely, and within predetermined time limits. Such enclosed fuses were subsequently mass-produced for uniformity and low cost.

Today, there are many types of enclosed fuses. The so called "small-dimension fuses" are normally associated with electronic circuits. Some large fuses find their way into electronic circuits, but these fuses are mainly confined to the power supplies that feed the electronic portions of the system.

There are two principal sizes for small-dimension fuses: those that measure $\frac{1}{4}$" in diameter by $1\frac{1}{4}$" long, and those with diameter of $\frac{13}{32}$" and are $1\frac{1}{2}$" long. Within these size limitations there is a great variety of fuse types to meet the requirements of most applications. Most fuses in the electronic circuits are of the smaller size.

Recently, even smaller fuses have gained popularity, but their electrical characteristics are limited by their dimensions. Typical subminiature fuses measure 0.15" in diameter by 0.300" in length with pigtails. Another type is 0.25" in diameter by 0.27" high, and is a plug-in type. Table 6.1 gives the details.

Table 6.1: Typical small dimension fuses

Ampere Range	Voltage Rating	Characteristics	Dimensions (inches)
$\frac{1}{500}$ to 2	250 or less	Fast Acting	$\frac{1}{4} \times 1$
$\frac{1}{500}$ to 3	250 or less	Fast Acting	$\frac{1}{4} \times 1\frac{1}{4}$
4 to 30	32 or less	Fast Acting	$\frac{1}{4} \times 1\frac{1}{4}$
$\frac{1}{4}$ to 20	250 or less	Fast Acting	$\frac{1}{4} \times 1\frac{1}{4}$
25 to 30	125 or less	Fast Acting	$\frac{1}{4} \times 1\frac{1}{4}$
$\frac{1}{100}$ to 1	250 or less	Time Delay	$\frac{1}{4} \times 1\frac{1}{4}$
$1\frac{2}{10}$ to $2\frac{8}{10}$	125 or less	Time Delay	$\frac{1}{4} \times 1\frac{1}{4}$
3 to 30	32 or less	Time Delay	$\frac{1}{4} \times 1\frac{1}{4}$
$1\frac{1}{4}$ to 2	250 or less	Time Delay	$\frac{1}{4} \times 1\frac{1}{4}$
3 to 7	125 or less	Time Delay	$\frac{1}{4} \times 1\frac{1}{4}$
1 to 15	250 or less	Fast Acting	$1\frac{3}{32} \times 1\frac{1}{2}$
20 to 30	125 or less	Fast Acting	$1\frac{3}{32} \times 1\frac{1}{2}$
1 to 30	250 or less	Fast Acting	$1\frac{3}{32} \times 1\frac{1}{2}$
$\frac{1}{10}$ to 10	250 or less	Time Delay	$1\frac{3}{32} \times 1\frac{1}{2}$
12 to 15	125 or less	Time Delay	$1\frac{3}{32} \times 1\frac{1}{2}$
20 to 30	32 or less	Time Delay	$1\frac{3}{32} \times 1\frac{1}{2}$
0 to 30	600 or less	Fast Acting	$1\frac{3}{32} \times 1\frac{1}{2}$

Some manufacturers use the old 3AG - 8AG fuse designations. Table 6.2 gives the details.

Table 6.2: Old fuse designations

Type	Size (in.)
1AG	$\frac{1}{4} \times \frac{5}{8}$
3AG	$\frac{1}{4} \times 1\frac{1}{4}$
4AG	$\frac{9}{32} \times 1\frac{1}{4}$
5AG	$1\frac{3}{32} \times 1\frac{1}{2}$
7AG	$\frac{1}{4} \times \frac{7}{8}$
8AG	$\frac{1}{4} \times 1$
9AG	$\frac{1}{4} \times 1\frac{7}{16}$

As stated earlier, the aim of a fuse is to interrupt current in the event of a fault that causes excessive current to flow. All electronic circuits, unless of the microwatt variety powered by a high impedance battery, must be protected by a fuse. A wide variety of fuses and fuse holders are available. Figure 6.3 shows some of the types of fuses and fuse holders used in electronic power supplies. Fuse holders are mostly mounted in positions where they can be accessed without having to open the equipment. Some fuse holders are mounted on chassis or printed circuit boards and are hard wired.

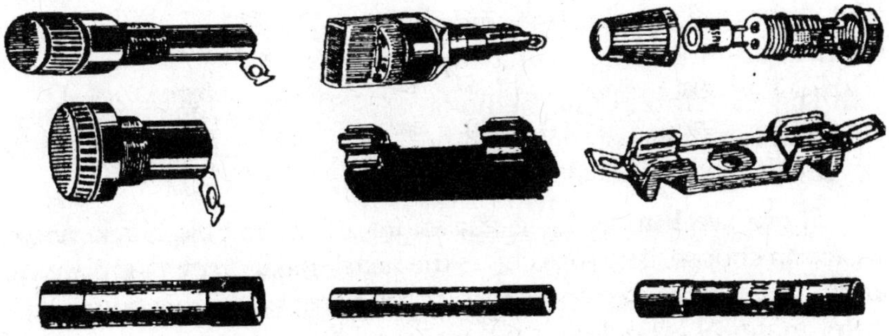

Figure 6.3: Fuses and fuse holders used in power supplies.

Fuses carry a voltage rating and a current rating. They should not be used beyond the specified voltage rating. For example, a fuse rated at 125 V is smaller in size. Such a fuse should not be used in a circuit in which 240 V could exist across a blown fuse because that can cause arcing. You can use a fuse of higher voltage ratings in a lower voltage circuit.

Slow-blow and Fast-blow Fuses

A quantity that is always quoted for a fuse is its nominal current rating. However, another important thing to be considered is the time taken for a fuse to blow. Semiconductor circuits need fuses that will act very quickly on short-circuit conditions. These are now described as super quick-acting fuses. If a fast blow or even a normal blow type of fuse is used in circuits in which there is a considerably high current at switching on times (as in the case of charging of capacitors in a power supply with capacitor-input filters or heating of cold filaments in case of vacuum tubes or incandescent lamps), it will tend to blow every time the device is turned on. Slow-blow fuses should be used in such circuits.

Fuse Mountings

The best fuse is no better than the mounting used to connect it into the circuit. A bad connection between the fuse terminal and its mounting upsets the thermal path so that the fuse cannot dissipate its internally generated heat properly. Such a bad connection will cause a higher-than-normal resistance thereby adding to the heat and inducing a runaway condition.

There are three basic methods of installation: open fuse-blocks with clips; panel mounted fuse-holders; and direct attachment of the pigtail lead-in by soldering or by binding with screws.

There are hundreds of variations of open fuse-blocks from which to choose. The fuse clip is the heart of this type of mounting and is usually made from a spring bronze or beryllium copper. The latter material is the better type.

Although panel mounted fuse holders are not available in as many varieties, there are plenty from which to choose. The most common such fuse holder has a rating of 30 amperes and 250 volts, which is adequate for the ¼" by 1¼" fuses that it accommodates. It fits into a ½" hole and is secured by means of a metal nut.

Indicating-lamp type fuse holders are popular. These fuse holders accept ¼" by 1¼ fuse and include a blown-fuse indication. A neon lamp in parallel with the fuse glows when the fuse opens or when there is no fuse in the holder. Such holders are available for circuits with voltages between 90 and 250.

Circuit Breakers

Heavy duty power supplies are often protected by fast-acting circuit breakers. Figure 6.4 shows the inside and outside views of such breakers.

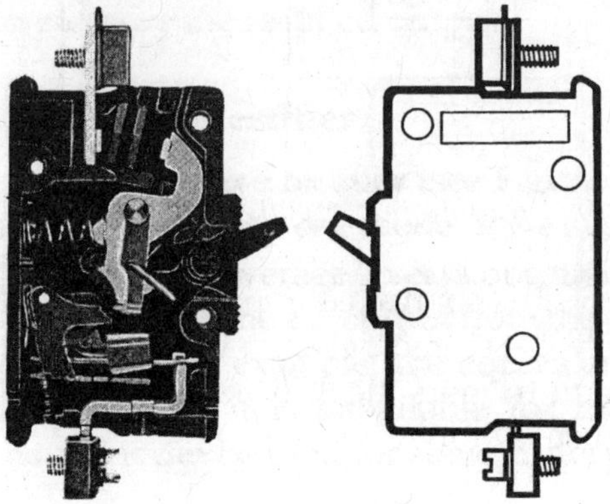

Figure 6.4: Internal and external views of a circuit breaker.

Circuit breakers provide two kinds of protection. In case of short circuits, an electromagnetically operated mechanism cuts off the supply fast. In case of mild, but sustained overloads. a thermal cutoff device actuates the circuit breaker to trip and switch off the incoming power. Circuit breakers are so designed that they cannot be held on, or reset till the overload is removed. If the fault situation has been corrected, the circuit breaker can be reset simply by pushing its knob upwards. The operation is much more reliable, easy and simple compared to replacement of a blown-off fuse.

Protection Against Power Supply Failures

If a power supply supplying to an electronic circuit fails, it can pass through high voltages to the load circuit and cause extensive damage to the load circuit. This excessive loading can cause further damage to the power supply also.

In many designs, electronic crowbar circuits are used to provide protection of the circuits from this kind of damage. The load circuit can be protected by using a *crowbar* circuit. The principle of operation of a crowbar circuit is shown in Figure 6.5: A practical circuit is shown in Figure 14.33.

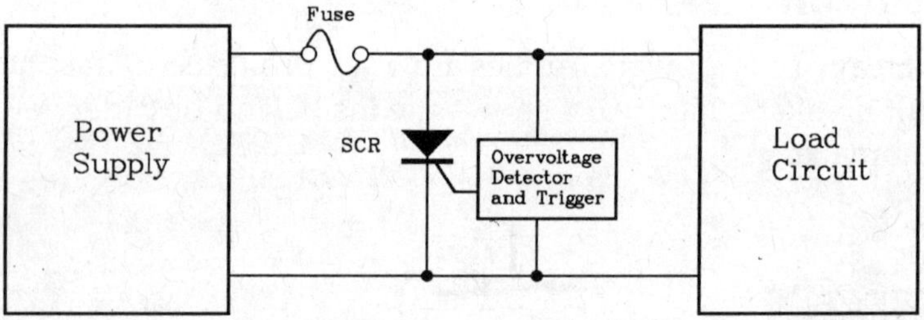

Figure 6.5: A crowbar protection circuit.

A high current capacity SCR is connected in parallel with the d.c. supply line. When the detector circuit detects a voltage higher than the preset limit, the circuit triggers the SCR. A large current flows through SCR and the fuse is blown off.

Protection From Transients

Fuses used in power supplies for semiconductor circuits will offer no protection against damaging transients that are of much shorter duration than what would blow a fuse. Bypass capacitors help in reducing the chances of damages that can be caused by transients. For more effective protection, spike suppression devices are used. They are described in Appendix D.

7

Unregulated Supplies

Unregulated electronic supplies are widely used to supply d.c. power to those appliances where the device is not very sensitive to voltage variations. Examples are radio receivers, battery eliminators, battery chargers, toys etc.

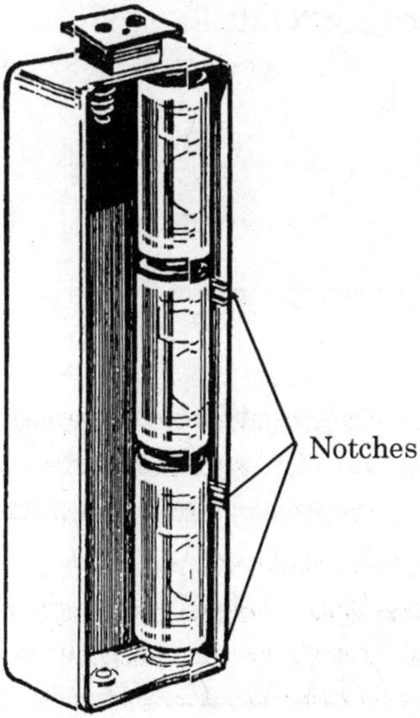

Figure 7.1: A battery box

The simplest type of unregulated supply is a battery. A battery's voltage goes down as the chemicals inside it get consumed. Its internal resistance also increases with usage and the drop in output voltage goes on becoming more and more with time and load.

For experimenters, a battery type of unregulated power supply with cells inside a battery box like the one shown in Figure 7.1 is most common. A battery box has a base plate of metal which can be inserted into notches provided in the box. This enables the box to be used as a 3 V, 6 V or 9 V supply by using 2 or 4 or 6 cells. Power is taken out from the 3-pin socket mounted on top of the box.

Unregulated Power Supplies

Figure 7.2 shows a simple half-wave rectifier circuit power supply for battery charging. No filter capacitor is needed because the battery itself works as a large capacity capacitor.

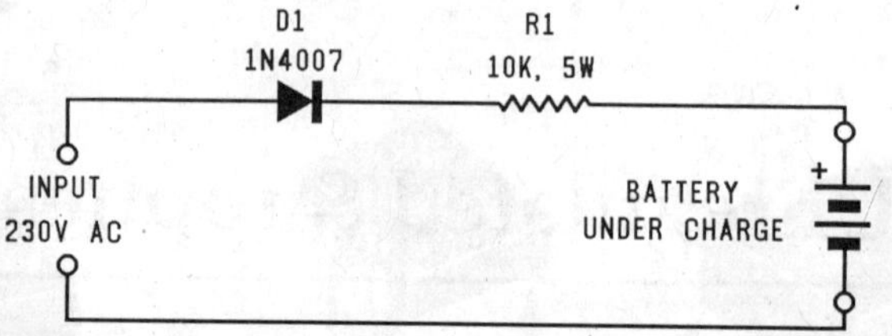

Figure 7.2:A simple battery charger.

Figure 7.3 shows two examples of unregulated type of d.c. power supplies. The upper one uses a silicon rectifier as a half-wave rectifier followed by an electrolytic capacitor working as a reservoir capacitor. The lower one uses a full-wave rectifier that is followed by a two-stage capacitor-resistor-capacitor filter. It also has an LED to indicate ON condition. Simple circuits like these are often used as battery eliminators for small radio receivers, cassette recorders and small d.c. operated toys.

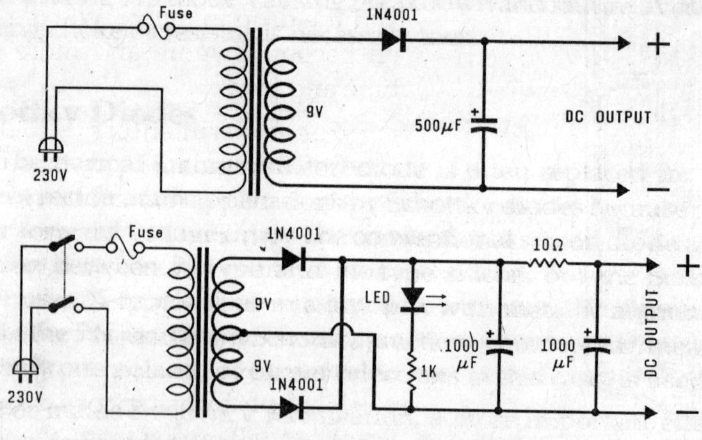

Figure 7.3: Simple unregulated d.c. supplies using half-wave and full-wave rectifiers.

Transient Problems

A common cause of trouble is transient voltages on the a.c. power lines. These are high-voltage short-duration voltages, often referred to as *spikes*. They come from distant lightning strokes, electric motors turning on and off, and so on. These spikes can temporarily increase the voltage seen by the rectifier to values much higher than the normal transformer voltages. Transients can cause unexpected, and often unexplained, loss of silicon rectifiers.

It is always wise to suppress line transients, and it can be easily done. Figure 7.4 shows one way. Capacitor C1 looks like 320,000 ohms at 50 Hz, but to a sharp transient (which has only high-frequency components), it is an effective bypass. C2 provides additional protection on the secondary side of the transformer. Its value should be 0.01 μF for a transformer voltage of 100 V or less, and 0.001 μF for high-voltage transformers. In many designs, rectifier diodes are also bypassed by low-value high-voltage capacitors.

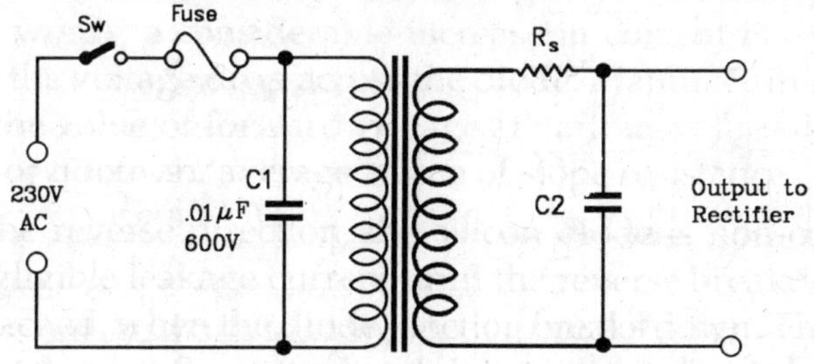

Figure 7.4: Transient suppression by using capacitors.

Figure 7.5 shows another transient-suppression method using selenium suppressor diodes. These diodes do not conduct unless the peak voltage becomes abnormally high. Then they conduct heavily and clip the transient peaks.

Transient voltages can go as high as twice the normal line voltage before the suppressor diodes clip the peaks. Capacitors cannot give perfect suppression either. Thus, it is a good idea to use power supply rectifiers rated at about twice the expected peak reverse voltage (PRV) to ensure safety against transients.

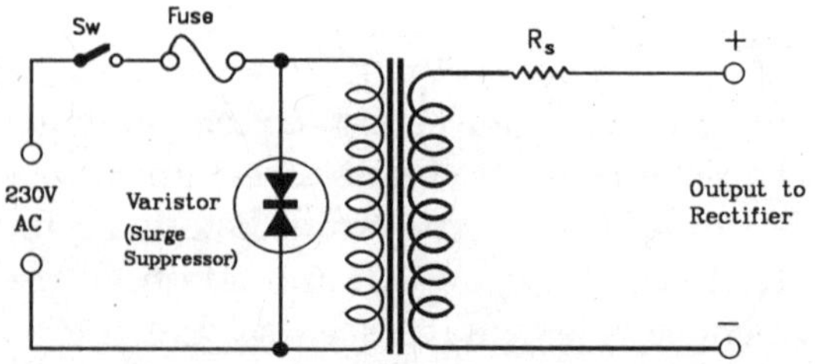

Figure 7.5: Transient suppression by using suppressor diodes.

Diodes in Series

When PIV rating of a single diode is not sufficient for due application, similar diodes may be connected in series and used as a single rectifier. For example, two 500 PRV diodes in series will withstand 1000 volts. When this is done, a resistor and a capacitor should be placed across each diode in the string as shown in Figure 7.6 to equalize PRV drops and guard against transient voltage spikes

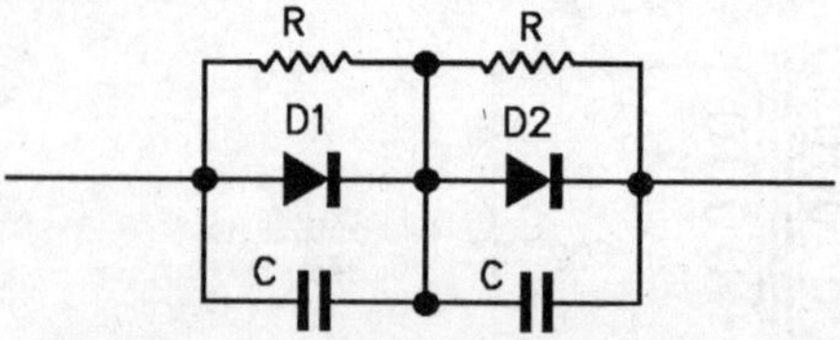

Figure 7.6: Diodes can be connected in series to get higher PRV ratings.

Even if the diodes are of the same type and have the same PRV rating, they may have widely different back resistances when they are cut off. The applied reverse voltage divides according to Ohm's law, and the diode with higher back resistance will have the higher voltage developed across it. That diode may break down. Once that happens, others may also fail one by one.

If we put a swamping resistor across each diode, R as shown in Figure 7.6, the resultant resistance across each diode will be almost the same, and the back voltage will divide almost equally. A good rule of thumb for resistor size is this: Multiply the PRV rating of the diode by 500 ohms. For example, a 500-PRV diode should be shunted by 500 × 500, or 250,000 ohms.

The change from forward conduction to high back resistance does not take place instantly in a silicon diode. Some diodes take longer than others to develop high back resistance. To protect the "fast" diodes in series string until all the diodes are properly cut-off, a 0.01 μF capacitor should be connected across each diode, as shown in Figure 7.6. A complete power supply circuit using this method of protection is shown in Figure 7.7.

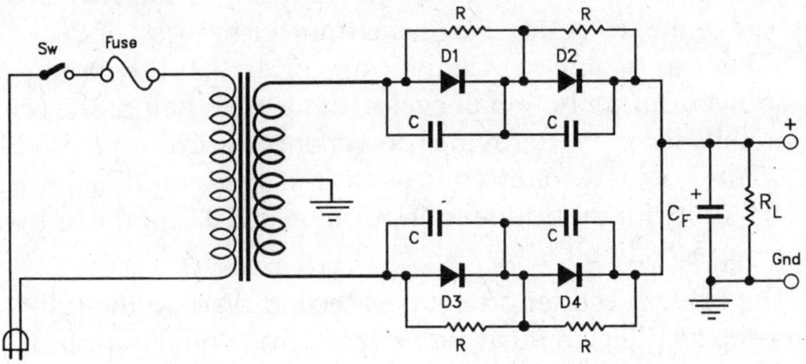

Figure 7.7: A high voltage power supply using diodes in series.

Diodes in Parallel

When a single diode of the needed current handling capability is not available, a number of similar diodes can be placed in parallel to increase current-handling capability. Equalizing resistors, as shown in Figure 7.8, should be added in series. Without these resistors, one diode may take most of the current. This happens because the forward resistances of similar diodes may not be equal when the diodes are conducting. The diode with least forward resistance passes maximum current. The series resistors should be selected to have about one volt drop at the expected peak current. However, so far as practicable, it is far better to use a single diode having larger current rating.

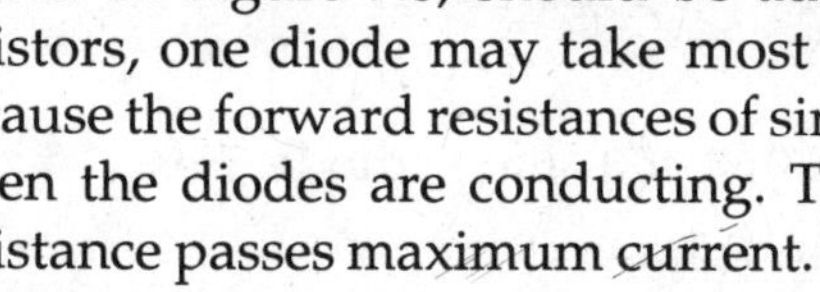

Figure 7.8: Diodes can be connected in parallel to get higher current capability.

Rectifiers Circuits – Pros and Cons

Comparing the full-wave center-tap rectifier circuit and the full-wave bridge-rectifier circuit, we can see that both circuits have almost the same rectifier requirement, since the center tap has half the number of rectifiers as the bridge. These rectifiers have twice the inverse voltage rating requirement of the bridge diodes, however. The diode current ratings are identical for the two circuits. The bridge makes better use of the transformer's secondary than the center-tap rectifier, since the transformer's full winding supplies power during both half cycles, while each half of the center-tap circuit's secondary provides power only during its positive half cycle. This is usually referred to as the *transformer utilization factor,* which is unity for the bridge configuration and 0.5 for the full-wave center-tapped circuit.

The bridge rectifier often takes second place to the full-wave center-tap rectifier circuit in high-current low-voltage applications. This is because the forward voltage drops in the bridge introduce a volt or more of additional loss, and thus more heat to be dissipated, than does the single diode drop of the full-wave rectifier.

The half-wave rectifier is not used much except in some low power applications. However, it does see considerable use in high-frequency switching power supplies in what are called *forward converter* and *flyback converter* topologies.

Rectifier Protection

The important specifications of a silicon diode are:

1. PIV, the peak inverse voltage
2. I_O, the average d.c. current rating
3. I_{REP}—The peak repetitive forward current
4. I_{SURGE}, a non-repetitive peak half sine wave of one-half cycle of the line frequency (10 ms for 50 Hz)
5. Switching speed
6. Power dissipation and thermal resistance

The first two specifications appear in most catalogues. IREP and ISURGE often are often not specified in the catalogues, but they are very important. Because the rectifier never allows current to flow more than half the time, when it does conduct it has to pass at least twice the average direct current. With a capacitor-input filter, the rectifier conducts much less than half the time, so that when it does conduct, it may pass as much as 10 to 20 times the average d.c. current, under certain conditions. This was shown in Figure 5.11.

After the capacitor is charged to the peak rectified voltage, a period of diode nonconduction elapses while the output voltage discharges through the load. As the voltage begins to rise on the next positive half-cycle, a point is reached where the rectified voltage equals the stored voltage in the capacitor. As the voltage rises beyond that point, the diode begins to supply current. The diode will continue to conduct until the waveform reaches the crest, as shown in Figure 5.11. Since the diode has only that short time in which it has to charge the capacitor with enough energy to provide power to the load for the non-conducting balance of the cycle, the current will be high. The larger the capacitor for a given load, the shorter the diode conduction time and the higher the peak repetitive current (IREP).

Current Inrush

When the supply is first turned on, the discharged filter-input capacitor looks like a dead short, and the rectifier passes a very heavy current. This current transient is called ISURGE. The maximum surge current rating for a diode is usually specified for a duration of one-half cycle at (50 Hz), or about 10 ms. Some form of surge protection is usually necessary to protect the diode until the input capacitor becomes nearly charged, unless the diodes used have a very high surge current rating (several hundred amperes).In most cases, the resistance of the secondary winding limits the surge current. If a manufacturer's data sheet is not available, an educated guess about diode's capability can be made by using these rules of thumb for silicon diodes commonly used in power supplies.

Rule 1: The maximum I_{REP} rating can be assumed to be approximately four times the maximum I_o rating.

Rule 2: The maximum I_{SURGE} rating can be assumed to be approximately 12 times the maximum I_o rating. (This figure should provide a reasonable safety factor. Silicon rectifiers with 750-mA d.c. ratings, for example, seldom have 1-cycle surge rating of less than 15 A or more.) From this you can see that the rectifier should be selected on the basis of I_{SURGE} and not on I_o ratings.

Although you can sometimes rely on the d.c. resistance of the transformer secondary to provide ample surge-current limiting, this is seldom true in high-voltage power supplies. Series resistors are often installed between the secondary and the rectifier strings or in the transformer's primary circuit, but these can be a deterrent to good voltage regulation.

Voltage Regulation

The output voltage of a power supply always decreases as more current is drawn, not only because of increased voltage drops in the transformer and filter chokes, but also because the output voltage at light loads tends to soar to the peak value of the transformer voltage as a result of charging the first capacitor. Proper filter design can eliminate the soaring effect. The change in output voltage with load is called voltage regulation, and is expressed as a percentage.

$$\text{Percent regulation} = \left(\frac{E1 - E2}{E2}\right) \times 100\,\% \qquad (7\text{-}1)$$

where

E1 = the no-load voltage

E2 = the full-load voltage

A low percentage indicates good regulation. A steady load such as that represented by a receiver, speech amplifier or unkeyed stages of a transmitter, does not require good regulation as long as the proper voltage is obtained under load conditions. The filter capacitors must have a voltage rating safe for the highest value to which the voltage will soar when the external load is removed.

A power supply will show more (higher) regulation with long-term changes in load resistance than with short temporary changes. The regulation with long-term changes is often called the *static* regulation, to distinguish it from the *dynamic* regulation (short temporary load changes). A load that varies at a syllabic or keyed rate, as represented by some audio and RF amplifiers, usually requires good dynamic regulation (15% or less) if distortion products are to be held to a low level. The dynamic regulation of a power supply can be improved by increasing the value of the output capacitor.

When essentially constant voltage regardless of current variation is required (for stabilizing an oscillator, for example), special voltage regulating circuits described in later chapters are used.

Bleeder Resistors

A bleeder resistor is a resistance connected across the output terminals of the power supply. Its functions are to discharge the filter capacitors as a safety measure when the power is turned off and to improve voltage regulation by providing a minimum load resistance. When voltage regulation is not of importance, the resistance may be as high as 100 Ω per volt of power supply output voltage. The resistance value to be used for voltage regulating purposes was discussed with reference to Figure 5.10. From the consideration of safety, the power rating of the resistor should be as conservative as possible, since a burned-out bleeder resistor is dangerous!

Ripple Voltage and Frequency

Pulsations at the output of the rectifier can be considered to be the result of an alternating current superimposed on a steady direct current (see Figures 5.9). From this view point, the filter may be considered to consist of shunt capacitors that short circuit the a.c. component while not interfering with the flow of the d.c. component. Series chokes will readily pass d.c. but will impede the flow of the a.c. component.

The alternating component is called *ripple*. The effectiveness of a filter can be expressed in terms of percent ripple, which is the ratio of the RMS value of the ripple to the d.c. value in percent.

$$\text{Percent ripple (RMS)} = \frac{E1}{E2} \times 100\ \% \quad \text{- - - - - - - - - - - - (7-2)}$$

where: E1 = the RMS value of ripple voltage
E2 = the steady d.c. voltage.

Any frequency multiplier or amplifier supply in a CW transmitter should have less than 5% ripple on the plate voltage. Bias supplies for linear amplifiers should have less than 1% ripple. VFOs, speech amplifiers and receivers may require ripple no greater than to 0.01%.

Ripple frequency refers to the frequency of the pulsations in the rectifier output waveform—the number of pulsations per second. The ripple frequency of half-wave rectifiers is the same as the line-supply frequency — 50 Hz with a 50-Hz supply. Since the output pulses are doubled with a full-wave rectifier, the ripple frequency is doubled—to 100 Hz with a 50-Hz supply.

The amount of filtering (values of inductance and capacitance) required to give adequate smoothing depends on the ripple frequency. More filtering is required as the ripple frequency is decreased.

Input Versus Output Voltage

The average output voltage of a capacitor-input filter is generally poorly regulated with load current variations. This is because the rectifier diodes conduct for only a small portion of the a.c. cycle to charge the filter capacitor to the peak value of the a.c. waveform (see Figure 5.11). When the instantaneous voltage of the a.c. passes its peak, the diode ceases to conduct. This forces the capacitor to support the load current until the a.c. voltage on the opposing diode in the bridge or full-wave rectifier is high enough to pick up the load and recharge the capacitor. For this reason, the diode currents are usually quite high.

Since the cyclic peak voltage of the capacitor filter output is determined by the peak of the input a.c. waveform, the minimum voltage and, therefore, the ripple amplitude, is determined by the amount of voltage discharge, or "droop" occurring in the capacitor while it is discharging and supporting the load. Obviously, the higher the load current, the proportionately greater the discharge, and therefore the lower is the average output voltage will be.

Although not exactly accurate, an easy way to determine the peak-to-peak ripple for a certain capacitor and load is to assume a constant load current. We can calculate the droop in the capacitor by using the relationship:

$$C \times E = I \times T \qquad \text{(7-3)}$$

where:

C	=	the capacitance in microfarads.
E	=	the voltage droop, or peak-to-peak ripple voltage
I	=	the load current in milliamperes
and t	=	the time between half-cycles of the rectified waveform, in milliseconds. For 50 Hz full-wave rectifiers, t is 9 ms.

As an example, let us assume that we need to determine the peak-to-peak ripple voltage at the d.c. output of a full-wave rectifier filter combination that produces 13.8 V d.c. and supplies a transceiver drawing 2.0 A. The filter capacitor in the power supply is 5000 μF Using the above relationship:

$$C \times E = I \times T$$

$$5000\ \mu F \times E = 2000\ mA \times 9\ ms$$

$$E = \frac{2000\ mA \times 9\ ms}{5000\ \mu f} \qquad = 3.6\ V_{P\text{-}P}$$

Obviously, this is too much ripple. A capacitor value of about 20,000 μF would be better suited for this application. If a linear regulator is used after this rectifier/filter combination, however, and the source voltage is raised to produce a d.c. voltage of about 20 V, the 5000 μF capacitor with its 3.6 V peak-to-peak ripple would work well, since the regulator would remove the ripple content before the output power was applied to the transceiver.

Choke-Input Filters

Choke-input filters have become less popular than they once were, because of the high surge current capability of silicon rectifiers. These filters provide the benefits of greatly improved output voltage stability over varying loads and low peak-current surges in the rectifiers. On the negative side, the choke is bulky and heavy, and the output voltage is lower than that of a capacitor input filter.

As long as the inductance of the choke is large enough to maintain a continuous current over the complete cycle of the input a.c. waveform, the filter output voltage will be the average value of the rectified output. The average d.c. value of a full-wave rectified sine wave is 0.637 times its peak voltage. Since the RMS a.c. voltage is 0.707 times the peak, the output of the choke-input filter will be (0.637/0.707), or 0.90 times the RMS a.c. voltage. For light loads, however, there may not be enough energy stored in the choke during the input waveform crest to allow continuous current over the full cycle. When this happens, the filter output voltage will rise as the filter assumes more and more of the characteristics of a capacitor-input filter.

Choke-input filters see extensive use in the energy-storage networks of the switch-mode regulators.

Dual Voltage Power Supplies

Many applications need a symmetrical supply of equal plus and minus voltages with respect to ground. This can be done by using a transformer with two similar windings, or, more easily using a bridge rectifier as shown in Figure 7.9.

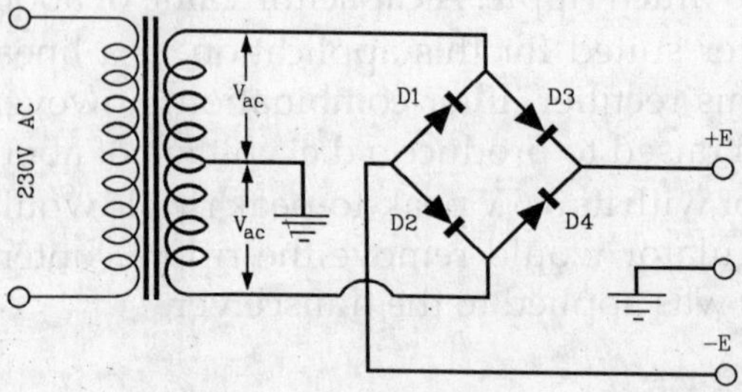

Figure 7.9: A symmetrical dual supply using a bridge rectifier

The center tap of the transformer's secondary is grounded. Diodes D3 and D4 therefore, work as fullwave rectifier with a center-tapped transformer as shown in Figure 7.3 and deliver a positive voltage +E with respect to ground. Diodes D1 and D2 also work in the same manner, but since the diode polarities are reversed, they produce a fullwave rectified negative voltage -E with respect to ground. Filter circuits can be added independently to get filtered d.c. output.

The same circuit can be used to deliver two levels of voltages simply by connecting the -E point to ground instead of the center point as shown in Figure 7.10. You now get two voltages E1 and a higher voltage E2 with respect to ground. Obviously, filter capacitors for E2 should have higher voltage ratings.

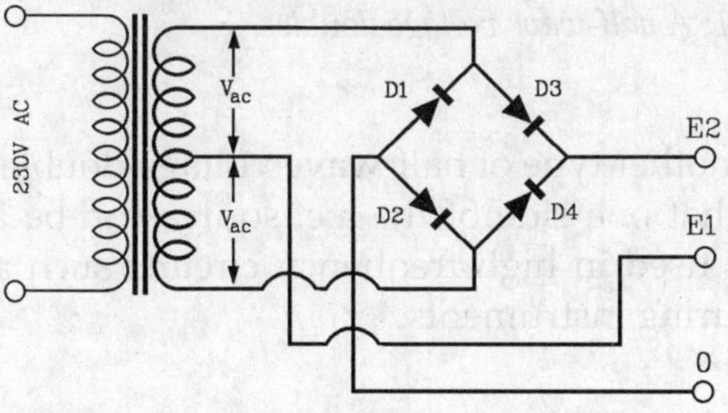

Figure 7.10: A two-level d.c. supply.

Voltage Multipliers

A voltage multiplier circuit is one which provides a d.c. output whose level is higher than the peak a.c. voltage from the transformer or other a.c. source. Such circuits are often used to get higher voltage to operate some low-current-drain device when only one source of a.c. voltage is available. Very often, good regulation is not needed.

Halfwave Doublers

The simplest form of circuit, the half-wave doubler, is shown in Figure 7.11. One end of the secondary is connected to the junction of two capacitors. Diode D1 rectifies the positive half-cycles of the a.c. voltage and charges capacitor C1 to the peak voltage of a.c. Diode D2 rectifies the negative half-cycles of the a.c. input and charges C2 to the peak voltage of a.c. The voltages of the two capacitors add up in series to provide a d.c. voltage of two times the peak value of a.c. Since both diodes work independently in this configuration, equalizing resistors are not needed.

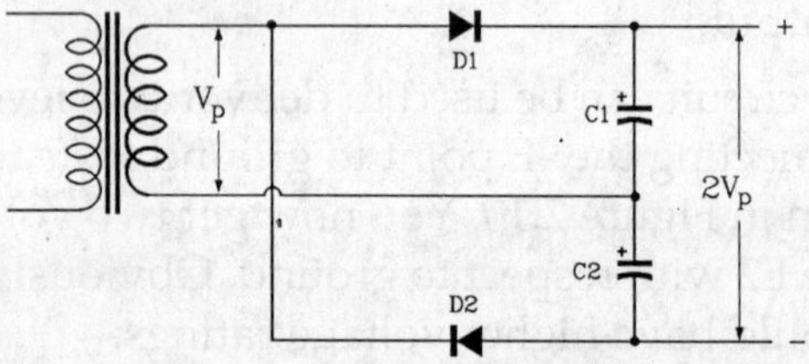

Figure 7.11: A half-wave voltage doubler.

Figure 7.12 shows another type of half-wave voltage doubler. The advantage here is that one side of the a.c. source can be at ground level. It is often used in high frequency circuits such as detector probes in measuring instruments.

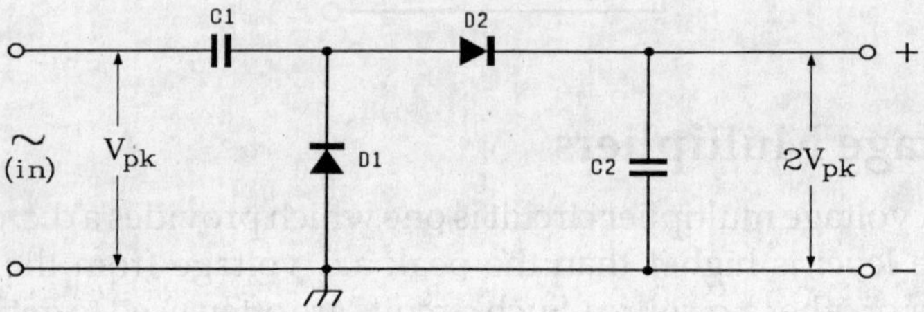

Figure 7.12: Half-wave voltage doubler with one line at ground level.

The configuration of Figure 7.11 has the advantage that each capacitor is subjected to only the a.c. peak voltage unlike the second capacitor in the circuit of Figure 7.12. But the a.c. output from the secondary of transformer is floating in this design, which may not always be convenient or possible. Many circuit designers prefer to keep one end of the a.c. input at the ground level as in Figure 7.12.

The working of this circuit is a little different. If we imagine this at the point of being switched on with all the capacitors uncharged and there is no load, then the first positive half-cycle will cause D2 to conduct, charging C1+C2 to the peak of the a.c. voltage. When the a.c. voltage reverses, D1 will conduct, and this will now discharge C1 and charge it in the opposite direction to the positive peak voltage, positive at the cathodes of D1. On the next positive cycle of a.c. the voltage across C1 and in the secondary will add up in series to charge C2 to a higher voltage. In course of a few cycles, C2 will be charged to twice the peak voltage of the a.c. This capacitor must therefore be rated for at least twice a.c. peak voltage working. C1 can have the ratings of peak voltage of the a.c. input. After this, voltages that are maintained will be equal to peak a.c. at the cathode of D1 and twice the peak a.c. at the cathode of D2.

Voltage regulation of a voltage doubler is very poor. This is mainly because the capacitors have a fairly large reactance even at the higher supply frequencies which are almost used along with multiplier circuits. Unlike the rectifier circuits for low-voltage supplies, there is no real lower limit to the output of a multiplier circuit on heavy load, and if the load resistance fluctuates considerably, some stabilization will be needed. Most multiplier circuits make use of waveforms derived from oscillators, and a control circuit for stabilization samples the output voltage and uses this to control the amplitude of the oscillator. As for any other rectifier-reservoir type of supply, the ripple amplitude increases as the d.c. output level drops.

Voltage Triplers

Voltage tripler and quadrupler circuits make use of the same principles of using a capacitor to charge by using current from one diode and then to superimpose this voltage on to another conducting diode along with an a.c. half-wave. A tripler circuit is shown in Figure 7.13. Diodes D1 and D2 and capacitors C1 and C2 form a voltage doubler of the type shown in Figure 7.12. Diode D3 and capacitor C3 simply rectify the negative half-cycle of the a.c. input. The two voltages add up to give thrice the peak voltage (V_P).

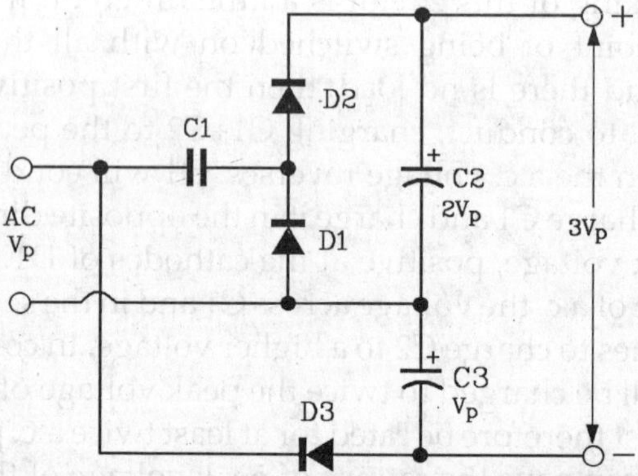

Figure 7.13: Voltage tripler with floating input.

This circuit has the disadvantage that it needs a floating a.c. supply. Many designers need one end of the input a.c. source to be at ground level. The tripler shown in Figure 7.14 is a circuit configuration in which one side of the input a.c. line is at ground level.

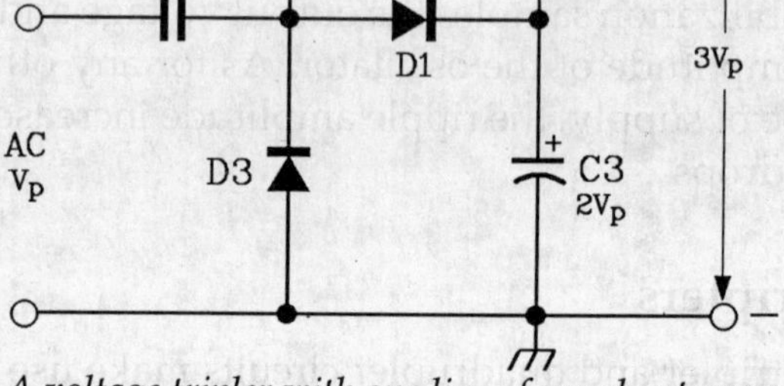

Figure 7.14: A voltage tripler with one line of supply at ground level.

Voltage Quadruplers

A quadrupler circuit is shown in Figure 7.15. This circuit is a combination of two voltage doubler circuits of the type shown in Figure 7.11. In this case, the a.c. input has to be a floating one and this can be a disadvantage in some situations..

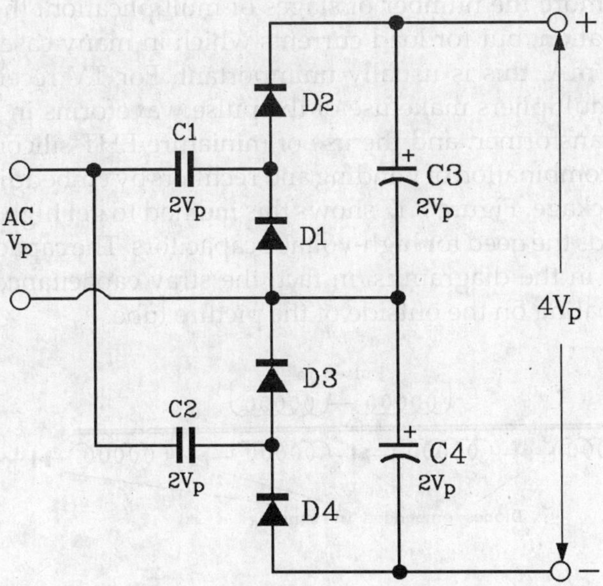

Figure 7.15: A voltage quadrupler with floating a.c. input.

Figure 7.16 shows a voltage quadrupler which is a combination of two circuits of the type shown in Figure 7.12. Here, one of the lines of a.c. input voltage can be connected to ground.

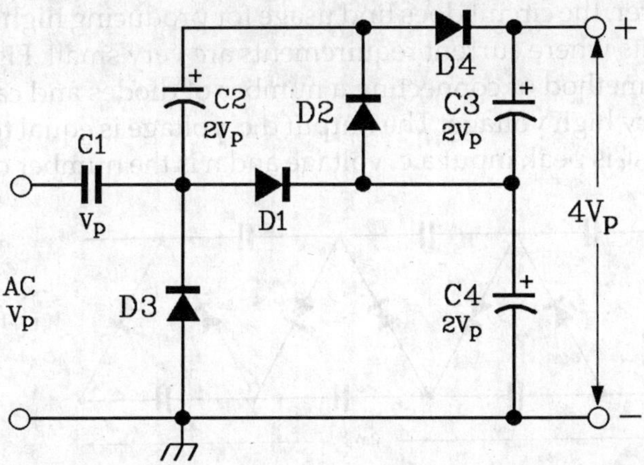

Figure 7.16: A voltage quadrupler with one side of input line grounded.

The more the number of stages of multiplication, the poorer the regulation, but for load currents which in many cases do not exceed 1 mA, this is usually unimportant. For TV receiver use, voltage multipliers make use of the pulse waveforms in the line-output transformer, and the use of miniature EHT silicon diodes allows a combination of winding and rectifiers by embedding them in one package. Figure 7.17 shows this method to get high voltage. This avoids the need for high-voltage capacitors. The capacitor that is shown in the diagram is, in fact, the stray capacitances of the carbon coating on the outside of the picture tube.

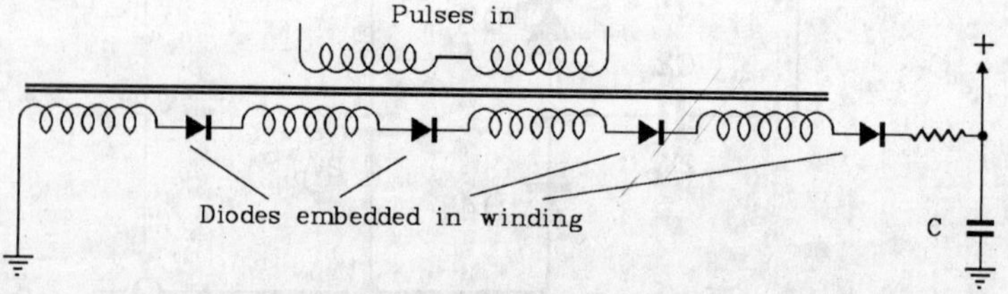

Figure 7.17: A high voltage supply with embedded diodes.

The principle of multiplication of output d.c. voltage used in voltage doubler and voltage quadrupler circuits described above can be extended to a number of stages to get higher and higher voltages. Though the voltage regulation goes on becoming poorer and poorer, the circuit does find usage for producing high voltages for circuits where current requirements are very small. Figure 7.18 shows a method of connecting a number of diodes and capacitors to get very high voltage. The output d.c. voltage is equal to $2nV_{pk}$, where V_{pk} is peak input a.c. voltage and n is the number of stages.

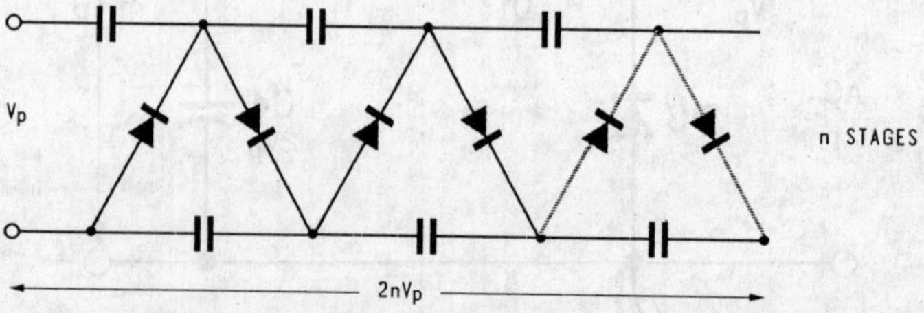

Figure 7.18: Principle of voltage multiplication.

A practical application of this type of multiplication is the ionizer power supply shown in Figure 7.19. Ions are positively or negatively charged molecules of the gases that make up the atmosphere and their concentration depends on the location and the prevailing weather conditions. It is believed that a preponderance of negative ions has a positive effect on physical well-being, while a preponderance of positive ions has a negative effect. The average concentration of ions of either polarity is normally fairly small, around 400 to 1500 ions per cc. of air, but in mountain resorts the concentration of negative ions is considerably higher. This may account, in part, for the nice and energetic feeling at these resorts.

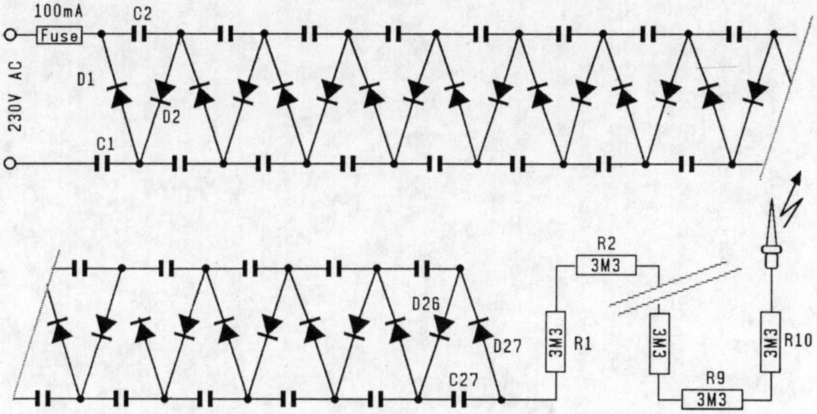

D1 ..... D27 = 1N4007 (1000V)
C1 ..... C27 = 33nF or 47nF 650V

Figure 7.19: A high voltage power supply using diodes in series

With mains supply of 230 V a.c. as input and 27 rectifying diodes, the circuit shown in Figure 7.19 produces a negative d.c. voltage of around 7.5 kV. The negative output terminal of the multiplier is connected to a pointed object such as a sewing needle. The electric field strength around a charged body is greatest where curvature is greatest, i.e., around sharp points. An intense field is thus produced at the tip of the needle, and electrons from the tip of the needle are 'sprayed' on to the surrounding air molecules, turning them into negative ions. These ions are then repelled by the negative charge on the needle point, and other gas molecules take up their place and are ionized in turn. This means that a constant 'wind' of negative ions emanates from the needle point.

As the needle must be exposed to the air on order to generate the ion stream it is necessary to limit the current that can flow in the event of the needle being inadvertently touched. This is achieved by inserting series resistors R1 through R10 in series *Under no circumstances should these resistors be omitted or bypassed, as this could result in a fatal electric shock.*

8

Voltage Stabilization

Many electronic circuits need stable d.c. voltages for their reliable operation. Maintenance of voltage (or current) at a fixed level is commonly referred to as *regulation* or *stabilization*. Stabilization is needed against two main causes of variation of output voltage in a power supply. The first one is changes in the incoming mains supply voltage (line voltage variations) and the second one is changes in load current (load current variations). A good stabilizer should keep the terminal output voltage of a power supply steady within specified limits despite the line voltage changes and load current variations from zero to a little more than full value.

Constant Voltage Sources

In Chapter 1 (see Figure 1.5) we explained that the output voltage available to a load falls because of internal resistance of the voltage source. In theory, an ideal source of constant voltage will be a generator which has zero internal resistance. With such a generator, the output voltage will remain same irrespective of load current. In practice, a battery of very low internal resistance can be taken as a constant voltage source as shown in Figure 8.1. The internal resistance of this battery is 0.01 ohms and the voltage loss is 0.01 V at 1 A and 0.1 V at 10 A of load current. The curve on the right shows that V_L is almost constant for loads from 0 A to 10 A.

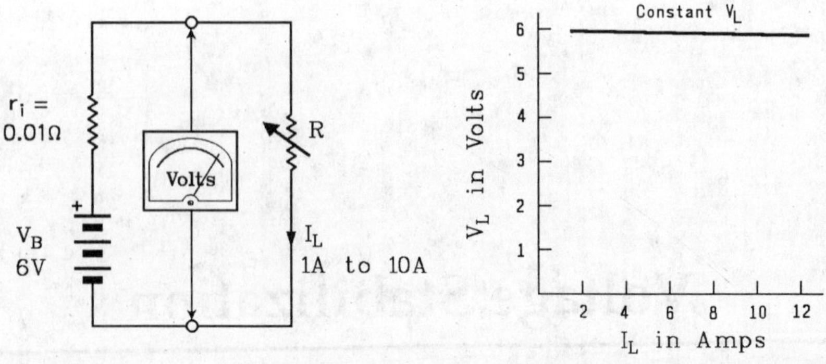

Figure 8.1: Constant voltage source.

Principles of Stabilization

In electronic circuits, fixed voltages are obtained by using stabilizing electronic circuits. Stabilizers, in general, operate by wasting power. This implies that the source d.c. voltage should be higher than the required stabilized voltage at the rated current. The main unregulated power supply therefore generates d.c. voltages which are higher than the regulated voltages needed. The excess voltage is dropped across a series resistive element. The current flowing through it causes power loss which generates heat. This resistive element should have sufficient heat dissipating capability.

Series Regulator

Figure 8.2 shows principle of operation of a series resistor type of regulator. A non-linear resistive element R_S (often a vacuum tube or a transistor) is connected in series with the load. The value of R is so adjusted electronically that while the input voltage E changes, or the load current I changes, the voltage drop across the load R_L remains constant. The excess voltage equal to (E-V) is dropped across the series resistance R_S resulting in a power loss of of (E-V)×I in R_S. This loss results in heating up of R_S and R_S should therefore have adequate heat dissipation capability.

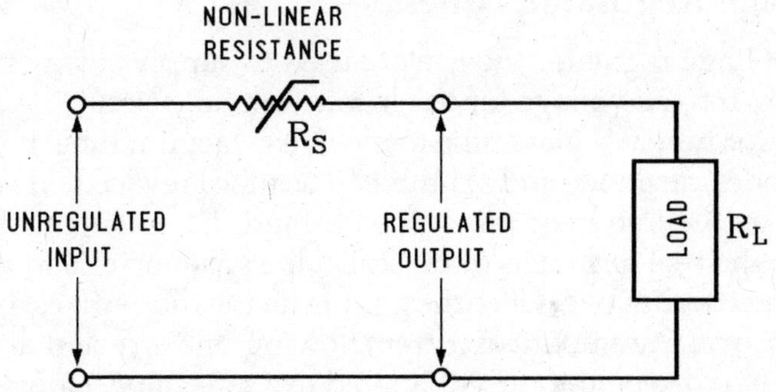

Figure 8.2: Principle of operation of a series type regulator.

Shunt Regulator

In Figure 8.3 the series resistor R_S has a fixed value and the non-linear resistor R is connected in parallel with the load. The value of R adjusts itself to keep the voltage drop across it, and hence the voltage across the load R_L, constant. As before, the difference between input and output voltages is lost in the series resistor R_S. The current through this series dropping resistor is now the sum of the currents in the load circuit and that in the shunt resistor R.

The heat dissipated in the non-linear resistor R is maximum when the load current is minimum. This non-linear shunt resistor R therefore, should also have adequate heat dissipating capability.

In practice, the non-linear resistor R is either a voltage regulator tube or a zener diode.

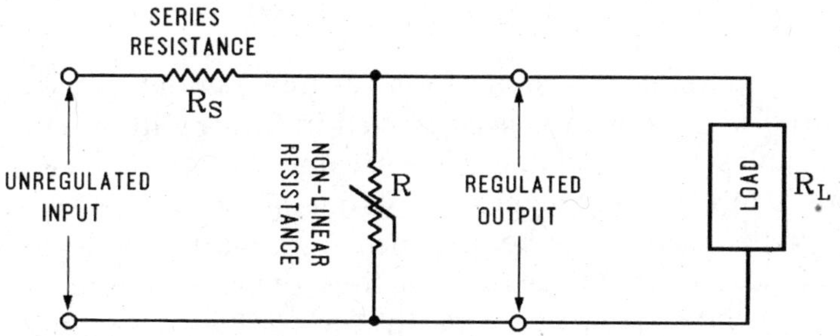

Figure 8.3: Principle of operation of a shunt regulator.

Voltage Regulator Tubes

Voltage regulator tubes were used to supply steady voltages of known fixed voltage for use in vacuum tube circuits. A voltage regulator tube is a glass tube made like a vacuum tube. It has two electrodes, an anode and a cathode. The tube is evacuated and then filled with some inert gas. When a high d.c. voltage is applied across the two terminals of this tube, the gas ionizes and the tube becomes conductive. Once the gas inside the tube strikes, it has no control over the amount of current flowing, and an external resistor is needed to keep the current limited to a safe value. However, the voltage across the tube remains constant over a moderately wide range of current through the tube.

The value of this steady voltage depends on type of gas that is filled inside the tube. Voltage regulator tubes like VR90, VR105 and VR150 were used to regulate voltages at 90 V, 105 V and 150 V respectively.

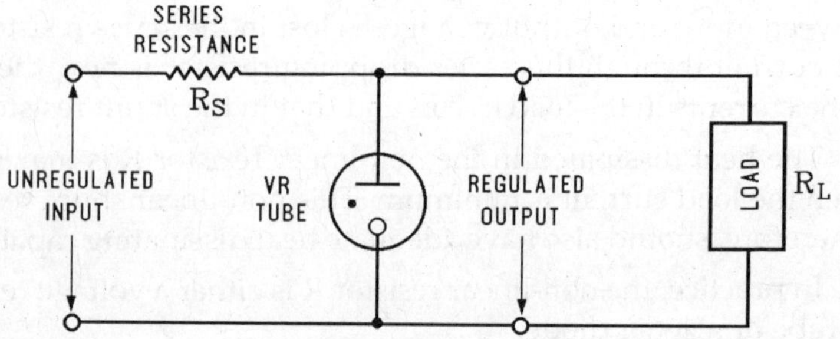

Figure 8.4: Voltage stabilizer with a VR type of tube.

The fundamental circuit for a gas tube regulator is shown in Figure 8.4. The VR tube is connected in series with a resistor R_S across the source of voltage. The starting voltage of a VR tube is about 30 to 40 percent higher than its operating voltage. The source voltage therefore, must be higher than the starting voltage of the tube. Once the gas inside the tube fires, the voltage drop across it falls and then it remains constant for a given range of current through the tube. This range is specified by the manufacturer.

The load is connected in parallel with the tube (see Figure 8.4). For stable operation, a minimum current of 5 to 10 mA is required to flow through the tube all the time. The maximum permissible current with most types is 40 mA. Consequently, the load current cannot exceed 30 to 35 mA, if the voltage is to be stabilized over a range from zero load current to maximum load current. A single VR tube may also be used to regulate the voltage to a load current of almost any value so long as the *variation* in the current does not exceed 30 to 35 mA.

If, for example, the average load current is 100 mA, a VR tube may be used to hold the voltage constant provided the current does not fall below 85 mA or rise above 115 mA.

The value of the limiting resistor R_S must be between that which on full load, allows the minimum tube current to flow, and that will just pass the maximum permissible tube current when there is no load current. The latter value is generally used. It is given by the equation:

$$R_S = \frac{E-V}{I} \qquad \text{(8-1)}$$

where R_S is the series current limiting resistor in ohms, E is the voltage of the source, V is the rated voltage across the regulator tube, and I is the maximum tube current in amperes (usually 40 mA or 0.04 A)

A capacitor is sometimes connected in parallel with the VR tube to filter out noise. The capacitor in shunt with a VR tube, if used, should be limited to 0.1 μF or less. Larger values may cause the tube's voltage-drop to oscillate between the operating and starting voltages. This is because the tube, the capacitor and the series resistor combination make a relaxation type of oscillator.

Two tubes may be used in series to give a higher regulated voltage than what is obtainable with one tube, and also to give two values of regulated voltages as shown in Figure 8.5. Regulation of the order of over one percent can be obtained with these regulator tubes when they are operated within their operating current range.

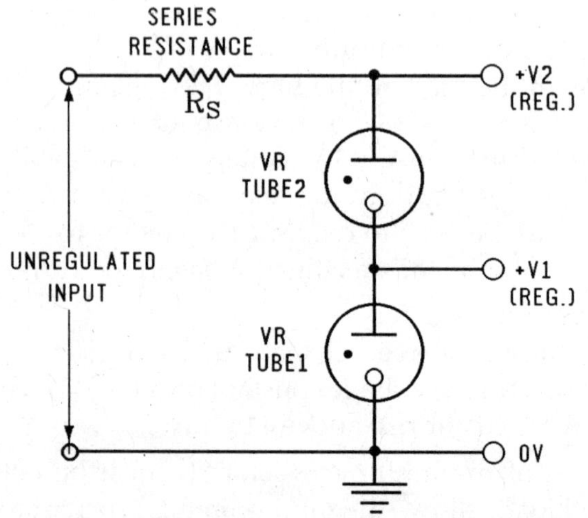

Figure 8.5: Two VR tubes in series give higher regulated voltages.

Zener Diodes

For supplying regulated voltages to semiconductor circuits zener diodes (named after Dr. Carl Zener) are used in place of VR tubes, much in the same way as VR tubes are used. A zener diode looks like any other semiconductor diode and conducts in the forward direction in the same way. However, in the reverse direction, it has different characteristics. It is designed to break down at a lower reverse d.c. voltage.

Figure 8.6 shows electrical characteristics of zener diodes. Compare these characteristics with that of a rectifier diode shown in Figure 4.3. The reverse breakdown voltage for a silicon rectifier diode is made as high as possible, subject to the requirements of the diode. Altering the doping of the silicon allows zener diodes to be made with a range of breakdown voltage levels. These levels are specified as 3V3, 4V7, 5V6, 6V8 and so on. The letter V is used in place of the decimal point to avoid any confusion due to misplaced decimal points or the use of a comma in place of a point. Thus, 3V3 and 4V7 represent zener diodes with reverse breakdown voltages of 3.3 V and 4.7 V respectively.

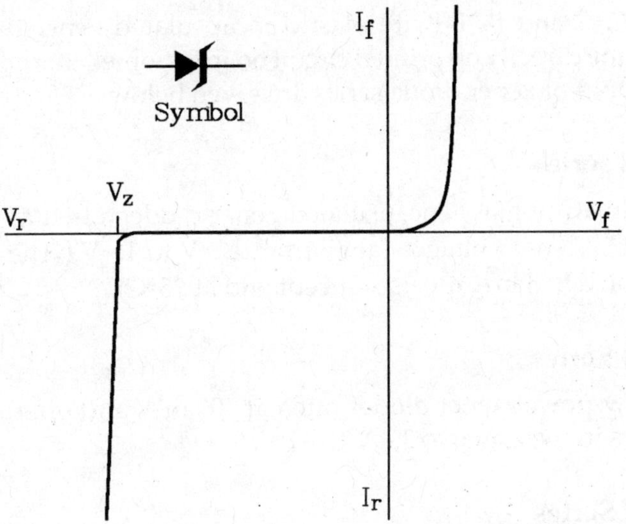

Figure 8.6: Characteristics of a zener diode and its symbol.

Once the diode junction breaks down, the zener diode has no control over the current through it. But the diode junction does not get irreparably damaged if the reverse current through it is kept limited to a safe value by external resistors or other components. This type of diode is used as a source of constant voltage.

Figure 8.6 also shows the symbol used for representing zener diodes in schematic diagrams. This symbol is similar to that for a silicon diode except that the cathode bar is replaced by a zig-zag "Z" shaped line to show its breakdown characteristics.

Zener diodes are available in a wide variety of zener voltages and power ratings. The voltages range from less than 2 volts to a few hundred volts, while the power ratings (power the diode dissipates) run from less than 0.25 watts to 50 watts. This gives a very wide range of choice for using zener diodes.

The zener voltage of zener diodes is specified at some specified test current and temperature because the zener voltage changes slightly with change in the current through the zener diode and with changes in temperature. The reverse breakdown voltage is also affected by variations in temperature.

BZ147 and BZ148 are plastic encapsulated zener diodes for mounting directly on printed circuit boards. Some other commonly used types of zener diode series are given below:

BZY88 series

Miniature glass encapsulated zener diodes rated at 500 mW (at 25°C). Zener voltages range from 2.7 V to 15 V.(The voltages are quoted at 5 mA reverse current and at 25°C).

BZX55 Series

Low-power zener diodes rated at 500 mW and offering zener voltages in the range of 2.4 V to 91 V.

BZX61 Series

Encapsulated alloy junction diodes rated at 1.3 W at 25°C ambient. Zener voltages range from 7.5 V to 72 V.

BZX85 Series

Medium-power glass-encapsulated zener diodes rated at 1.3W and offering zener voltages in the range of 5.1 V to 62 V.

BZY93 Series

High power zener diodes in stud mounting encapsulation. Rated at 20 W for ambient temperatures up to 75°C and zener voltages ranging from 9.1 V to 75 V.

BZY97 Series

Medium power wire-ended diodes rated at 1.5 W and offering zener voltages in the range of 9.1 V to 37 V.

1N5333 Series

Plastic encapsulated zener diodes rated at 5 W. Zener voltages range from 3.3 V to 24 V.

Using Zener Diodes

In the breakdown condition, the voltage across the zener diode is almost constant as shown in Figure 8.7. The current I_Z through the zener diode is given by the equation:

$$I_Z = \frac{E - V_Z}{R_S} \quad \text{-----------------------(8-2)}$$

where I_Z is the current through the series resistor Rs and through the zener diode, E is the unregulated input voltage and V_Z is the regulated output voltage.

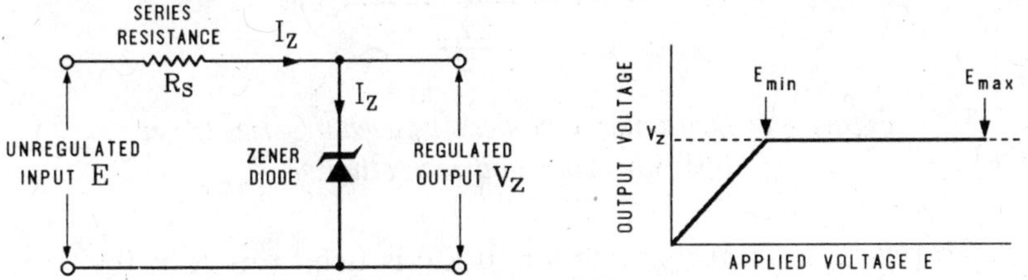

Figure 8.7: Principle of voltage stabilization with a zener diode.

When the zener diode is conducting, a power equal to $V_Z \times I_Z$ watts is dissipated inside the zener diode, where V_Z is in volts and current I_Z is in amperes. This value should be within the power dissipation ratings of the zener diode used.

Two zener diodes can be connected in series to get higher regulated voltages, or two values of regulated voltages as shown in Figure 8.8. Since the current through the two series connected diodes is the same, the power dissipated in each zener ($V_Z \times I_Z$) depends on its zener voltage.

The ability of a zener diode to stabilize a voltage in an external load is dependent upon the conducting impedance of the diodes, which can be as low as one ohm in a low-voltage high-power diode to as high as a thousand ohms in a low-power high voltage diode. Detailed characteristics of some zener diode types are given in Appendix F.

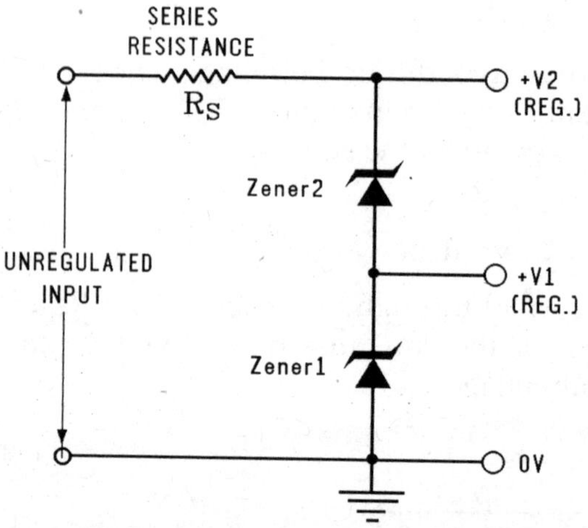

Figure 8.8: Two zener diodes can be used in series to get two values of stabilized voltages.

In practical circuits, a zener diode is used much in the same way as a gaseous regulator tube is used. A typical circuit is shown in Figure 8.9. Note that the cathode of the zener diode is connected to the positive side of the supply. This is just the reverse of the way we connect rectifier diodes because zener diodes are used in their reverse breakdown state.

As explained earlier, a resistor R_S is connected in series to drop the excess voltage. The current through this series resistor is the sum of the currents I_Z through the zener diode and I_L through the load. Hence, maximum dissipation of heat occurs in the series resistance when the input voltage E is maximum. On the other hand, maximum heat dissipation occurs in the zener diode when the load happens to be disconnected ($I_L = 0$) and all the current passes through the zener diode(s).

The value of series resistor R_S is determined by the load requirements. If R_S is too large the diode will be unable to regulate at large values of load current I_L, which is the current through R_L. If R_S is too small, the diode dissipation rating may be exceeded at low values of I_L.

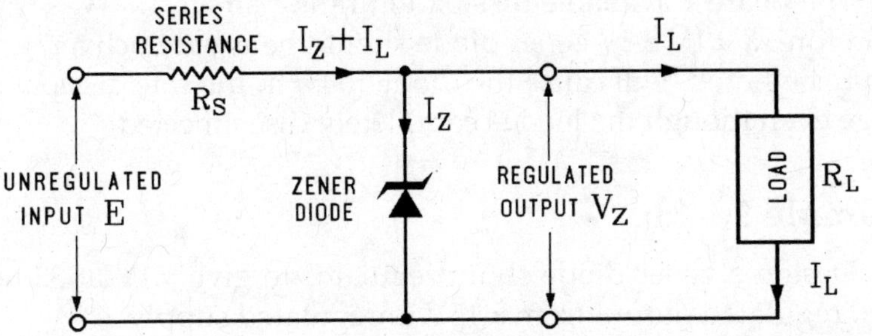

Figure 8.9: Zener regulated supply design..

In practical designs, a current of about ten percent of full load current I_L is allowed to flow through the zener diode to keep it conducting (the minimum current in low current circuits is about 5 mA). That makes the total current through the series resistance R_S to $1.1I_L$. This assures diode regulation under maximum load. The optimum value of R_S can be calculated by the equation:

$$R_S = \frac{E_{DC\,(min)} - E_Z}{1.1\,I_{L\,(max)}} \qquad (8\text{-}3)$$

When R_S is known, the maximum dissipation P_D in the zener diode may be determined by:

$$P_D = \left[\frac{E_{DC\,(max)} - E_Z}{R_S} - I_{L\,(min)}\right] E_Z$$

Example 1

A 12-volt source is to supply a circuit requiring 9 volts. The load current varies between 200 and 350 mA.

$E_Z = 9.1\text{V}$ (nearest available value)

$$R_S = \frac{12-9.1}{1.1 \times 0.35} = \frac{2.9}{0.385} = 7.5 \text{ ohms}$$

$$P_D = \left[\frac{12-9.1}{7.5} - 0.2\right] \times 9.1 = 0.185 \times 9.1 = 1.7 \text{ W}$$

The nearest available dissipation rating above 1.7 W is 5 W. Therefore, a 9.1V 5-W zener diode should be used. Such a rating, it may be noted, will cause the diode to be in the safe dissipation range even though the load is completely disconnected.

Example 2

Design a zener diode shunt regulator to give 9.1V at 33 mA max. regulated output from a 15V unregulated supply

Voltage to be dropped across Rs	15-9.1 = 5.9V
Zener diode current	5 mA (min)
Total current through R_S	33+5 = 38 mA
Series resistance R_S	5.9/38 = 155 ohms
Nearest practical value for R_S	147 ohms
Power dissipated in R_S	5.9 × 38 = 224 mW
Power dissipated in zener diode	9.1 × 5 = 45.5 mW
Maximum power dissipated in zener	9.1 × 38 = 346 mW

An important thing to be noted about this type of regulators where the regulating element is connected in shunt (parallel) with the load is that they are inherently protected against short circuits on the load side. If a short occurs across the regulated supply output terminals and continues for a long time, the maximum damage that can occur is the burn out of the series resistor R_S. The circuit is ready again as soon as the short is removed.

Low Voltage Stabilization

Zener diodes with zener voltages below 3V are not easily available. In many cases, the forward voltage drop across a rectifier diode (about 0.6 V at low currents) makes it possible to use it to serve as a low-voltage zener diode. Figure 8.10 shows a simple shunt regulated supply for giving a stabilized voltage of about 1.4V using rectifier diodes. Note the polarities of the diodes used in this case. These diodes are conducting in the forward direction.

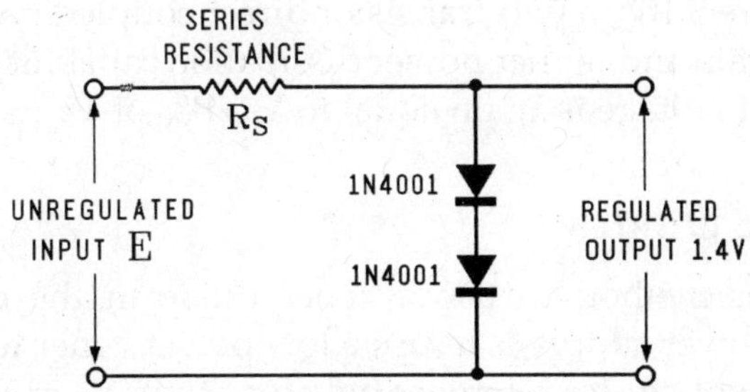

Figure 8.10: Using rectifier diodes to get low stabilized voltages.

Amplified Zener

Considerable amount of heat gets dissipated in the zener diodes when regulating voltages in high current circuits. High power zener diodes may not be readily available in the required voltage range. However, it is possible to combine a low-power zener diode with a high power transistor to simulate a high-power zener diode. Two such combinations are shown in Figure 8.11.

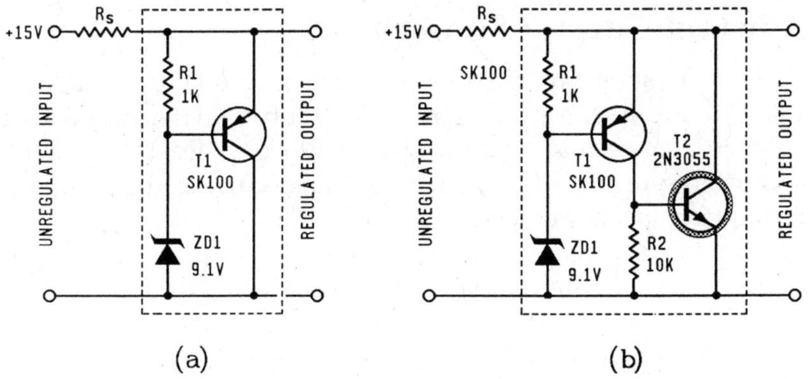

Figure 8.11: Amplified Zeners.

In Figure 8.11a, the excess voltage above zener voltage is dropped across R1. If this voltage exceeds the V_{be} of T1 (about 0.6V for silicon transistors), transistor T1 conducts and that current through T1 causes additional voltage drop across Rs. The circuit stabilizes when the regulated output voltage stands at V_Z+V_{be}.

In Figure 8.10b, a two-transistor direct-coupled pair is used to get higher gain and higher power dissipation capability. The regulated output voltage is again equal to V_Z+B_{be} of T1.

Reference Diodes

Another method to use a zener diode in the design of a stabilized power supply is to use a low power zener to generate a stable fixed voltage for reference only for comparison with another voltage. These designs are described in the next chapter.

When a zener diode is used to generate a fixed voltage for reference only, it is often called a *reference diode*. However, the zener voltage of an ordinary type of zener diode changes a little with the amount of current through it and also with temperature. Special reference diodes are made in which these variations are much smaller. Their reference voltages are quoted for a particular value of zener current. This current is often supplied from a constant current source to get a highly stable voltage reference source.

Constant Current Sources

In a constant voltage source (see Figure 8.1), the available voltage remains same irrespective of load current. That implies that the internal resistance of the source is zero, or at least very small. In constant current sources, it is the load current that remains constant irrespective of load resistance variations. An ideal current source will have infinite internal resistance. In practice, we can imagine that a constant current source is a voltage source with a very high resistance in series with it.

In Figure 8.12, a voltage source of 1000 volts has internal resistance of one megohm. If the output terminals are shorted together, the current that will flow is 1000V/1MΩ = 1 mA.

If a load resistance R_L of 100k is connected externally, the load current becomes 1000/1.1MΩ = 0.91 mA, which is approximately the same as 1 mA. In other words, the load current remains almost 1 mA even when the load resistance varies from 0 to 100k ohms.

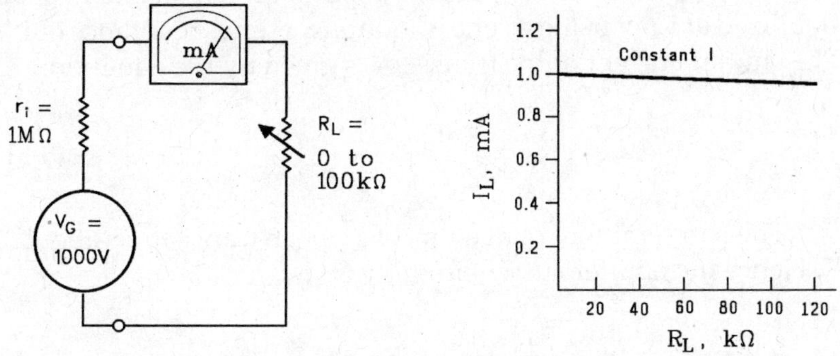

Figure 8.12: Constant current source.

It is obvious from the above example that a constant current source should have a high voltage source, which is not always convenient in practice. Other methods have therefore been evolved to simulate a constant current sources. One method makes use of the high collector impedance of a transistor.

The collector current of a transistor depends only on its base current and is not affected by changes in collector voltages. In other words, the collector impedance of a transistor is very high. Using this characteristics of transistors, two circuits, one using an n-p-n transistor, and the other using a p-n-p transistor, are shown in Figure 8.13. Both circuits operate on a positive power supply.

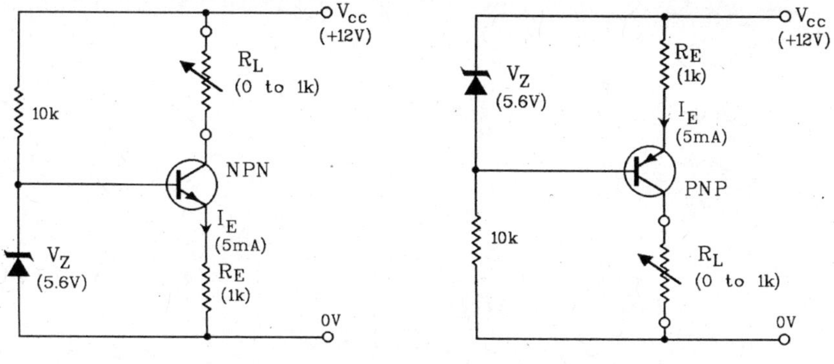

Figure 8.13: Constant current sources using transistors.

In Figure 8.13, the base of the transistor receives a fixed voltage stabilized at 5.6 V using a zener diode and a series resistance of 10k. The emitter current of the transistor is given by the equation:

$$I_E = \frac{V_Z - V_{BE}}{R_E} \quad \text{------------------} \quad (8\text{-}3)$$

The base-emitter voltage V_{BE} of a silicon transistor is about 0.6 V. Hence the emitter current in this case is:

$$I_E = \frac{5.6 - 0.6}{1000} = 5 \text{ mA}$$

If the current gain of the transistor is high, the collector current will be almost equal to the emitter current and will remain constant irrespective of load resistance variations, if they are within its range. For higher current gain, or for getting higher constant currents, a Darlington transistor can be used.

In schematic diagrams, a constant current generator is shown by a symbol shown in Figure 8.14.

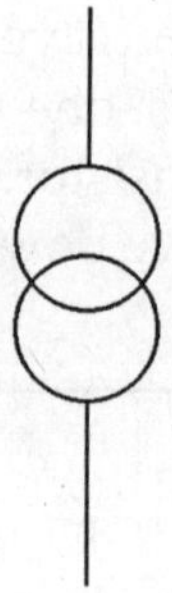

Figure 8.14: Symbol for representing a constant current source.

9

Electronic Regulators

Many electronic circuits need highly stable d.c. voltages that should that should not deviate more than, say, ± 5% under all working conditions. Very stable voltages can be obtained by adding an electronic regulator at the output of an unregulated supply as was shown in Figure 1.7.

Maintenance of a fixed level of voltage (or current) is commonly referred to as *regulation* or *stabilization*. Nowadays, most of the stabilizers use integrated circuits, but this is one method out of many, and some other methods have peculiar advantages for special purposes. In order to make effective use of stabilization, it is essential to understand the principles of their working.

Stabilization is needed against two causes of variation of output. The first one is changes in the mains supply voltage (often referred to as *line voltage variations*). This happens when heavy loads such as lifts, refrigerators and motors are switched on or off suddenly, generating a high voltage transient on the line. A good stabilizer should not pass any of these surges to the d.c. output, and in some cases, something more than simple stabilization is needed. The other cause is the *load current variation*, which can happen due to charging of capacitors, or by changes in the current requirement of the load circuit. A good stabilizer should keep the terminal output voltage steady (within specified limits) despite current variations between full rated load and zero load.

Electronic Stabilization

Stabilizers, in general, operate by wasting power. We have seen that the d.c. output voltage of a rectifier-filter type of power supply will fall to a low value at the rated load current, and this is the maximum value to which the output can be stabilized. In fact, allowing for ripples and for the effect of the stabilizer itself, a still lower value is the one at which the output can be stabilized. This implies that the amount of waste can be considerable. For example, a 5V d.c. output may need a supply of 8 to 10 V. All this leads to fairly generous voltage levels at the input to a stabilizer, and this in turn, can present heat dissipation problems.

Series Regulators

The principle of operation of series regulators was shown in Figure 8.2. The non-linear element in the regulator is usually a silicon power transistor which can dissipate the power wasted in it. A simple circuit using a zener diode and an n-p-n transistor is shown in Figure 9.1. The two circuits are identical electrically. The one shown on left is easy to understand, while the one on the right is the more conventional way of drawing series regulators in schematic diagrams..

Resistor R1 and zener diode ZD1 form a stable voltage supply of the type shown in Figures 8.7 and 8.9. Transistor T1 is connected as an emitter follower. Its base is maintained at a fixed voltage

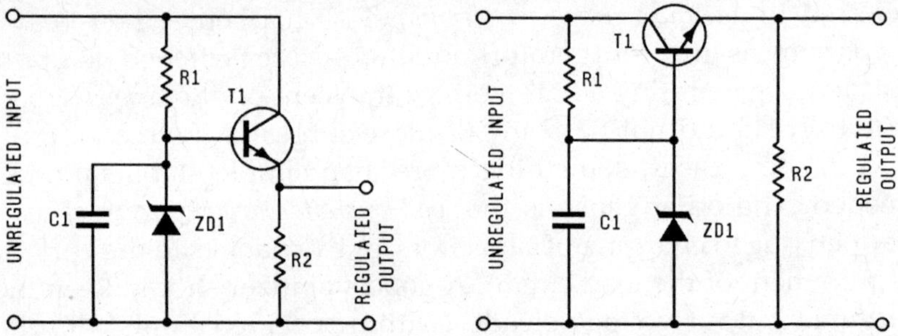

Figure 9.1: Simple regulated supply using a zener diode and a transistor.

equal to the zener voltage of the diode ZD1. Hence, the output voltage at the emitter of the transistor is held at a steady voltage which 0.6 V (V_{be} of a silicon transistor) less than the zener voltage of ZD1. Resistor R2 is not necessary but is a help in quickly establishing the circuit voltages. A capacitor is often connected across the zener diode to filter out noise generated inside the diode.

In this type of circuit, power is dissipated in the transistor only when the load is drawing current. This is different from the shunt regulator. The output voltage is stabilized against both: input voltage variations and load current variations. The maximum variation of output voltage will be equal to the variation of base-to-emitter voltage for the range of current that is being used. This may amount to no more than about 0.2 V.

For a closer control of output voltage a comparator-amplifier method is used. The principle of working of this type of voltage regulator is shown in Figure 9.2. A standard d.c. voltage (usually provided by a zener diode and denoted as a *reference voltage* or V_{ref}, and a fraction of the output voltage are compared. The algebraic difference of these two voltages, called the *error signal* is amplified, and the amplified error signal is used to control the conductivity of the series-pass transistor in such a way so as to reduce the error voltage at the input of the comparator. The system tries to maintain the output voltage at a steady level decided by circuit components. The higher the gain of the amplifier, the better is the stabilization.

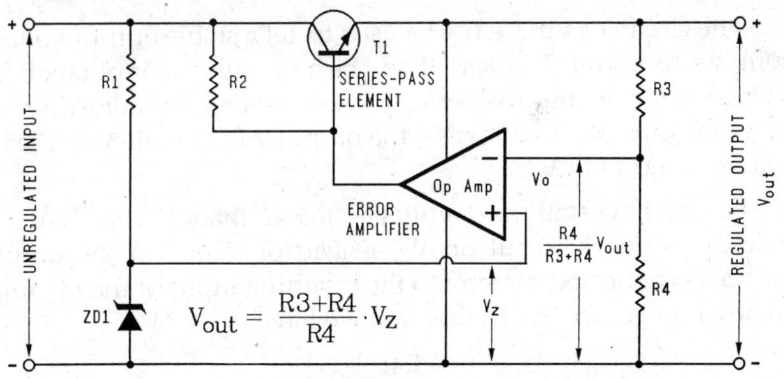

Figure 9.2: Principle of operation of series regulated power supplies.

In Figure 9.2, an operational amplifier is used for doing the task of comparison. An operational amplifier, (often referred to as an op amp), is a differential amplifier with two inputs and a very high gain. Here, the op amp is used to compare the two input signals and amplify the error signal which is the algebraic difference of the two input signals. For more information on op amps see Author's book OP AMPS MADE SIMPLE by BPB Publications.

In this circuit, the fixed reference voltage V_Z is applied to the non-inverting (+) input of the op amp while a fraction of the output voltage, as decided by R3 and R4, is applied to the inverting input (-) of the op amp. The amplified output controls the conductivity of the series-pass transistor. If the output voltage, for any reason, happens to rise, the output of the op amp goes down. This reduces the base drive to the series-pass transistor thereby reducing its conductivity and reduce the output voltage. The reverse happens when the output voltage happens to fall. The output is thus held steady within very close limits.

The voltage that is compared with the zener voltage depends on the ratio of R3 and R4 and is equal to Vz when the error is very small. Hence:

$$V_Z = V_{out} \cdot \frac{R4}{R3+R4}$$

or

$$V_{out} = \frac{R3+R4}{R4} \cdot V_Z \quad = \left(1 + \frac{R3}{R4}\right) \cdot V_Z$$

The circuit can therefore be used to get a stable output voltage from V_Z to about 3 V less than the unregulated input voltage (voltage drop across the series-pass transistor) by adjusting the ratio of R3 and R4. In this case, the output voltage cannot be set to a value lower than V_Z.

For getting stabilized voltages lower than V_Z, the voltage dividing network is put on V_Z instead of V_{out} and the output voltage is connected directly to the inverting input of the op amp. as shown in Figure 9.3. In this case we have:

$$V_{out} = V_Z \times \frac{R4}{R3+R4}$$

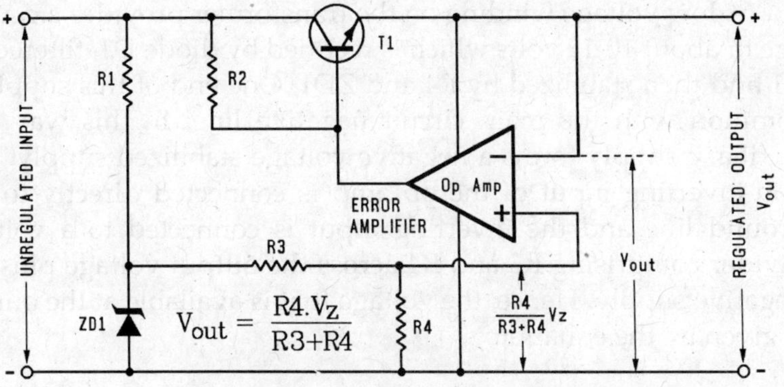

Figure 9.3: Low voltage stabilizer using an op amp.

With this circuit, the output voltage can be set from a maximum of V_Z to a low value. The output voltage can, however, not be reduced down to zero because the op amp itself does need a some voltage for its operation. If you want to be able to set the voltage from zero to a high value, a separate negative voltage regulated supply is needed for providing reference voltage. This supply is not required to deliver any power and a simple zener stabilized supply as shown in Figure 8.9 is sufficient. Figure 9.4 shows a circuit arrangement for getting voltages down to zero.

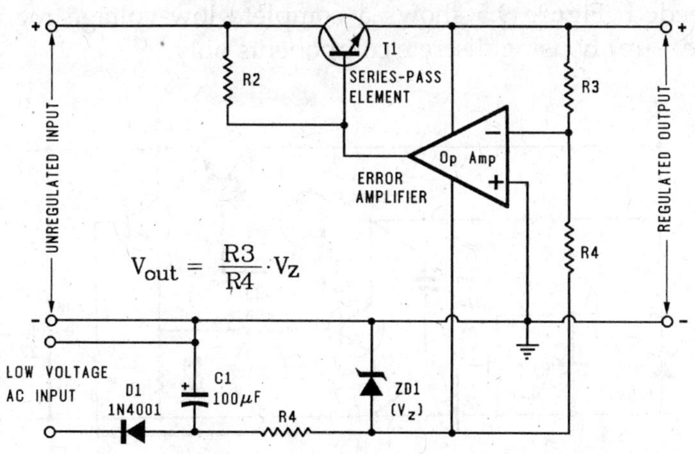

Figure 9.4: Regulated supply adjustable down to zero volts.

A low voltage winding on the transformer provides a.c. voltage of about 10-12 volts which is rectified by diode D1, filtered by C1 and then stabilized by R4 and ZD1. One end of this supply is common with the main circuit negative line. In this way this auxiliary supply forms a negative voltage stabilized supply. The non-inverting input of the op amp is connected directly to the ground line and the inverting input is connected to a voltage divider comprising R3 and R4 across the output voltage plus the negative supply. Hence, the voltage that is available at the output is given by the equation:

$$V_{out} = (1+\frac{R3}{R4})\,V_Z - V_Z \quad = \frac{R3}{R4}\,.V_Z$$

If R4 is a fixed value, the output voltage becomes directly proportional to the value of R3 and can be adjusted from zero to a value about 2 to 3 volts lower than the unregulated supply voltage, allowing for a drop in the output series transistor..

Though in the above examples we have shown an op amp for amplification of the error signal, it is possible to make a complete regulated power supply using discrete components. In fact, in some cases it is advantageous because sometimes integrated circuits are not available in the high voltage rating at which the output is needed. Figure 9.5 shows a complete low-voltage regulated power supply using discrete components only.

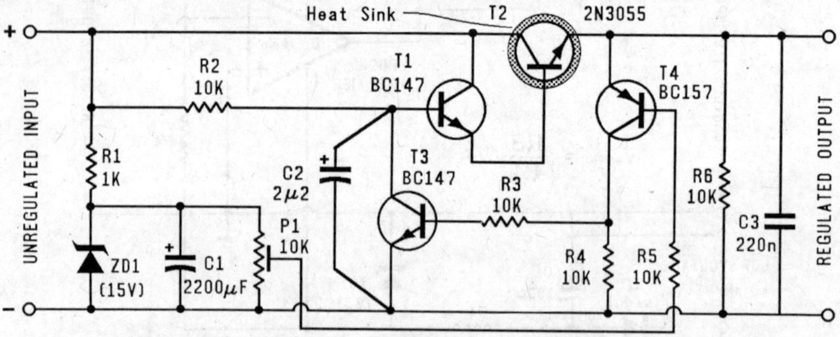

Figure 9.5: Low voltage regulated supply using discrete components.

R1, ZD1, and C1 form a low-power zener stabilized reference voltage supply of the type shown in Figure 8.7. A fraction of this output voltage, as set by the wiper position of P1, is applied to the base of transistor T4. Transistors T1 and T2 are connected in a Darlington configuration to form a high-gain d.c. amplifier. Transistor T2 is mounted on a heat sink for dissipating the heat generated by power wasted in it. Resistor R2 provides a forward bias to T1 and T2.. The forward bias current can be shunted to ground wholly or partially depending on the conductivity of T3.

If the output voltage happens to rise, transistor T4 conducts more. This makes T3 conduct more and bypass the forward bias current to the series-pass element comprising T1 and T2. The output voltage therefore falls. The reverse happens when the output voltage tends to fall. The output voltage is therefore held steady under all conditions of line voltage and load current variations.

Series Pass Transistors

The series pass transistors shown in Figures 9.1 to 9.5 may not be able to pass high currents needed by the load. If a high power transistor is used, the output of the operational amplifier may not be sufficient to provide needed drive current to its base. Two solutions to this problem are shown in Figure 9.6.

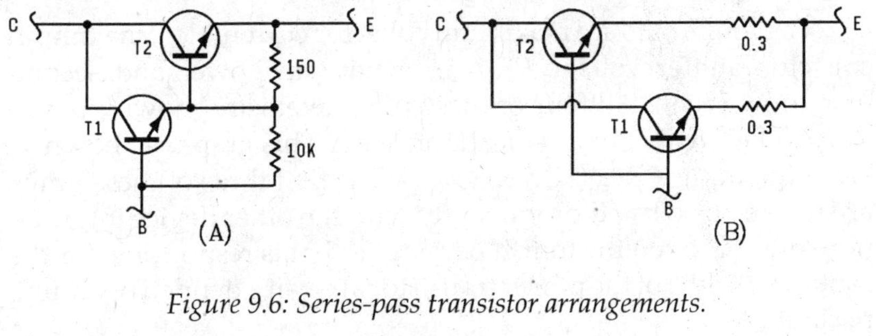

Figure 9.6: Series-pass transistor arrangements.

The Darlington arrangement shown in Figure 9.6A multiplies the beta of the pass transistor, thereby extending the control range of the error amplifier.

If the Darlington connection is implemented with discrete transistors, resistors across the base-emitter junctions, as shown in Figure 9.7A, may be necessary to prevent collector-to-base leakage currents in T1 from being amplified and turning on the transistor pair . These resistors are contained within the envelope of a monolithic Darlington device.

When a single-pass transistor is not available to handle the current required from a regulator, the current-handling capability of the series-pass element may be increased by connecting two or more pass transistors in parallel. Figure 9.6B shows the method of connecting these pass transistors in parallel. The resistances in the emitter leads of each transistor are necessary to equalize the currents. They should have adequate heat dissipating capability.

Transistor Ratings

When bipolar (n-p-n, p-n-p) power transistors are used in applications in which they are called upon the handle power on a continuous basis, rather than switching, there are four parameters that must be examined to see if any maximum limits are being exceeded. Operation of the transistor outside these limits can easily result in the failure of the device. Unfortunately, not all are aware of all these pass parameters. Yet, for a transistor to provide reliable operation, the circuit designer must not allow his power supply or amplifier to cause overstress.

The four limits are maximum collector current (I_C), maximum collector-emitter voltage (V_{CEO}), maximum power and second breakdown (I_{SB}). All four of these parameters are shown graphically in Figure 9.7 for a typical transistor. This graph is known as Safe Operating ARea (SOAR) graph. The first three of these limits are usually also listed prominently with the other device information, but it is often the fourth parameter that is responsible for the "sudden death" of the power transistor after an extended operating period.

The maximum current limit of the transistor (I_C max) is usually the current limit for fusing the wire connected to the emitter, rather than anything pertaining to the transistor chip itself. When this limit is exceeded, the bond wire can melt and open circuit the emitter. On the operating curve, this limit is shown as a horizontal line extending out from the Y-axis (zero volts between collector and emitter) and ending at the voltage point where the constant power limit begins.

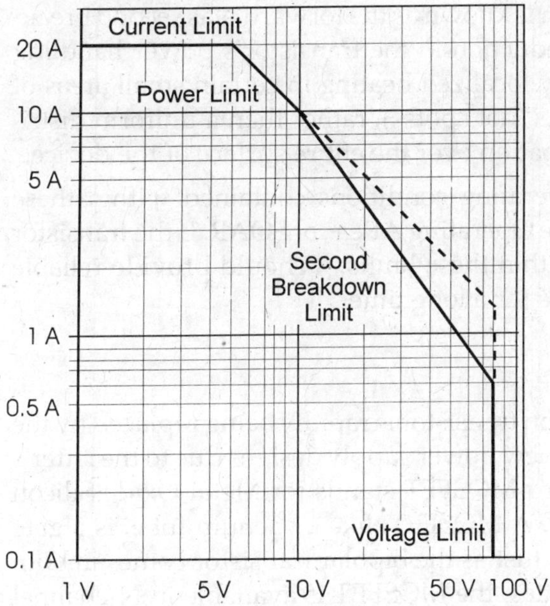

Figure 9.7: Typical graph of Safe Operating Area.

The maximum collector-emitter voltage limit of the transistor ($V_{CE\ max}$) is the point at which the transistor can no longer stand off the voltage between collector and emitter.

With increasing collector-emitter voltage drop at maximum collector current, a point is reached where the power in the transistor will cause the junction temperature to rise to a level where the device leakage current rapidly increases and begins to dominate. In this region, the product of the voltage drop and the current would be constant and represent the maximum power (Pt) rating for the transistor, that is, as the voltage drop continues to increase, the collector current must decrease to maintain the power dissipation at a constant value.

With most of the higher voltage rated transistors, a point is reached on the constant power portion of the curve whereby, with further increased voltage drop, the maximum power rating is not constant, but decreases as the collector to emitter voltage increases. This decrease in power handling capability continues until the maximum voltage limit is reached.

This special region is known as the forward bias second breakdown (FBSB) area. Reduction in the transistor's power handling capability is caused by localized heating in certain small areas of the transistor junction ("hot spots"), rather than a uniform distribution of power dissipation over the entire surface of the device.

The region of operating conditions contained within these curves is called the Safe Operating ARea, or SOAR. If the transistor is always operated within these limits, it should provide reliable and continuous service for a long time.

MOSFET Transistors

The bipolar junction transistor is rapidly being replaced by the MOSFET transistor in new power supply designs due to the latter's ease of drive. The word MOSFET stands for Metal-Oxide Silicon Field Effect Transistor. A MOSFET, like a vacuum tube, is a gate voltage driven device. Just as the bipolar transistor comes in both n-p-n and p-n-p varieties, the MOSFET is available in N-channel and P-channel types, with the N-channel being more popular of the two. The N-channel MOSFET is equivalent to the n-p-n bipolar, and the P-channel is equivalent to the pnp bipolar transistor.

There are some considerations that should be observed when using a MOSFET as a linear regulator series-pass transistor. Several volts of gate drive are needed in order to start conduction of the device, as opposed to less than one volt for the bipolar. However, no current is drawn from the driving circuit. MOSFETs are inherently very high frequency devices, and will readily oscillate with stray circuit capacitances. In order to prevent oscillation in the transistor and surrounding circuits, it is common practice to insert a small resistor of about 100 ohms directly in series with the gate of the series-pass transistor to reduce the gate circuit Q.

Regulation

The type of regulators we have described above are linear regulators. Tightness or "stiffness" of regulation of a linear regulator depends on the gain of the error amplifier and the ratio of the output scaling resistors. In any regulator, the output is cleanest and regulation stiffest at the point where the sampling network or error amplifier is connected. If heavy load current is drawn through long leads, the voltage drop in the leads can degrade the regulation at the load, To combat this effect, the feedback connection to the error amplifier can be made directly to the load. This technique, called *remote sensing,* moves the point of best regulation to the load by bringing the connecting loads inside the feedback loop. This is shown in Figure 9.8.

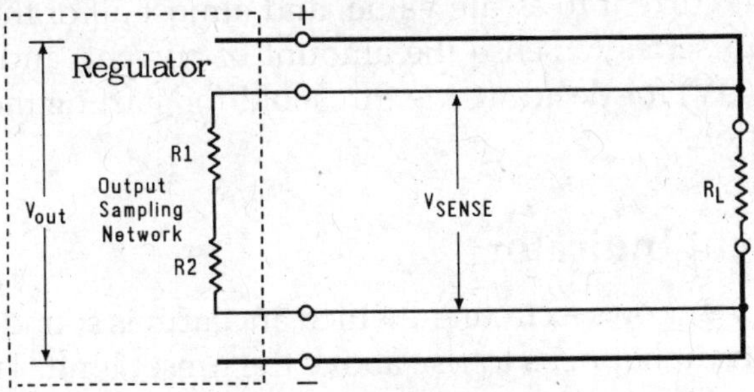

Figure 9.8: Remote sensing to overcome voltage drop in connecting wires.

Input Versus Output Voltage

In a series regulator, the pass-transistor power dissipation is directly proportional to the load current and input/output voltage differential. The series pass element can be located in either leg of the supply. Either n-p-n or p-n-p devices can be used, depending on the ground polarity of the unregulated input.

The differential between the input and output voltages is a design tradeoff. If the input voltage from the rectifiers and filter is only slightly higher than the required output voltage, there will be minimal voltage drop across the series pass transistor resulting in minimal thermal dissipation and high power-supply efficiency. The supply will have less capability to provide regulated power in the event of power line brownout and other reduced line voltage conditions, however. Conversely, a higher input voltage will provide operation over a wider range of input voltage, but at the expense of increased heat dissipation in the series-pass element.

Overcurrent Protection

None of the series-pass element type of regulators described above have a protection against shorts in the load circuits. If the output terminals happen to be shorted, a heavy current will flow and burn out the series-pass transistor, and possibly a few other components also. Several techniques have been used to limit the short-circuit current to a safe value, and almost all of them use a resistance in series to sense the amount of current, and the V_{BE} (approx. 0.65V) for deciding the threshold for starting the current limiting action.

Overcurrent Indicator

Figure 9.9 shows a circuit in which an alarm is sounded when the load current happens to rise above the preset limit. Transistor T1 and piezo-electric buzzer form an alarm circuit that sounds when load current exceeds a certain value.

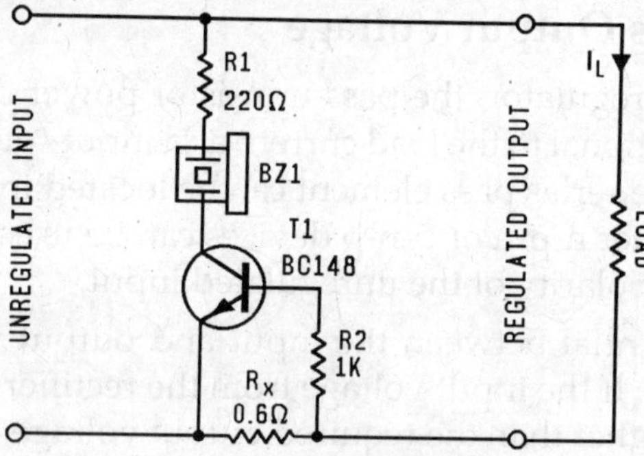

Figure 9.9: Overload current alarm.

In this circuit, all the load current passes through resistance R_X and develops a voltage equal to $I_L R_X$ across it. So long as this voltage is less than V_{BE} of T1, nothing happens because transistor T1 does not conduct. When the load current exceeds V_{BE}/R_X, T1 conducts and sounds a high-pitched alarm. For a particular value of maximum load current I_L the value of R_X can be calculated by:

$$R_X = 0.65/I_L$$

Current Limiting

As stated earlier, damage to a pass transistor can occur when the load current exceeds the safe amount. Figure 9.10 illustrates a simple current-limiter circuit that will protect T1.

All of the load current is routed through R1. A voltage difference will exist across R1, the value of which will depend on the exact load current at a given time. When the load current exceeds a predetermined safe value, the voltage drop across R1 will forward bias T2 and cause it to conduct. Because T2 is a silicon transistor, the voltage drop across R1 must exceed 0.6V to turn T2 on. This being the case, R1 is chosen for a value that provides a drop of 0.6 V when the maximum safe load current is drawn. In this instance, the drop will be 0.6 V when I_L exceeds 0.5 A. Resistance R2 protects the base-emitter junction of T2 from current spikes and from destruction of T2 in the event T1 fails under short-circuit conditions.

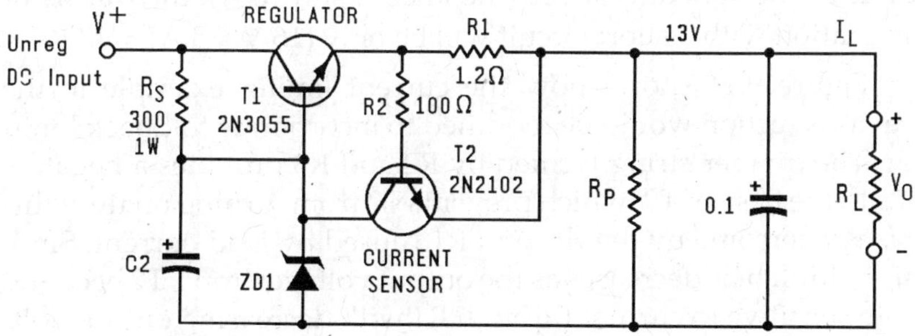

Figure 9.10: Overload protection for a regulated supply.

When T2 turns on, some of the base current through R_S flows through T2. thereby depriving T1 of some of its base current. This action, depending upon the amount of T1 base current at a precise moment, cuts off T1 conduction to some degree, thus limiting the current through it.

Since all the load current passes through R1, it should have adequate heat dissipation capability. Capacitor C2, as stated earlier, bypasses noise generated inside a zener diode because of the internal avalanche action.

Foldback Current Limiting

Under short-circuit conditions, a constant-current type current limiter must still withstand the full source voltage and limited short circuit current simultaneously, which can impose a very high power dissipation or second breakdown stress on the series pass transistor. For example, a 12-V regulator with current limiting set for 10A and having a source of 16 V will have a dissipation of 40 W,[(16-12 V) × 10A] at the point of current limiting. Its dissipation will rise to 160 W (16 V × 10A) under short-circuit conditions.

A modification of the limiter circuit can cause the regulated output current to decrease with decreasing load resistance after the overcurrent knee. With the output shorted, the output current is only a fraction of the knee current value, which protects the series pass transistor from excessive dissipation and failure. Using the previous example of the 12-V, 10-A regulator, if the short-circuit current is designed to be 3A (the knee is still 10 A), the transistor dissipation with a short circuit will be only (16 V × 3 A) = 48 W.

Figure 9.11 shows how the current limiter example in the previous section would be modified to incorporate foldback limiting. The divider string formed by R2 and R3 provides a negative bias to the base of T2, which prevents T2 from turning on until this bias is overcome by the drop in R1 caused by load current. Since this holdoff bias decreases as the output voltage drops, T2 becomes more sensitive to current through R1 with decreasing output voltage. This is shown in Figure 9.12.

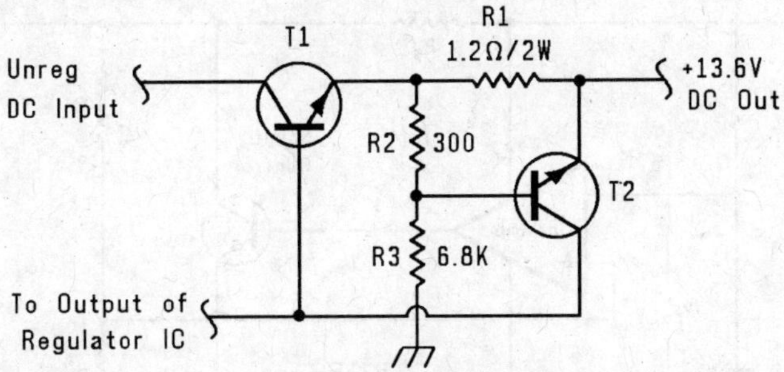

Figure 9.11: Circuit for foldback current limiting.

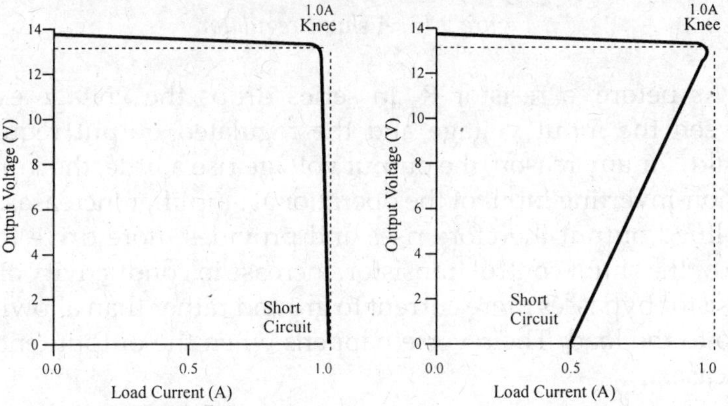

Figure 9.12: Current limiting; normal on left and foldback on right.

The circuit is designed by first calculating the value of R1 for short-circuit current. For example, if 0.5 A is chosen, the value for R1 is simply 0.6 V/ 0.5 A = 1.2 Ω.

Shunt Regulators

The principle of working of a shunt type of regulator was described with reference to Figure 8.9. A practical circuit using a zener diode for providing reference voltage, an operational amplifier as an error amplifier, and a high power transistor as the regulating element connected in shunt. Note the difference in the input connections of this circuit with circuit in Figure 9.3.

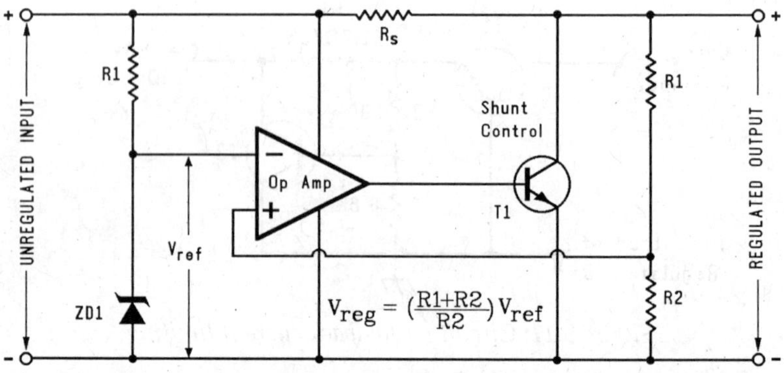

Figure 9.13: A shunt regulator.

As before, a resistor R_S in series drops the voltage excess between the input voltage and the regulated output required. Should, for any reason, the output voltage rise a little, the input at the non-inverting input of the operational amplifier increases. The amplified output therefore rises and provides more drive to the base of the shunt control transistor. Increase in conductivity of this transistor bypasses more current to ground rather than allowing it to go to the load. The reverse happens when the output tends to decrease.

Shunt regulators are, by nature, protected against overloads. The short-circuit current is limited by R_S. It may burn out during prolonged shorts.

Crowbar Protection

If the series-pass transistor in a regulated power supply happens to get shorted, the output voltage will soar high and damage the components in the load. To prevent such an eventuality, a crow bar circuit is used which blows the fuse when such a thing happens and protects the power supply as well as the load circuit from further damage. The basic principle of working of crow bar circuits was explained with reference to Figure 6.5. Practical crowbar circuits for protection are shown in Chapter 14.

10

IC Regulators

We have seen that it is possible to make an electronic regulated power supply using discrete components only as well as using operational amplifiers with other discrete components. To make the job of power supply design easy, several ready-made ICs (integrated circuits) containing all the essential circuitry are available in the market. Specifications, electrical characteristics, pin connections and methods of using some of the popular ones are described in this chapter.

CA3085 Series Regulators

Positive Voltage Regulators for regulated voltages from 1.7 V to 46 V at currents up to 100 mA.

CA3085 and CA3085A are silicon monolithic integrated circuits designed for service as voltage regulators at output voltages ranging from 1.7 to 46 volts at currents up to 100 milliamperes.

These types are supplied in 8-lead TO-5 package and the TO-5 with dual-in-line formatted leads. The CA3085 is also supplied in the 8-lead dual-in-line plastic package. Pin connections for the same are shown in Figure 10.1.

The CA3085A is unilaterally interchangeable with the CA3085.

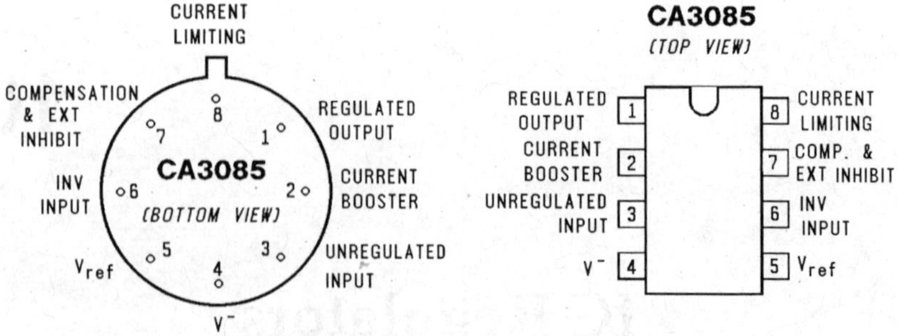

Figure 10.1: Pin connections for CA3085 voltage regulator IC.

A block diagram of CA3085 Series is shown in Figure 10.2. The diagram shows the connecting terminals that provide access to the regulator circuit components. The voltage regulator provides important features such as frequency compensation, short-circuit protection, temperature compensated reference voltage, current limiting and booster input. These devices are useful in a wide range of applications for regulating high current, switching, shunt, and positive and negative voltages. They are also applicable for current and dual tracking regulations.

The CA3085A and CA3085B have output current capabilities up to 100 mA, and the CA3085 up to 12 mA without the use of external pass transistors. The CA3085 Series has an unregulated input voltage ranging from 7.5 to 30 V (CA3085), 7.5 to 40 V (CA3085A), and 7.5 to 50 V (CA3085B), and a minimum regulated output voltage of 26 V (CA3085), 36 V (CA3085A), and 46 V (CA3085B). This is summarized in the following table.

Type	V_{IN} Range (V)	V_{OUT} Range (V)	Max. I_{OUT} (mA)	Max. Load Regulation (%V_{out})
CA3085	7.5 to 30	1.8 to 26	12	0.1
CA3085A	7.5 to 40	1.8 to 36	100	0.15
CA3085B	7.5 to 50	1.8 to 46	100	0.15

Features:

- ☐ Up to 100 mA output current
- ☐ Input and output short-circuit protection
- ☐ Load and line regulation : 0.025%
- ☐ Pin compatible with LM100 Series
- ☐ Adjustable output voltage
- ☐ Low noise

Applications

- ☐ Shunt voltage regulator
- ☐ Current regulator
- ☐ Switching voltage regulator
- ☐ High-current voltage regulator
- ☐ Combination positive and negative voltage regulator
- ☐ Dual tracking regulator

Absolute Maximum Rating Values $T_A = 25°C$

Power Dissipation Without Heat Sink	With Heat Sink (TO-5 Only)
Up to $T_A = 55°C$ --------630 mW	Up to $T_C = 55°C$ = ------1.6 W
Above $T_A = 55°C$ ---- derate @ 6.67 mW/°C	Above $T_C = 55°C$ - derate @ 16.7 mW/°C

Fundamentally, the circuit shown in Figure 10.2 consists of a frequency-compensated error amplifier which compares an internally generated reference voltage with a sample of the output voltage and controls the series-pass amplifier to regulate the output. The starting circuit assures stable latch-on of the voltage reference circuitry. The current limiting portion of the circuit is an operational feature that protects the IC in the event of overload.

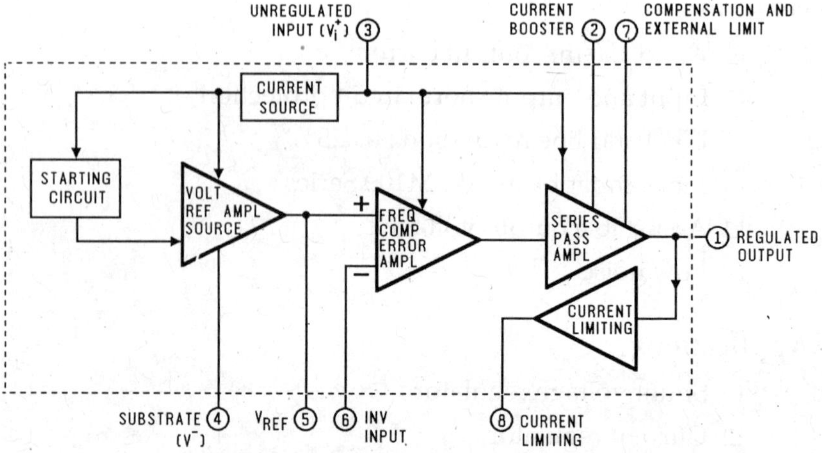

Figure 10.2: Block diagram of CA3085 voltage regulator.

The IC has a built-in reference voltage source of about 1.5 V. Terminal 5 provides this voltage as a source of stable reference voltage for auxiliary use. A current of about 250 microamperes can be supplied to an external circuit without significantly disturbing the reference- voltage stability. If necessary, filtering of the inherent noise of the reference-voltage circuit can be accomplished by connecting a suitable bypass capacitor between terminals 5 and 4.

Terminal 6 ("inverting input" in accordance with operational amplifier terminology) is the input through which a sample of output regulated voltage is applied.

The collector of the series-pass output transistor is brought out at terminal 2 ("current booster") to provide base drive to an external p-n-p transistor; this method is one method of regulating currents greater than 100 milliamperes.

Because the voltage regulator is essentially an operational amplifier having considerable feedback, frequency compensation is necessary in some circuits to prevent oscillations. Terminal 7 is provided for external frequency compensation; it can also be used to "inhibit" (strobe, squelch, pulse, key) the operation of the series-pass amplifier.

Application Hints

A Simple Voltage Regulator

Figure 10.3 shows the schematic diagram of a simple power supply using the CA3085. The a.c. supply voltage is stepped down by a transformer, full-wave rectified by the diode bridge, and smoothed by a large electrolytic capacitor C1 to provide unregulated d.c. to the CA3085 regulator circuit. Frequency compensation of the error amplifier is provided by capacitor C2. Capacitor C3 bypasses residual noise in the reference-voltage source, and thus decreases the incremental noise-voltage in the regulator circuit output.

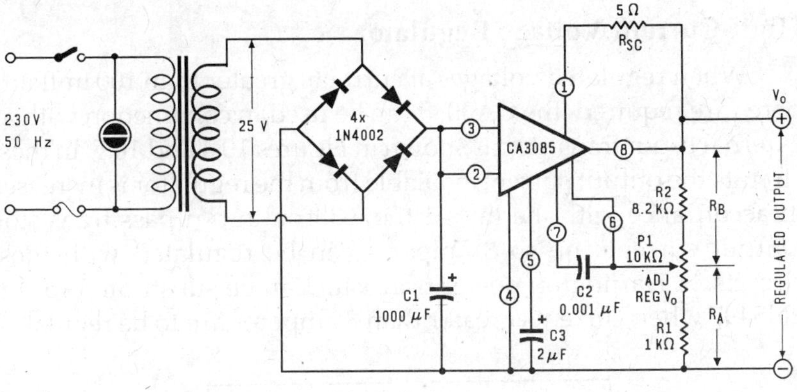

Figure 10.3: Basic power supply using CA3085.

Because the open-loop gain of the error amplifier is very high (greater than 1000), the output voltage may be directly calculated from the following expression:

$$V_O = \frac{R_B + R_A}{R_A} \cdot V_{ref} \quad \text{-------------------- (10-1)}$$

In the circuit shown in Figure 10.3, the output voltage can be adjusted from 7 to 20 volts by varying P1. The maximum output current is determined by R_{SC}.

The line and load regulation for this circuit are approximately 0.25 percent of the output voltage. The maximum output current under short-circuit conditions is limited by R_{SC}. The short-circuit current is determined by :

$$I_{SC} = \frac{V_{BE}}{R_{SC}} = \frac{0.6}{R_{SC}} \text{ amperes} \quad \text{-------------- (10-2)}$$

When this circuit is used to provide high output currents at low output voltages, care must be exercised to avoid excessive IC dissipation. In the circuit shown in Figure 10.3, this dissipation control can be accomplished by using a step down transformer with lower output voltages (a reduction to V_i), or by using a dropping a resistor between the rectifier and the CA3085 regulator.

High-Current Voltage Regulator

When regulated voltages at currents greater than 100 milliamperes are required, the CA3085 can be used in conjunction with an external n-p-n transistor as shown in Figures 10.4 and 10.5. In these circuits the output current available from the regulator is increased in accordance with the h_{FE} of the external n-p-n pass-transistor. Output currents up to 8 amperes can be regulated with these circuits. A Darlington power transistor can be substituted for the 2N5497 when currents greater than 8 amperes are to be regulated.

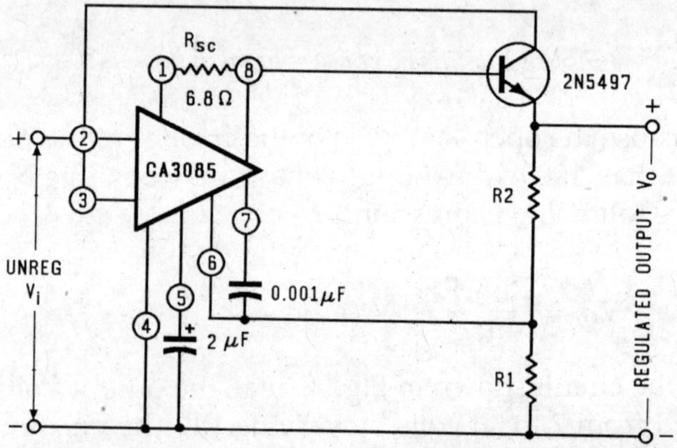

Figure 10.4: Power supply with simplified short-circuit protection.

A simplified method of short circuit protection is used in the circuit of Figure 10.4. The variable resistor R_{SC} serves two purposes: (1) it can be adjusted to optimize the base drive requirements (hFE) of the particular 2N5497 transistor being used, and (2) in the event of a short-circuit in the regulated output voltage the base drive current in the 2N5497 will increase, thereby increasing the voltage drop across R_{SC}. As this voltage drop increases the short-circuit protection system within the CA3085 correspondingly reduces the output current available at terminal 8. It should be noted that the degree of short-circuit protection depends on the value of R_{SC}, i.e., design compromise is required in choosing the value of R_{SC} to provide the desired base drive for the 2N5497 while maintaining the desired short-circuit protection.

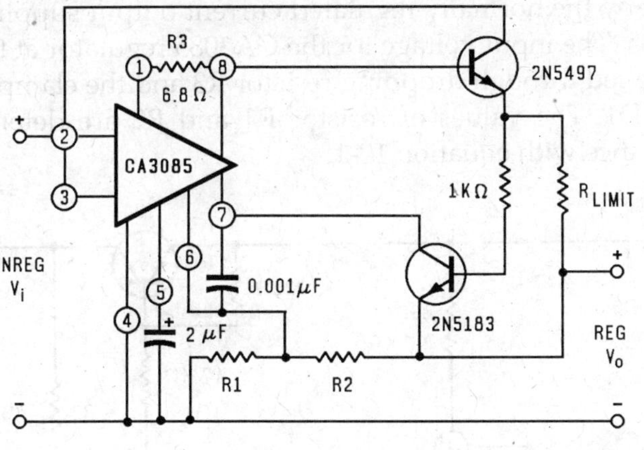

Figure 10.5: Power supply with auxiliary short-circuit protection.

Figure 10.5 shows an alternate circuit in which an additional transistor 2N5183 and two resistors have been added as an auxiliary short-circuit protection feature. Resistor R3 is used to establish the desired base drive for the 2N5497, as described above. Resistor R_{LIMIT} now controls the short-circuit output current because, in the event of a short-circuit, the voltage drop developed across its terminals increases sufficiently to increase the base drive to the 2N5183 transistor. This increase in the base drive results in the reduced output from the CA3085 because collector current flow in the 2N5183 diverts base drive from the Darlington output stage of CA3085 (inside the IC) through terminal 7. The load regulation of this circuit is typically 0.025 percent/volt change in the input voltage.

High Voltage Regulator

Figure 10.6 shows a circuit that uses the CA3085 as a voltage reference and regulator control device for high voltage power supplies in which the voltages to be regulated are well above the input-voltage ratings of the CA3085-series circuits. The external transistors T1 and T2 require voltage ratings in excess of the maximum input voltage to be regulated. Series-pass transistor T2 is controlled by the collector current of T1, which in turn, is controlled by the normally regulated current output supplied by the CA3085. The input voltage for the CA3085 regulator at terminal 3 is supplied through dropping resistor R3 and the clamping zener diode D1. The values of resistor R1 and R2 are determined in accordance with equation 10-1.

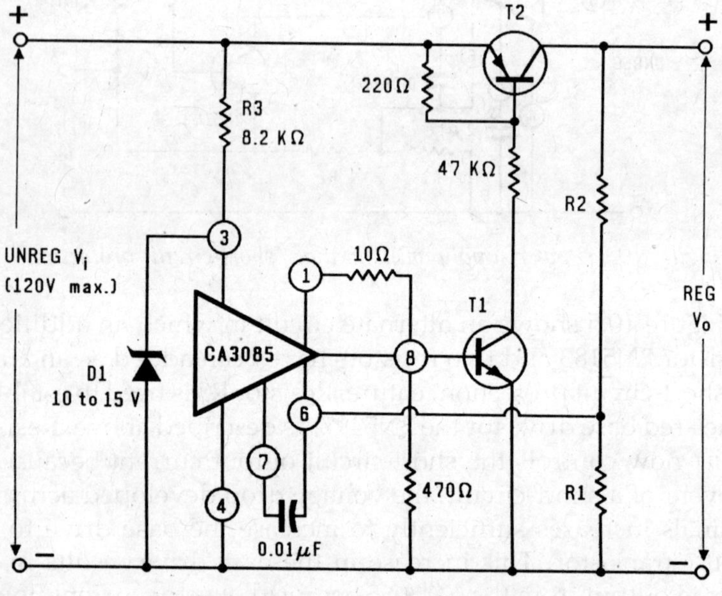

Figure 10.6: A high voltage regulator.

High-Output-Current Voltage Regulator With Foldback Current Limiting.

In high-current voltage regulators employing constant current limiting (e.g., Figures 10.4 and 10.5), it is possible to develop excessive dissipation in the series-pass transistor when a short-circuit develops across the output terminals. This situation can be avoided by the use of the "foldback" current-limiting circuitry as shown in Figure 10.7. In this circuit, terminal 8 of the CA3085 senses the output voltage, and terminal 1 is tied to a tap on a voltage-divider network connected between the emitter of the pass-transistor T3 and ground. The current-foldback-trip point is established by the value of resistor R_{SC}.

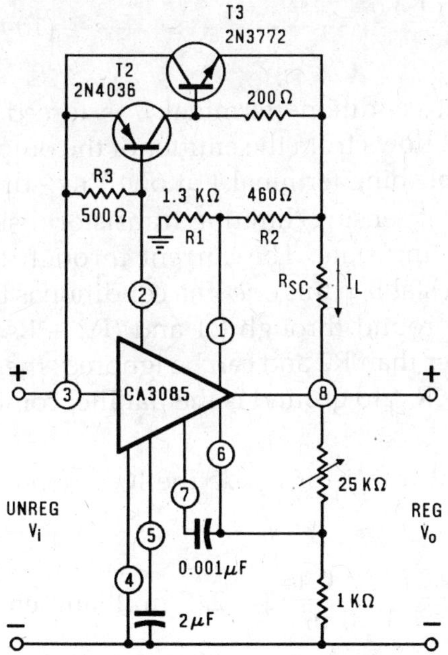

Figure 10.7: High output current voltage regulator with "foldback" current limiting.

The protective tripping is accomplished by forward biasing the current sensing/limiting transistor inside CA3085. Conditions for tripping circuit operation are defined by the following expression:

V_{BE} (of internal transistor)

$$= \text{(voltage at terminal 1)} - \text{(output voltage)}$$

$$= \left[(V_O + I_L R_{SC}) \times \frac{R1}{R1 + R2}\right] - V_O \text{ - - - - - - - - (10-3)}$$

If $\frac{R1}{R1 + R2} = K$, then

$$V_{BE} = (V_O + I_L R_{SC}) K - V_O$$

$$= KV_O + KI_L R_{SC} - V_O$$

And therefore

$$R_{SC} = \frac{V_O + V_{BE} - KV_O}{K I_L} \text{ - - - - - - - - - - - - - - - - - (10-4)}$$

Under load short-circuit conditions terminal 8 is forced to ground potential and current flows from the emitter of the output transistor inside the IC, establishing terminal 1 at one V_{BE} - drop ($\cong$ 0.7V) above and the current sensing/limiting transistor inside the IC in a partially conducting state. The current through the output transistor in the IC to establish this one V_{BE} condition is the sum of currents flowing to ground through R1 and (R2 + R_{SC}). Normally R_{SC} is much smaller than R2 and can be ignored; therefore the equivalent resistance R_{eq} to ground is the parallel combination of R1 and R2.

The current in the output transistor inside the IC CA3085 is given by:

$$I = \frac{V_{BE}}{R_{eq}} = \frac{V_{BE}}{\frac{R1R2}{R1 + R2}} = \frac{0.7\,(1.3 + 0.46)}{1.3 \times 0.46} = 2.06 \text{ milliamperes}$$

This current provides a voltage between terminals 2 and 3 as follows:

$$V_{2-3} = I \times 250 \text{ ohms} = 2.06 \times 10^{-3} \times 250 = 0.515 \text{ volt}$$

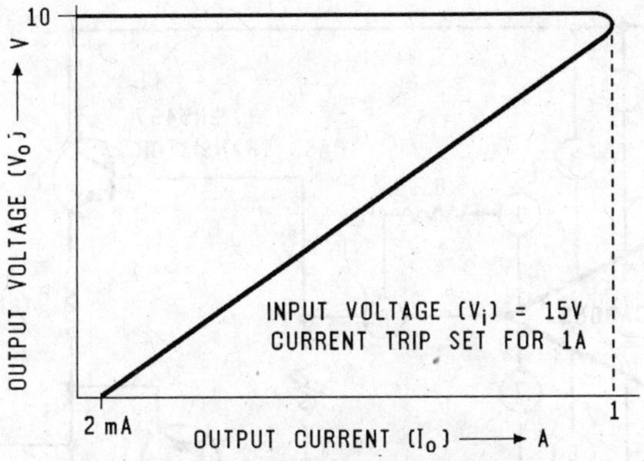

Figure 10.8: Typical foldback current limiting characteristic for circuit of Figure 10.7.

The effective resistance between terminals 2 and 3 is 250 ohms because the external 500 Ω resistor R3 is in parallel with the internal 500 Ω resistor inside the IC. It should be understood that the $V_{2\text{-}3}$ potential of 0.515 volt is insufficient to maintain the external p-n-p transistor T2 in conduction, and, therefore, T3 has no base drive. Thus, the output current is reduced to zero by the protective circuitry. Figure 10.8 shows the foldback characteristic typical of the circuit of Figure 10.7.

High-Output-Current Voltage Regulator Using Auxiliary Transistor For Foldback Current Limiting

An alternative method of providing foldback current limiting is shown in Figure 10.9. The operation of this circuit is similar to that in Figure 10.7 except that the foldback transistor T2 is external to the CA3085 to permit added flexibility in protection circuit design.

Under low-load conditions T2 is effectively reversed biased by a small amount, depending upon the values of R3 and R4. As the load current increases, the voltage drop R_{TRIP} increases, thereby raising the voltage at the base of T1, and T2 starts to conduct. As T2 becomes increasingly conductive it diverts base current from transistors inside CA3085, and thus reduces the base drive to the

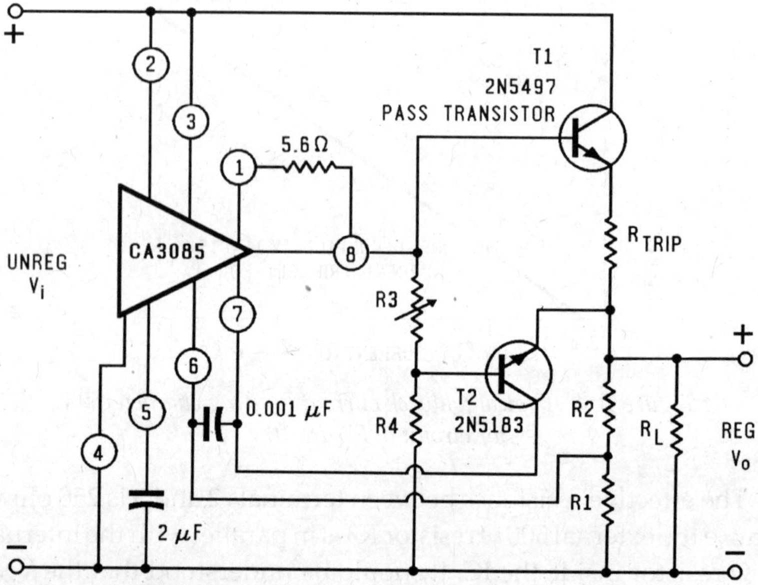

Figure 10.9: High-output-current voltage regulator using auxiliary transistor to provide foldback current limiting.

external pass-transistor T1 with a consequent reduction in the output voltage. The point at which current-limiting occurs, I_{TRIP}, is calculated as follows:

V_{BE} (T1) = voltage at terminal 8 - output voltage V_o (assuming a low value for R_{TRIP})

$$V_{BE}\,(T2) = \text{voltage at terminal } 8 \cdot \left(\frac{R4}{R3 + R4}\right) - V_o$$

$$= \left[V_o + I_L\,R_{TRIP} + V_{BE(T1)}\right]\left[\frac{R4}{R3 + R4}\right] - V_o$$

If $\dfrac{R4}{R3 + R4} = K$, then the trip current is given by:

$$I_{TRIP} = \frac{V_{BE(T2)} - K\left[V_o + V_{BE(T1)}\right] + V_o}{K\,R_{TRIP}} \text{-----------(10-4)}$$

In the circuit of Figure 10.7, the load current goes to zero when a short circuit occurs. In the circuit of Figure 10.9 the load current is significantly reduced but does not go to zero. The value of I_{SC} is computed as follows:

$$V_{BE(T2)} + \left[\frac{V_{BE(T2)}}{R2} + I_{B(T2)}\right] R1 = V_{BE(T1)} + I_{SC} R_{TRIP}$$

That gives

$$I_{SC} = \frac{V_{BE(T2)} + \left[\frac{V_{BE(T2)}}{R2} + I_{B(T2)}\right] R1 - V_{BE(T1)}}{R_{TRIP}} \quad \text{---- (10-5)}$$

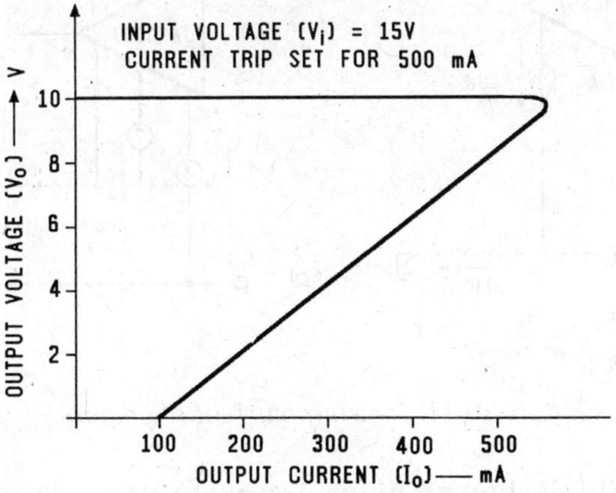

Figure 10.10: Typical foldback current limiting characteristic for circuit of Figure 10.9.

Figure 10.10 shows that the transfer characteristic of the load current is essentially linear between the "trip point" and the "short-circuit" point.

Current Regulators

The CA3085 series of regulators provide a constant source or sink current. A regulated-current supply capable of delivering up to 100 milliamperes is shown in Figure 10.11A. The regulated load current is controlled by R1 because the current flowing through this resistor must establish a voltage difference between terminals 6 and 4 that is equal to the internal reference voltage developed between terminals 5 and 4. The actual regulated current, reg I_L, is the sum of quiescent regulator current and the current through R1. That is:

$$\text{Reg } I_L = I_{quiescent} + I_{R1}$$

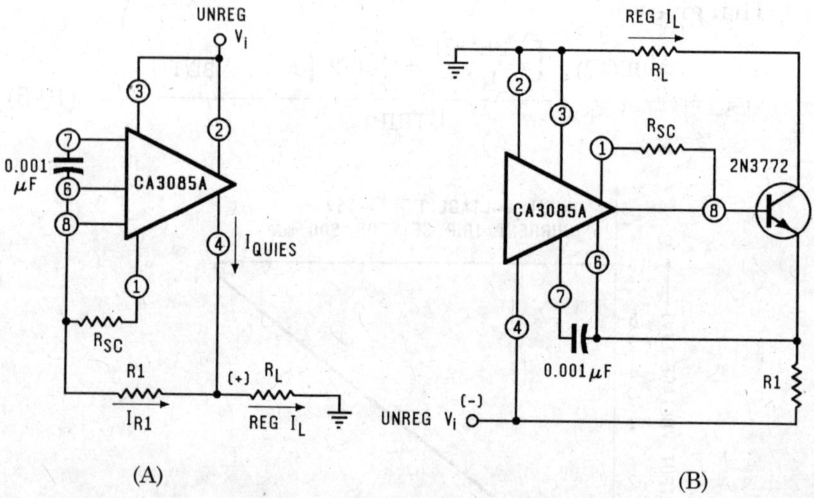

Figure 10.11: Constant current regulators.

Figure 10.11B shows a high current regulator circuit using the CA3085 in conjunction with an external n-p-n transistor to regulate currents up to 3 amperes. In this circuit the quiescent regulator current does not flow through the load and the output current can be directly programmed by R1. That is:

$$\text{Reg } I_L = \frac{V_{ref}}{R1}$$

With this regulator current between 1 milliamperes and 3 amperes can be programmed directly. At currents below 1 milliampere, inaccuracies may occur as a result of leakage in the external transistor.

CA723 Type Voltage Regulators

For Regulated Output Voltages from 2 V to 37 V at Output Currents up to 150 mA Without External Pass Transistor

Integrated circuits types CA723 and CA723C are silicon monolithic integrated circuits designed for service as voltage regulators at output voltages ranging from 2 to 37 volts at currents up to 150 milliamperes.

Each type includes a temperature compensated reference amplifier, an error amplifier, a power series-pass transistor, and a current limiting circuit. They also provide independently accessible inputs for adjustable current limiting and remote shut down and, in addition, feature low standby current drain, low temperature drift, and high ripple rejection.

The CA723 and 723C may be used with positive and negative power supplies in a wide variety of series, shunt, switching, and floating regulator applications. They can provide regulation at load currents greater than 150 milliamperes and in excess of 10 amperes with the use of suitable n-p-n or p-n-p external pass transistors.

The CA723 and CA723C are supplied in the 10-lead TO-5 type ceramic package (T suffix), and 14-lead dual-in-line plastic package (E suffix), and are direct replacements for industry types 723, 723C, µA723 and µA723C in packages with similar terminal arrangements. They are also available in chip form (H suffix). Pin connections for the metal case type and the plastic dual-in-line type packages shown in Figure 10.12.

All types are rated for operation over the full military temperature range of -55°C. to +125°C

To make it easy to understand the working of the CA723 types of regulators, a block diagram of the circuit inside the package is shown in Figure 10.13. It can be seen that the package contains all essential elements to make a good voltage regulator. Availability of pins for external connections make this IC versatile.

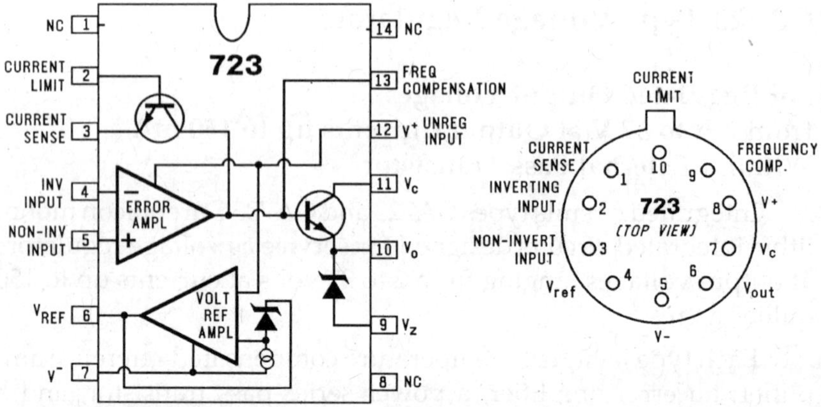

Figure 10.12: Pin connections for two types of packages of CA723

Features

- ☐ Up to 150 mA output current
- ☐ Positive and negative voltage regulation
- ☐ Regulation in excess of 10 A with suitable pass transistors
- ☐ Input and output short-circuit protection
- ☐ Load and line regulation: 0.03%
- ☐ Direct replacement for 723 and 723C industry types
- ☐ Adjustable voltage: 2 to 37 V
- ☐ Suitable for switching regulators

Applications

- ☐ Series and shunt voltage regulator
- ☐ Floating regulator
- ☐ Switching voltage regulator
- ☐ High-current voltage regulator
- ☐ Temperature controller

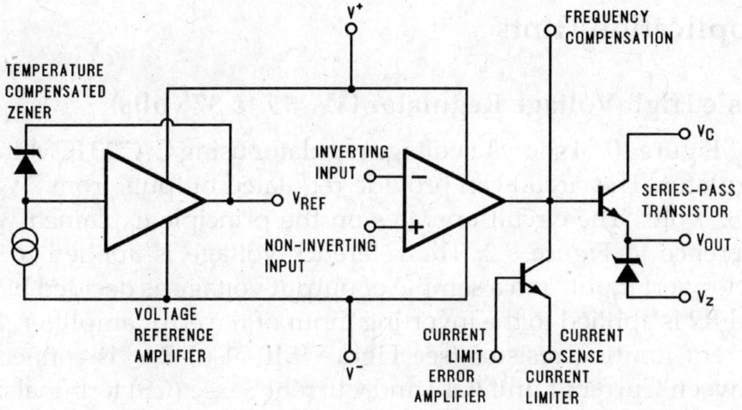

Figure 10.13: Block diagram of circuit inside a CA723 integrated circuit.

Maximum Ratings

DC SUPPLY VOLTAGE (Between V^+ and V^- Terminals)	40	V
PULSE VOLTAGE FOR 50 ms PULSE WIDTH (Between V^+ and V^- Terminals)	50	V
DIFFERENTIAL INPUT–OUTPUT VOLTAGE	40	V
DIFFERENTIAL INPUT VOLTAGE Between inverting and non-inverting inputs) Between non-inverting input and V^-	 ± 5 8	 V V
CURRENT FROM ZENER DIODE TERMINAL (Vz)	25	mA
CURRENT FROM VOLTAGE REF. TERMINAL	15	mA
DEVICE DISSIPATION (Up to T_A = 25°C) CA723T, CA723CT CA732E, CA723CE	 800 1000	 mW mW
DEVICE DISSIPATION (Above T_A = 25°C) CA723T, CA723CT Derate linearly) CA732E, CA723CE Derate linearly	 6.3 8.3	 mW/°C mW/°C
AMBIENT TEMPERATURE RANGE Operating Storage	 -55 to +125 -65 to +150	 °C °C
LEAD TEMPERATURE (During soldering) for 10 seconds max.	+265	°C

Application Hints

Basic High Voltage Regulator (V_O = 7 to 37 volts)

Figure 10.14 shows a voltage regulator using CA723 IC voltage regulator. The circuit can provide regulated outputs from 7 volts to 37 volts. The circuit operates on the principle explained with reference to Figure 9.2. The reference voltage is applied to the non-invert input, and a sample of output voltage as decided by R1 and R2 is applied to the inverting input of the error amplifier. The current limiting resistor (see Figure 9.10 also) R_{SC} is connected between Current Limit (CL) and Current Sense (CS) terminals.

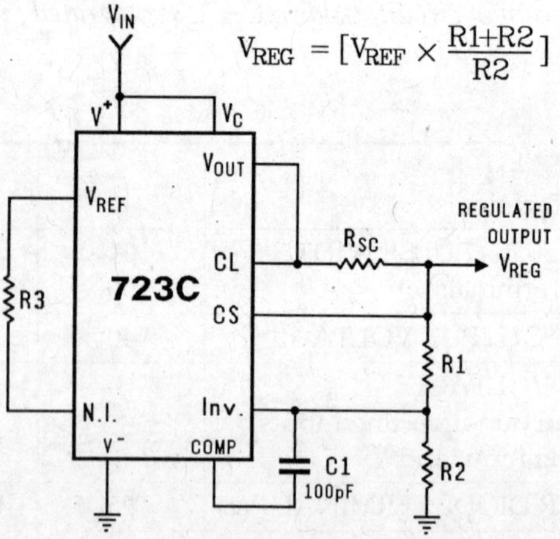

Figure 10.14: Basic high voltage regulator using CA723 IC.

Typical circuit performance data of this circuit is given below:

Regulated output voltage	15 V
Line regulation (ΔV_i = 3V)	1.5 mV
Load regulation (ΔI_L = 50 mA)	4.5 mV

Note that the reference voltage V_{REF} is connected to the non-inverting input through a resistor R3. For minimum temperature drift. The value of R3 is given by:

$$R3 = \frac{R1\,R2}{R1 + R2}$$

R3 can be eliminated for minimum component count.

An auxiliary output is also available from V_Z output terminal which is equal to the output voltage minus the zener voltage (about 7 V) of zener inside the IC. The maximum current that can be drawn from this terminal is limited to 25 mA.

Basic Low Voltage Regulator (V_O = 2 to 7 volts)

The circuit of a low voltage output regulator using CA723 is shown in Figure 10.15. This circuit works on the principle explained with reference to Figure 9.3 in which a fraction of reference voltage, as decided by R1 and R2 is applied to the non-inverting input, and the full output voltage is applied to the inverting input of the error amplifier inside the IC.

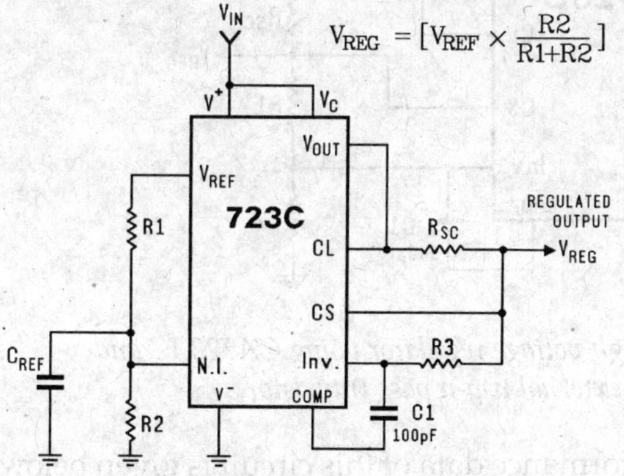

Figure 10.15: Basic low voltage regulator using CA723 IC.

Typical circuit performance data of this circuit is given below:

Regulated output voltage	5 V
Line regulation (ΔV_i = 3V)	0.5 mV
Load regulation (ΔI_L = 50 mA)	1.5 mV

Note that the regulated output (V_{REG}) is connected to the inverting input through a resistor R3 for minimizing temperature drift. For minimum temperature drift, the value of R3 is given by:

$$R3 = \frac{R1\,R2}{R1 + R2}$$

R3 can be eliminated for minimum component count.

Positive Voltage Regulator With External n-p-n Pass Transistor

A positive voltage regulator using an external n-p-n series-pass transistor is shown in Figure 10.16. The circuit is very similar to that shown in Figure 10.14 except that an n-p-n transistor 2N3054 (or 2N5294) is used to amplify the regulated output current.

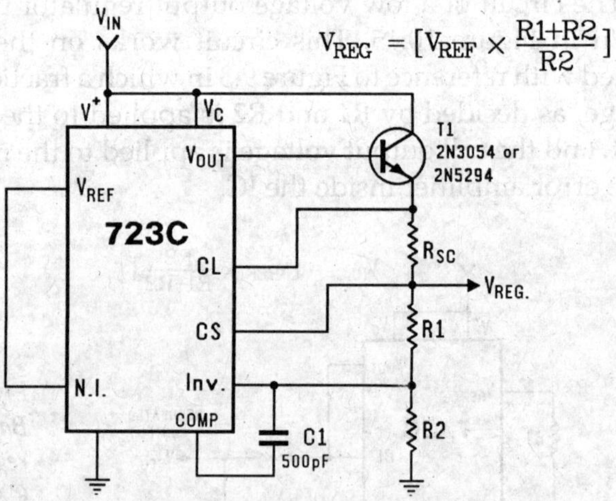

Figure 10.16: High voltage regulator using CA723 IC and an external n-p-n pass transistor.

Typical circuit performance data of this circuit is given below:

Regulated output voltage	15 V
Line regulation (ΔV_i = 3V)	1.5 mV
Load regulation (ΔI_L = 1 A)	15 mV

Positive Voltage Regulator With External-p-n-p Pass Transistor

A positive voltage regulator using an external p-n-p series-pass transistor is shown in Figure 10.17. The circuit is very similar to that shown in Figure 10.14 except that an external p-n-p transistor 2N3054 (or 2N5956, or 2N6108) is used to amplify the regulated output current.

The method used for connecting a p-n-p type of transistor is a little different. Instead of connecting the base of the transistor to get the drive from the regulated output of the IC, the base is connected across a resistance R3. All the d.c. current input to the regulator IC is passes through this resistance R3 and the voltage drop across it is used to control the conduction of the external p-n-p transistor.

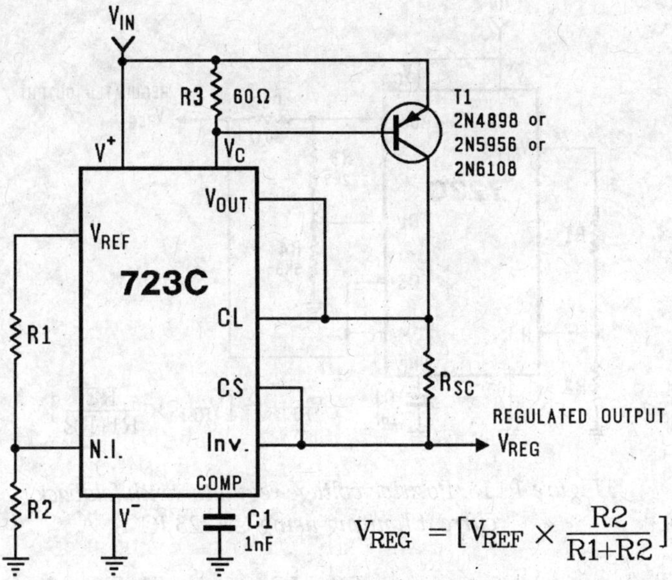

Figure 10.17: High voltage regulator using CA723 IC and an external p-n-p pass transistor.

As the load current drawn from the regulated output increases, the input current to the IC also increases, and so does the voltage drop across R3. That makes T1 conduct more and supply the increased current demand.

Typical circuit performance data of this circuit is given below:

Regulated output voltage	5 V
Line regulation (ΔV_i = 3V)	0.5 mV
Load regulation (ΔI_L = 1 A)	5 mV

Positive Voltage Regulator With Foldback Current Limiting

Figure 10.18 shows the circuit of a voltage regulator using CA723. This circuit provides foldback current limiting. The principle of working of this circuit is same as explained with reference to Figure 9.11 and 9.12.

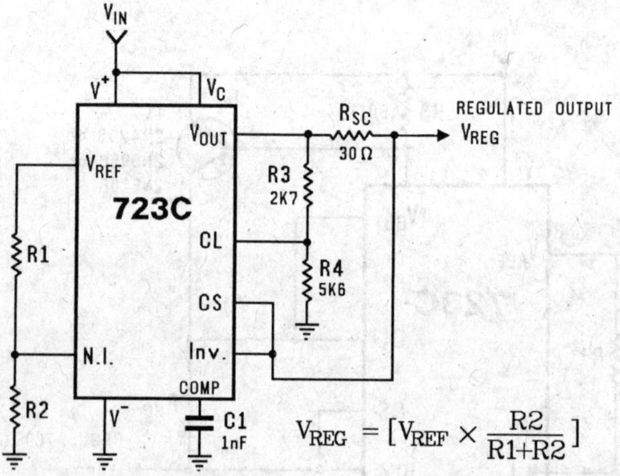

$$V_{REG} = [V_{REF} \times \frac{R2}{R1+R2}]$$

Figure 10.18: Positive voltage regulator with foldback current limiting using CA723 IC

The non-inverting input of the IC is fed with a fixed reference voltage as decided by R1 and R2 while the inverting input is connected to the full regulated output of the IC. The voltage between Current Sense (CS) and Current Limit (CL) terminals depends of the voltage drop across R_{SC} as well as on the voltage across R4 as decided by the voltage divider R3 and R4. This results in foldback type of current limiting.

Typical circuit performance data of this circuit is given below:

Regulated output voltage	+5 V
Line regulation (ΔV_i = 3V)	0.5 mV
Load regulation (ΔI_L = 10 mA)	1 mV
Short circuit current	20 mA

Positive Floating Voltage Regulator With Foldback Current Limiting

The working of circuit shown in Figure 10.19 is similar to those described above. However, to the voltage supply to the IC is kept limited to 12 V by zener diode D1. The floating output enables the circuit to be used in high voltage circuits.

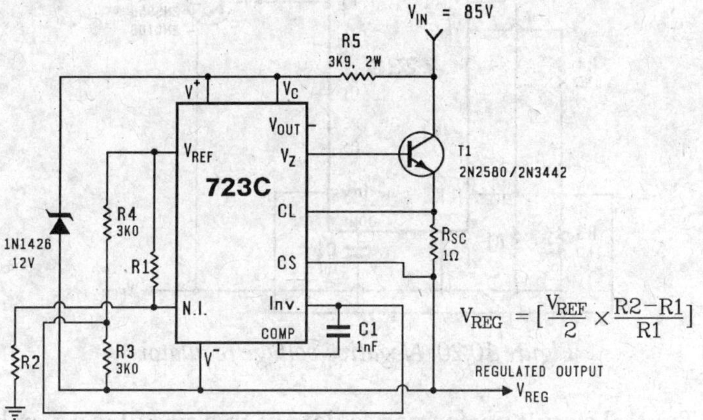

Figure 10.19: Positive floating voltage regulator with foldback current limiting

Typical circuit performance data of this circuit is given below:

Regulated output voltage	+50 V
Line regulation (ΔV_i = 20V)	15 mV
Load regulation (ΔI_L = 50 mA)	20 mV

Negative Voltage Regulator

The CA723 IC can also be used as a negative voltage regulator as shown in Figure 10.20. The circuit uses an external p-n-p transistor the base of which is connected to the V_Z output of the IC to provide an offset voltage. In this case, the non-inverting input is fed a sample of the regulated output voltage as decided by R2 and R1. The inverting input is fed with a fraction of the internally generated reference voltage as decided by R4 and R3.

For applications using TO-5 package, and where V_Z is required, an external 6.2 V zener diode should be connected in series with the Vo terminal (Terminal 6).

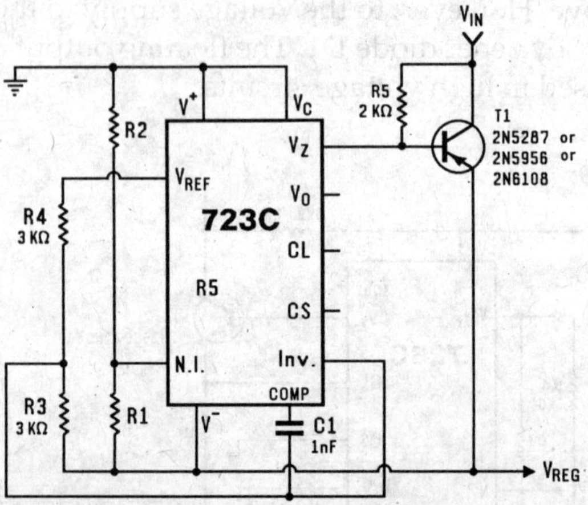

Figure 10.20: Negative voltage regulator.

Typical circuit performance data of this circuit is given below:

Regulated output voltage	-15 V
Line regulation (ΔV_i = 3V)	1 mV
Load regulation (ΔI_L = 100 mA)	2 mV

Shunt Regulator

Figure 10.21 shows a shunt regulator using a CA723 IC. The working of this circuit is similar to that of shunt regulator shown in Figure 9.13. Output available from Vz terminal of the 723 drives the base of an external n-p-n amplifier. An increase in output tends to increase the base drive current to T1 and that makes it to shunt more of the load current and keep the output stable.

Typical circuit performance data of this circuit is given below:

Regulated output voltage	+5 V
Line regulation (ΔV_i = 10V)	0.5 mV
Load regulation (ΔI_L = 100 mA)	1.5 mV

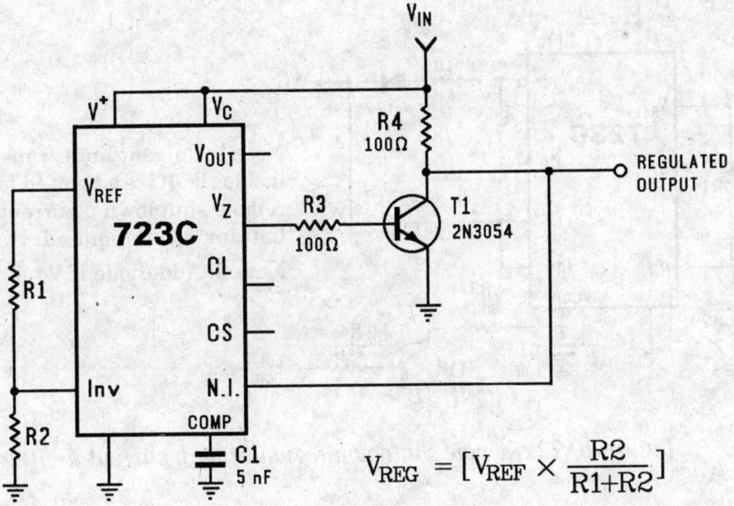

$$V_{REG} = [V_{REF} \times \frac{R2}{R1+R2}]$$

Figure 10.21: A shunt regulator.

Remote Shutdown Regulator

Figure 10.22 shows the circuit of a positive regulator using CA723 which can be shut down from a remote location by a logic positive voltage. The circuit works as a positive voltage regulator as explained above with reference to Figure 10.15 with short-circuit current limiting. When a positive signal is applied to the base of control transistor T1, it conducts and grounds the frequency compensation terminal of the IC. This stops drive to the internal output transistor and the regulated output falls to zero. Diode D1 is needed only if the output voltage exceeds 10 V.

The input is compatible with logic circuits where a positive voltage is 1 and a zero voltage is zero.

Typical circuit performance data of this circuit is given below:

Regulated output voltage	+5 V
Line regulation (ΔV_i = 3V)	0.5 mV
Load regulation (ΔI_L = 50 mA)	1.5 mV

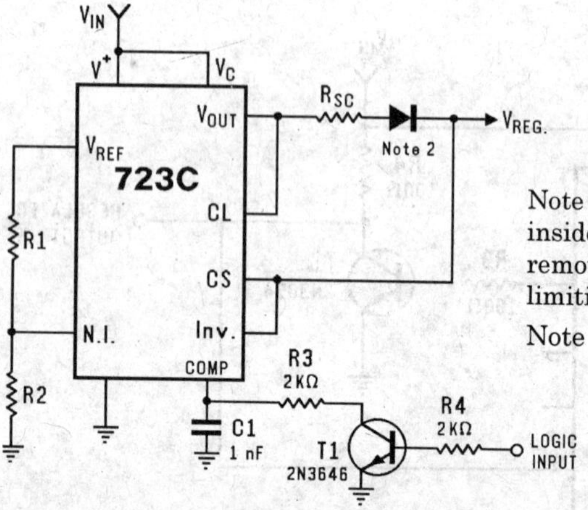

Note 1: Current limit transistor inside the IC can be used for remote shutdown if current limiting is not required.

Note 2. Add diode if Vout > 10V

Figure 10.22: Remote shutdown regulator with current limiting.

Useful Formulae

The short circuit current in all applications of 723 is given by the formula:

$$I_{LIMIT} = V_{SENSE}/R_{SC}$$

For foldback current limiting, the KNEE current and the short-circuit current can be calculated as follows:

$$I_{KNEE} = \left[\frac{V_{OUT}\ R3}{R_{SC}\ R4} + \frac{V_{SENSE}\ (R3 + R4)}{R_{SC}\ R4}\right]$$

$$I_{SHORT\ CKT} = \left[\frac{V_{SENSE}}{R_{SC}} \times \frac{R3 + R4}{R4}\right]$$

In the positive floating regulator shown in Figure 10.19, outputs from +4 volts to +250 volts can be calculated by the formula:

$$V_{OUT} = \left[\frac{V_{REF}}{2} \times \frac{R2 - R1}{R1}\right]; \quad R3 = R4$$

11

Fixed Voltage Regulators

The modern trend in regulators is toward the use of fixed voltage regulators. In most cases, they have three terminals only (a connection for unregulated d.c. input, regulated d.c. output and ground). These devices are commonly referred to as *three-terminal regulators*. Inside each regulator is a voltage reference, a high-gain error amplifier, temperature-compensated voltage sensing resistors and a pass element. Many currently available units have thermal shut-down, over-voltage protection and current feedback, making them virtually destruction proof. Fixed voltage regulators are available in a wide range of voltage and current ratings.

It is easy to see why regulators of this sort are so popular when you consider the low price and the number of individual components they can replace. Fixed voltage regulators are available in several different package styles, depending on current ratings. Low current (100 mA) devices frequently use the plastic TO-92 and DIP-style cases. TO-220 packages are popular in the 1.5 A range, and TO-3 cases house the larger 3-A and 5-A devices.

Three-terminal regulators are available as positive or negative types. In most cases, a positive regulator is used to regulate a positive voltage and a negative regulator to regulate a negative voltage. Depending on the system ground requirements, however, each regulator type may be used to regulate the "opposite" voltage as illustrated in Figure 11.1.

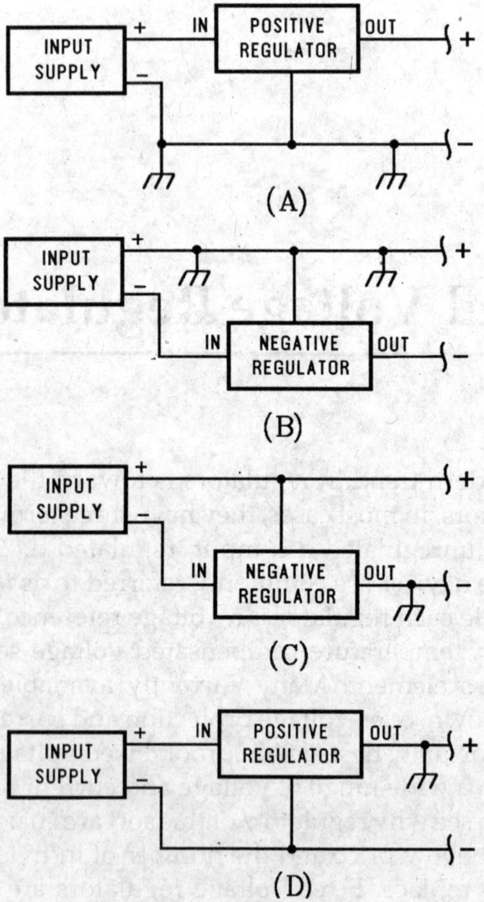

Figure 11.1: Using fixed-voltage regulators.

Figure 11.1A and B illustrate how the regulators are used in the conventional mode. Several regulators can be used with a common-input supply to deliver different voltages with a common ground. Negative regulators may be used in the same manner. If no other common supplies operate off the input supply to the regulator, the circuit of Figure 11.1C and 11.1D may be used to regulate positive voltages with a negative regulator and vice versa. In these configurations, the input supply is floated; neither side of the input is tied to the system ground.

Manufacturers have adopted a system of family members to classify three-terminal regulators in terms of supply polarity, output current and regulated voltage. For example, National uses the number LM7805C to describe a positive 5-V, 1.5-A regulator; the compatible unit from Texas Instruments is a UA7805KC. In the same way, LM7812C describes a 12-V regulator of similar characteristics. LM7905C denotes a negative 5-V, 1.5-A device. There are many such families with widely varied ratings available from manufacturers. Fixed-voltage regulators are available with output ratings in most common values between 5 V and 28 V. Other families include devices that can be adjusted from 1.25 to 50 V.

Regulator Specifications

When choosing a three-terminal regulator for a given application, the most important specifications to consider are device output voltage, output current, input-to-output differential voltage, line regulation, load regulation and power dissipation. Output voltage and current requirements are determined by the load with which the supply will ultimately be used.

Input-to-output differential voltage is one of the most important fixed-voltage regulator specifications to consider when designing a supply. The differential value (the difference between the voltage applied to the input terminal and the voltage on the output terminal) must be within a specified range. The minimum differential value, usually about 2.5 V, is called the *dropout* voltage. If the differential value is less than the dropout voltage, no regulation will take place. At the other end of the scale, the maximum input-output differential voltage is generally about 40 V. If this differential value is exceeded, device failure may occur.

Increases in either output current or differential voltage produce proportional increases in device power consumption. By employing a safety feature called current *foldback*, some manufacturers ensure that maximum dissipation will never be exceeded in normal operation. Figure 11.2 shows the relationship between output current, input-output differential and current limiting for a fixed voltage regulator nominally rated for 1.5-A output current.

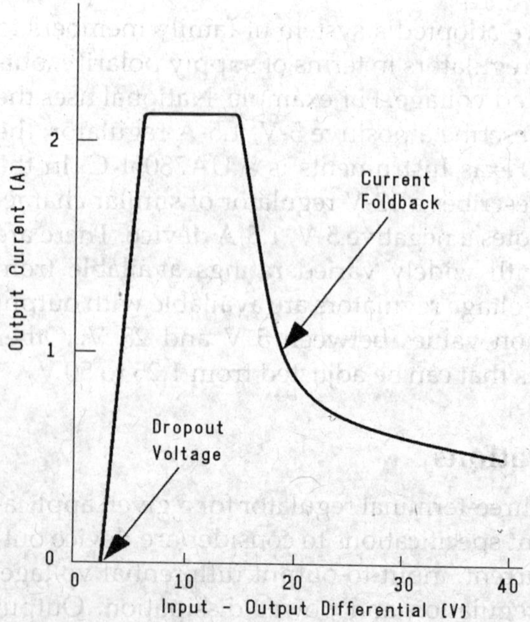

Figure 11.2. Effect of input-output differential voltage on three-terminal regulator current.

Maximum output current is available with differential voltages ranging from about 2.5 V (dropout voltage) to 12 V. Above 12V, the output current decreases, limiting the device dissipation to a safe value. If the output terminals are accidentally short circuited, the input-output differential will rise, causing current foldback, and thus preventing the power supply components from being over stressed. This protective feature makes three-terminal regulators particularly attractive in simple power supplies.

When designing a power supply around a particular three-terminal regulator, input-output voltage characteristics of the regulator should play a major role in selecting the transformer-secondary and filter components values. The unregulated voltage applied to the input of the three-terminal device should be higher than the dropout voltage, yet low enough that the regulator does not go into current limiting caused by an excessive differential voltage. If, for example, the regulated output voltage of the device shown in Figure 11.2 was 12, then unregulated volts between 14.5 and 24 volts would be acceptable if maximum output current is desired.

In use, all but the lowest current regulators generally require an adequate external heat sink because they may be called on to dissipate a fair amount of power. Also, because the regulator chip contains a high-gain error amplifier, bypassing of the input and output leads is essential for stable operation.

Most manufacturers recommend bypassing the input and output directly at the leads where they protrude through the heat sink. Solid tantalum capacitors are usually recommended because of their good high-frequency capabilities.

External capacitors used with IC regulators may discharge through the IC junction under certain circuit conditions, and high current discharges can harm ICs. Look at the regulator data sheet to see whether protection diodes are needed; what diodes to use and how to place them in any particular application.

In addition to fixed-output-voltage ICs, high-current adjustable voltage regulators are also available. These ICs require little more than an external potentiometer for an adjustable output range from 5 to 24 V at up to 5 A. They are not very expensive and that makes them attractive for test-bench power supplies. A very popular low current adjustable output voltage three terminal regulator, the LM317, is shown in Figure 11.3. It develops a steady 1.25V reference (V_{REF}), between the output and adjustment terminals. By installing R1 between these terminals, a constant current, I_1, is developed governed by the equation:

$$I_1 = \frac{V_{REF}}{R1} \qquad (11\text{-}1)$$

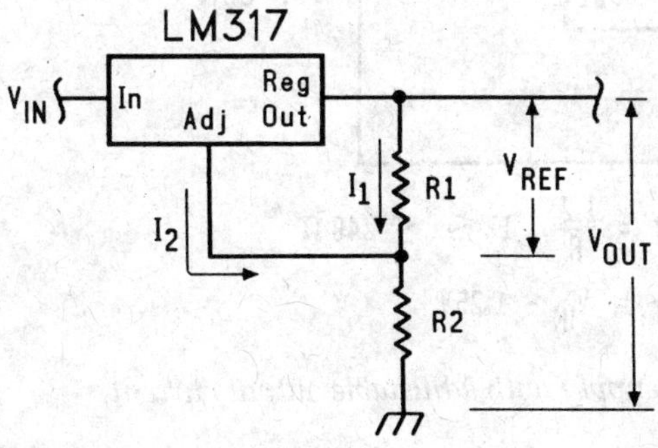

Figure 11.3: Adjustable output voltage power supply.

Both I_1 and a 100-μA error current, I_2 flow through R2, resulting in output voltage V_{OUT}. This V_{OUT} can be calculated using the equation:

$$V_{OUT} = V_{REF}\left[1 + \frac{R2}{R1}\right] + I_2 \times R2 \text{ - - - - - - - - } (11\text{-}5)$$

In most cases, the error current I_2 is very small in comparison to I_1 and the product $I_2 \times R2$ can be neglected. That simplifies calculations.

Any voltage between 1.2 and 37 V may be obtained with a 40-V input by changing the ratio of R2 to R1. At lower output voltages, however, the available current will be limited by the power dissipation of the regulator.

Constant Current Source

Figure 11.4 shows one of many flexible applications for the LM317. By adding only one resistor with the regulator, the voltage regulator can be changed into a constant current source capable of charging NiCd batteries, for example. Design equations are given in the Figure. The same precautions should be taken with adjustable regulators as with fixed voltage ones. Proper heat sinking and lead bypassing are essential for proper circuit operation. This has been described in more details in the next chapter.

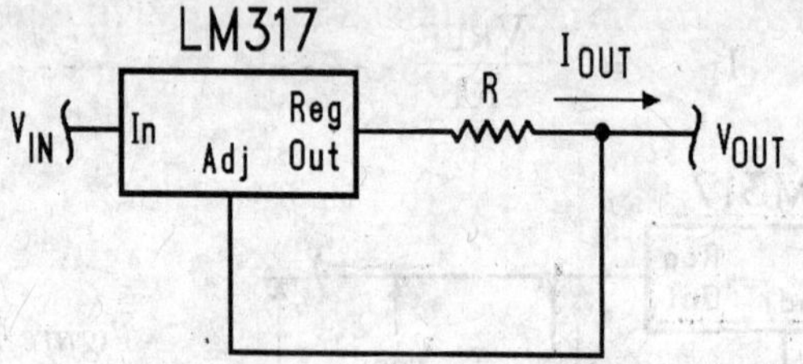

$$I_{OUT} = \frac{1.2}{R}, \quad 12 \leq R \leq 240\ \Omega$$

$$V_{OUT} = V_{IN} - 1.25V$$

Figure 11.4: Power supply with adjustable output current.

Increasing Regulator Output Current

When the maximum output current from an IC voltage regulator is insufficient to operate the load, discrete power transistors may be connected to increase the current handling capability of the IC. Figure 11.5 shows one method for boosting the output current of a positive regulator using an n-p-n type of transistor; although the same technology can be applied to negative regulators by using p-n-p transistors.

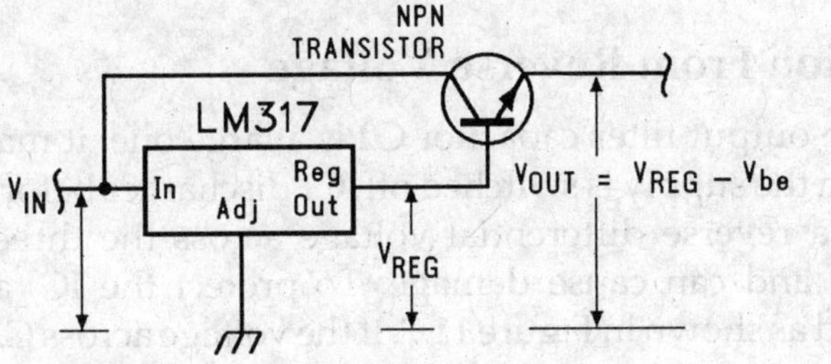

Figure 11.5: Boosting the output using an n-p-n transistor.

In Figure 11.5 an n-p-n transistor is connected as an emitter follower, multiplying the output current capacity by the transistor beta. The shortcoming of this approach is that the base-emitter junction is not inside the feed-back loop. The result is that the output voltage is reduced by the base-emitter drop in the externally connected transistor, and the load regulation is degraded by variations in this drop.

The circuit in Figure 11.6 has a p-n-p transistor "wrapped around" a positive voltage regulator. The regulator draws current through the base-emitter junction of the transistor, causing the transistor to conduct. The IC output voltage is unchanged by the transistor because the collector is connected directly to the IC output (sense point). Any increase in output voltage is detected by the IC regulator, which shuts off its internal pass transistor, and this stops the boost-transistor base current. An n-p-n transistor can be used similarly with a negative voltage regulator. The output voltage is not affected by the V_{BE} of the external transistor.

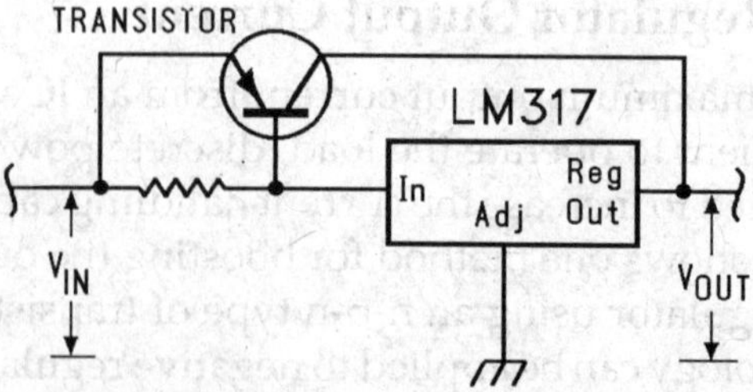

Figure 11.6: Boosting the output using a p-n-p transistor.

Protection From Reverse Voltage

If the output filter capacitor C2 is a large one, it may happen that when the supply is switched off, C1 discharges before C2. That will put a reverse differential voltage across the three-terminal regulator and can cause damage. To protect the IC, a diode is connected as shown in Figure 11.7. If the voltage across C2 is higher than that across C1, capacitor C2 will discharge through D1 and clamp the reverse differential voltage to about 0.6 V.

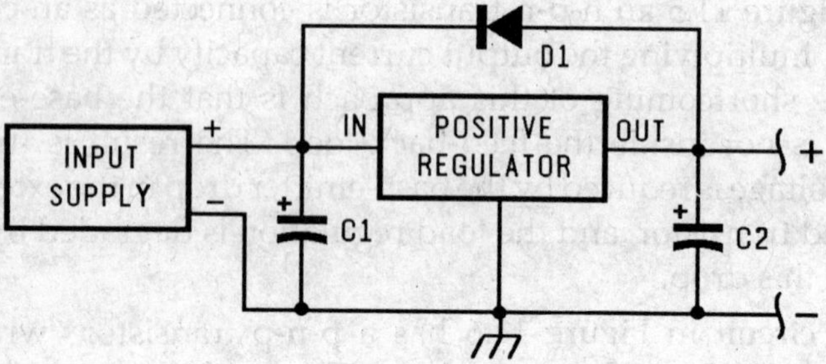

Figure 11.7: Protecting a regulator from reverse voltages.

12

Three-Terminal Regulators

Fixed voltage regulators mostly come in three-terminal packages in which one terminal is for input, one terminal for output and one terminal for ground. Due to easy availability, low cost, convenience of operation and host of other practical advantages, the trend today is to have localized regulation for each sub-system of the main system. The earlier practice of using single point regulation suffered not only from unequal voltage drops in the impedances of the line connecting the main regulator to various sub-systems, it also resulted in unwanted coupling between different subsystems constituting the main system. The concept of localized regulation using 3-terminal regulators is free from all these drawbacks.

In their basic operational mode, 3-terminal regulators require virtually no external components. They are available in both fixed voltage as well as adjustable voltage types with current ratings of 100 mA, 500 mA, 1.0 amp and 3.0 amps.

Technical information and application data on some of the popular types of 3-terminal positive voltage regulators and 3-terminal negative voltage regulators is given in this chapter. This data will enable the user to make use of the available devices to best of his advantage.

LM 78XX Series Positive Voltage Regulators

The LM 78XX series of three-terminal regulators is available with several fixed output voltages making them useful in a wide range of applications. One of these is local on-card regulation, eliminating the distribution problems associated with single point regulation. The voltages available allow these regulators to be used in logic systems, instrumentation, and other electronic equipment.

Although designed primarily as fixed voltage regulators these devices can be used with external components to obtain adjustable voltages and currents. Considerable effort was expended to make the LM 78XX series of regulators easy to use and minimize the number of external components. It is not necessary to bypass the output, although this does improve transient response. Input bypassing is needed only if the regulator is located far from the filter capacitor of the power supply. Current limiting is included to limit the peak output current to a safe value. Safe operating area (SOAR) protection for the output transistor is provided to limit internal power dissipation. If internal power dissipation becomes too high for the heat sinking provided, the thermal shutdown circuit takes over thereby preventing the IC from over-heating.

LM 78XXC

The LM 78XXC series is available in an aluminium TO-3 package which will allow over 1.0 A load current if adequate heat sinking is provided.

Features

- Output current in excess of 1 A
- Internal thermal overload protection
- No external components required
- Output transistor safe area protection
- Internal short circuit limit
- Available in aluminium TO-3 package

LM 78LXX

The LM 78LXX series is available in the metal three lead TO-39 (H) and the plastic TO-92 (Z) package . With adequate heat sinking the regulator can deliver 100 mA output current.

Features

- □ Output voltage tolerance of ± 5% (LM 78LXXAC) and ± 10% (LM 78LXXC) over the temperature range
- □ Output current of 100 mA
- □ Internal thermal overload protection
- □ Output transistor safe area protection
- □ Internal short circuit current limit
- □ Available in plastic TO-92 and metal TO-39 low profile packages

LM 78MXX

The LM 78MXX series is available in the plastic TO-202 package. This package allows these regulators to deliver over 0.5 A if adequate heat sinking is provided.

Features

- □ Output current in excess of 0.5 A
- □ Internal thermal overload protection
- □ No external components required
- □ Output transistor safe area protection
- □ Internal short circuit current limit
- □ Available in plastic TO-202 package
- □ Special circuitry allows start-up even if output is pulled to negative voltage (± supplies)

Pin connections for these devices are shown in Figures 12.1.

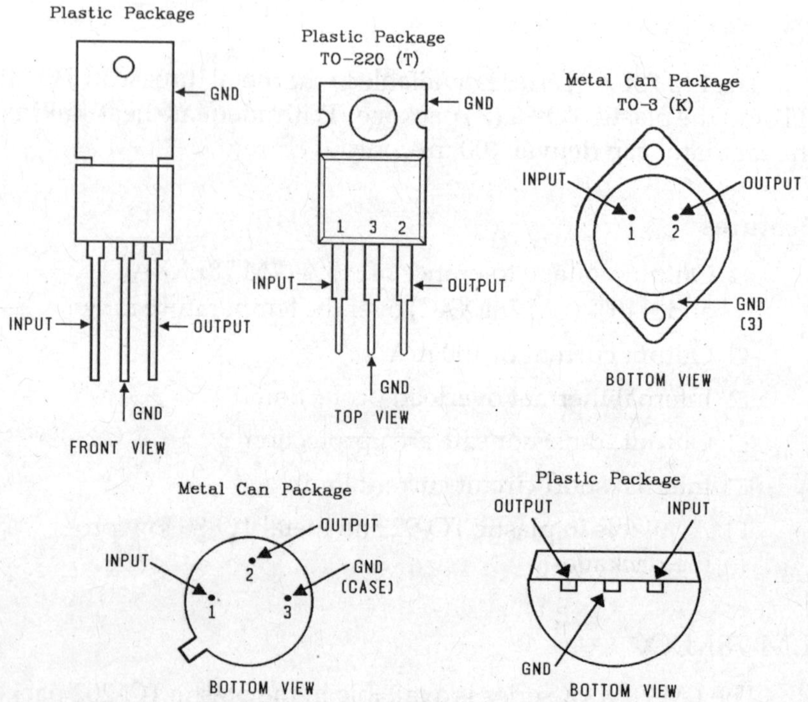

Figure 12.1: Packages for LM78XX series voltage regulators

Electrical Characteristics

Type	V_{out} (V)	I_{out} (A)			V_{in} (V)	
		78XXC	78LXX	78MXX	min.	max.
7805	5	1	0.1	0.5	7.5	20
7806	6	1	0.1	0.5	8.6	21
7808	8	1	0.1	0.5	10.6	23
7810	10	1	0.1	0.5	12.7	25
7812	12	1	0.1	0.5	14.8	27
7815	15	1	0.1	0.5	18	30
7818	18	1	0.1	0.5	21	33
7824	24	1	0.1	0.5	27.3	38

Application Hints

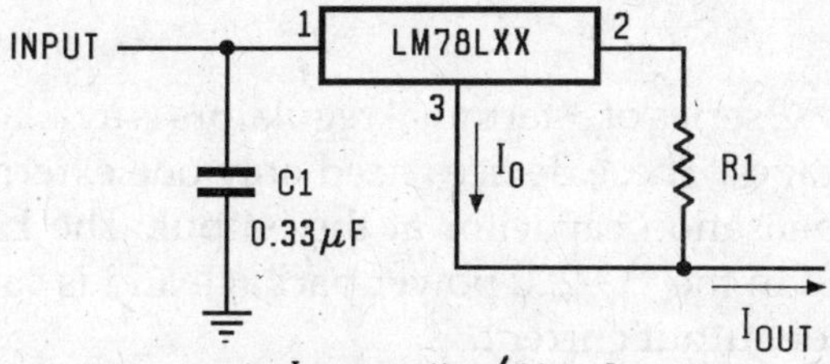

$$I_{OUT} = V_{23}/R1 + I_0$$

$$\Delta I_0 = 1.5 \text{ mA over line and load changes}$$

Figure 12.2: Current regulator using LM78 series regulator.

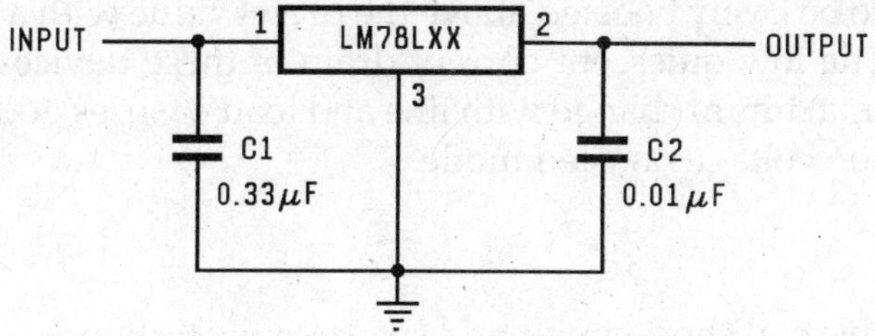

C2 required if regulator located far from the power supply filter

Figure 12.3: Fixed output regulator using LM78XX series regulators.

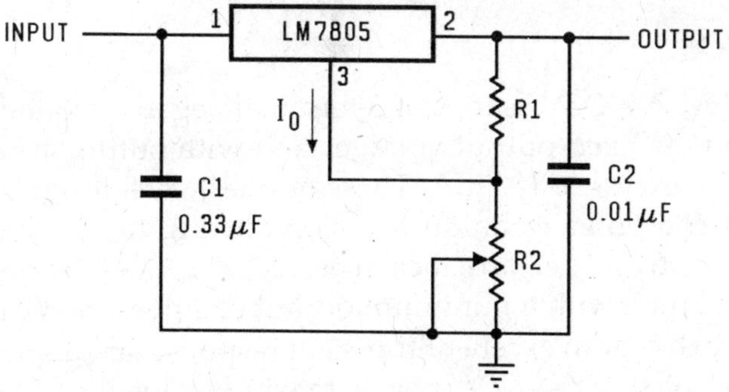

$$V_{OUT} = 5V + (5V/R1 + I_0)\, R2$$

Figure 12.4: Adjustable output regulator using LM78XX series regulators.

LM 79XX Series Negative Voltage Regulators

LM 79XXC

The LM 79XXC series of 3-terminal regulators is available with fixed output voltages. These devices need only one external component – a compensation capacitor at the output. The LM 79XX series is packaged in the TO-220 power package and is capable of supplying 1.5 A of output current.

These regulators employ internal current limiting, safe area protection and thermal shutdown for protection against all overload conditions.

Low ground pin current of the LM 79XXC series allows output voltages to be easily boosted above the preset value with a resistor divider. The low quiescent current drain of these devices with a specified maximum change with line and load ensures good regulation in the voltage boosted mode.

Features

- ☐ Thermal short circuit and safe area protection
- ☐ High ripple rejection
- ☐ 1.5 A output current
- ☐ 4% preset output voltage

LM 79LXX

The LM 79LXXAC series of 3-terminal negative voltage regulators features fixed output voltages and with output current capabilities in excess of 100 mA. These devices were designed using the latest computer techniques for optimizing the packaged IC thermal/electrical performance. The LM 79LXXAC series, even when combined with a minimum output compensation capacitor of 0.1 μF, exhibits an excellent transient response, a maximum line regulation of 0.07% V_O/V, and a maximum load regulation of 0.01% V_O/mA.

The LM 79LXXAC series also includes, as self-protection circuitry: safe operating area circuitry for output transistor, power dissipation limiting, a temperature independent short circuit current limit for peak output current limiting, and a thermal shutdown circuit to prevent excessive junction temperatures. Although designed primarily as fixed voltage regulators, these devices may be combined with simple external circuitry for boosted and / or adjustable voltages and currents. The LM LXXAC series is available in the 3-lead TO-92 package.

Features

- ☐ Preset output voltage error is less than ± 5% overload, line and temperature.
- ☐ Specified at an output current of 100 mA
- ☐ Easily compensated with a small 0.1 μF output capacitor
- ☐ Internal short-circuit thermal and safe operating area (SOAR) protection
- ☐ Easily adjustable to higher output voltages
- ☐ Maximum line regulation less than 0.07% V_{out}/V
- ☐ Maximum load regulation less than 0.01% V_{out}/mA
- ☐ TO-92 package

LM 79MXX

The LM 79MXX series of 3-terminal regulators is available with fixed output voltages and these devices need only one external component – a compensation capacitor at the output. The LM 79MXX series is packaged in the TO-202 power package and TO-39 metal can and is capable of supplying 0.5 A output current.

These regulators employ internal current limiting safe area protection and thermal shutdown for protection against virtually all overload conditions.

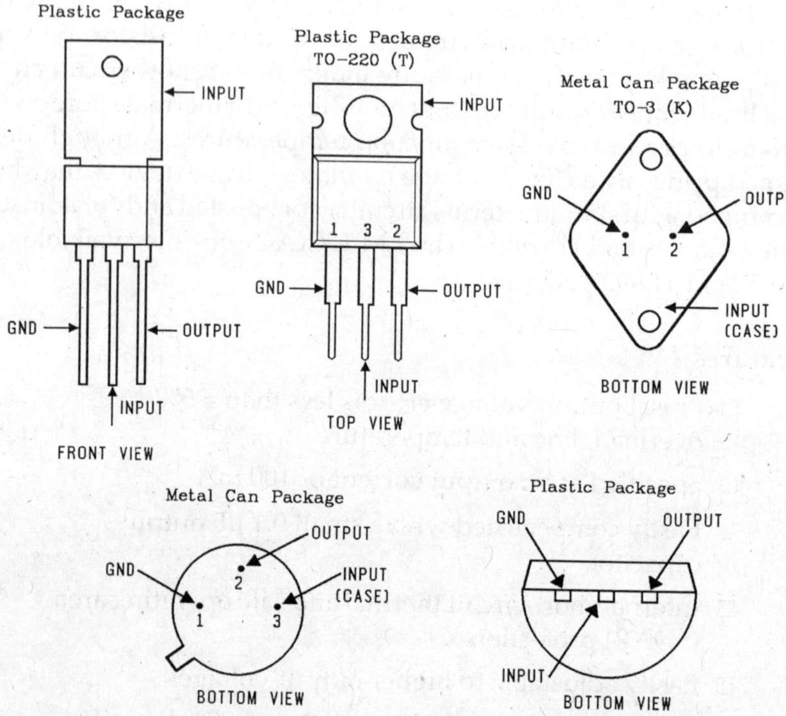

Figure 12.5: Packages for LM79XX series voltage regulators.

Low ground pin current of the LM 79MXX series allows output voltages to have easily boosted above the preset value with a resistor divider. This low quiescent current drain of these devices with a specified maximum change with line and load ensures good regulation in the voltage boosted mode.

Features

- ☐ Thermal, short-circuit and safe area protection
- ☐ High ripple rejection
- ☐ 0.5 A output current
- ☐ 4% preset output voltage

Electrical Characteristics

Type	V_{out} (V)	I_{out} (A)			Vin (V)	
		79XXC	79LXX	79MXX	min.	max.
7905	-5	1.5	0.1	0.5	-20	-7.5
7905,2	-5.2	1.5	-	-	-20.5	-7.5
7906	-6	1.5	-	0.5	-21	-8
7908	-8	1.5	-	0.5	-23	-10.5
7909	-9	1.5	-	-	-24	-11.5
7912	-12	1.5	0.1	0.5	-27	-14.5
7915	-15	1.5	0.1	0.5	-30	-18
7918	-18	1.5	0.1	-	-33	-21
7924	-24	1.5	0.1	0.5	-38	-24

Application Hints

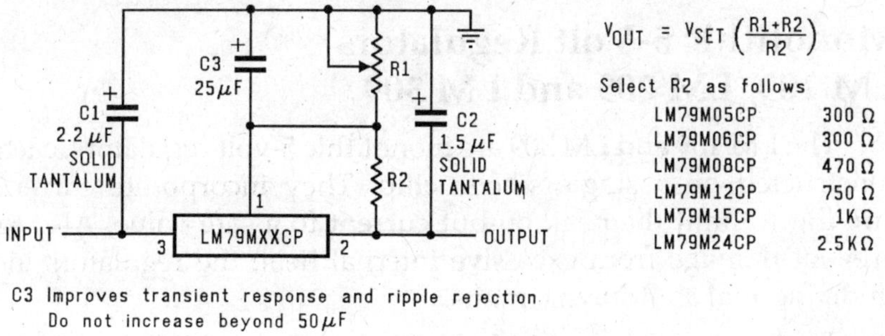

Figure 12.6: Variable output regulator using LM79XX series regulators.

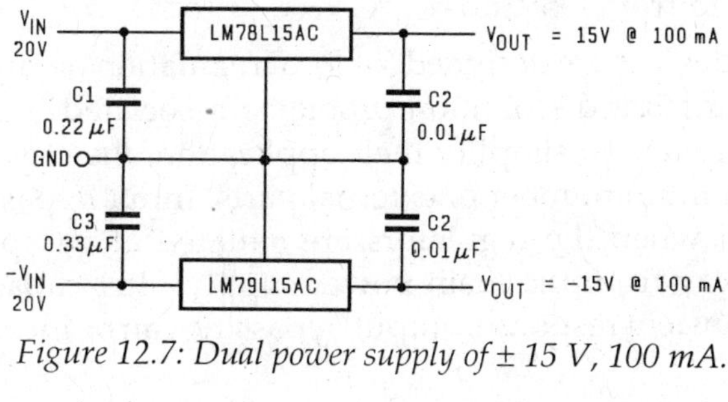

Figure 12.7: Dual power supply of ± 15 V, 100 mA.

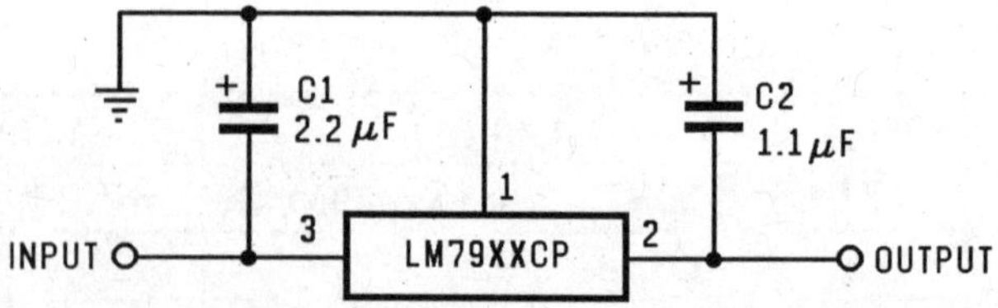

C1 is required if regulator is separated from the filter capacitor by more than 3". For value given, capacitor must be solid tantalum. A 25 μF aluminium electrolytic may be substituted.

C2 is required for stability. For value given capacitor must be solid tantalum. A 25 μF aluminium electrolytic may be substituted. Value may be increased without limit.

For output capacitance in excess of 100 μF, a high current diode from input to output (1N4001, etc.) will protect the regulator from momentary input shorts.

Figure 12.8: Fixed voltage regulator using LM79 series regulators.

Monolithic 5-Volt Regulators LM 109, LM 209 and LM 309

The LM 109 and LM 309 are monolithic 5-volt regulators, each constructed on a single silicon chip. They incorporate current limiting to limit the peak output current to a safe value. Also to prevent damage from excessive internal heat, the regulators include thermal shutdown circuits.

The military version, LM 109 operates from -55°C to +150°C. The LM 309, the commercial version, operates from 0°C to +125°C, The LM 209 is the same as the LM 109 except that its performance is guaranteed from -25°C to +85°C.

These devices are designed for local regulation on digital logic cards, eliminating distribution problems associated with single-point regulation. To simplify their application, the regulators require a minimum number of external parts. Input bypassing may be required when the regulators are situated at an appreciable distance away from the main power supply filter capacitors. To improve transient response, output bypassing can be incorporated.

The LM 109, LM 209 and LM 309 are available in the solid kovar TO-5 header and the TO-3 power package. The TO-5 package can handle currents in excess of 200 mA with adequate heat sinking. The TO-3 package can handle currents in excess of 1 A. Pin connections for these regulators are shown in Figure 12.9.

The regulators may be set to regulate voltages above 5 volts.

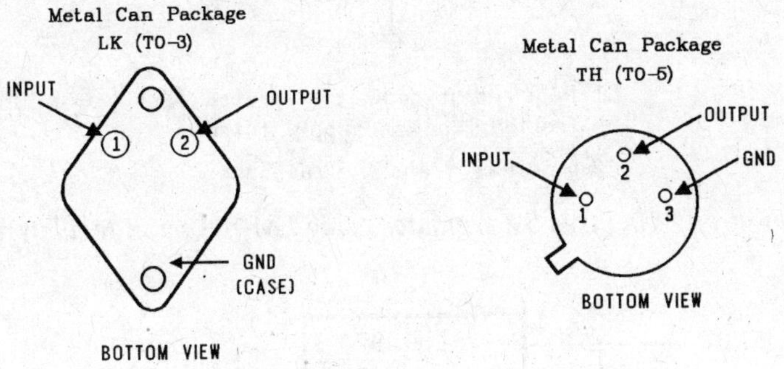

Figure 12.9: LM109 series regulators.

Features

- ☐ Output current in excess of 1 A
- ☐ Internal thermal overload protection
- ☐ No external components required
- ☐ Specified compatible, worst case, with TTL and DTL

ABSOLUTE MAXIMUM RATINGS

Input Voltage				35 V
Power Dissipation				Internally Limited
Operating Junction Temperature	LM 109	-55°C	to	+150°C
	LM 209	-25°C	to	+150°C
	LM 309	0°C	to	+125°C
Storage Temperature		-65°C	to	+150°C
Lead Temperature (soldering 10s)				+300°C

Application Hints

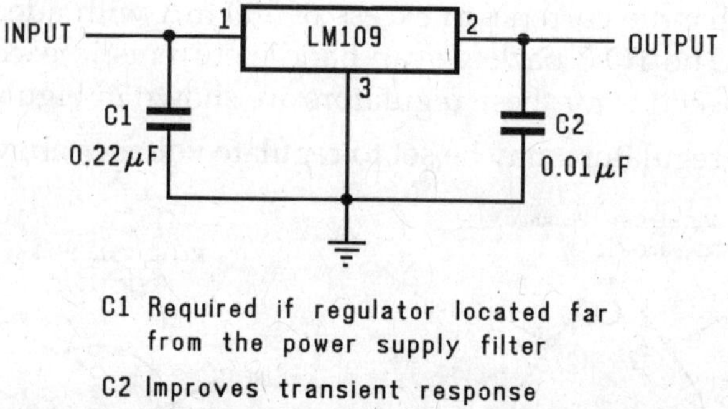

Figure 12.10: Fixed 5V regulator using LM 109 series regulators.

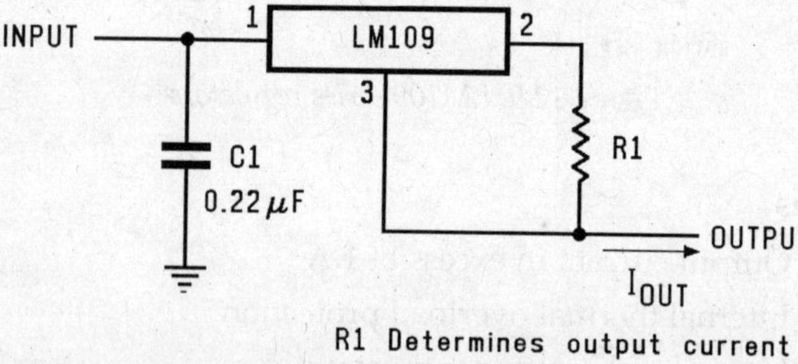

Figure 12.11: Current regulator using LM109 series regulators.

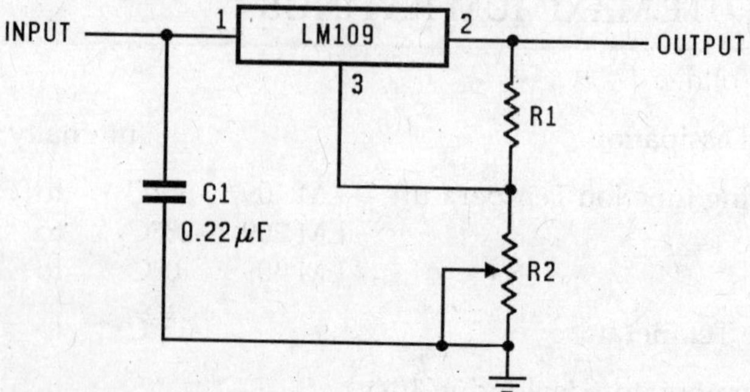

Figure 12.12: Adjustable output voltage regulator using LM109 series.

5-V, 3-A Positive Regulators LM 123, LM 223 and LM 323

The LM 123 is a three-terminal positive regulator with a preset 5 V output and a load driving capability of 3 amperes. New circuit design and processing techniques are used to provide the high output current without sacrificing the regulation characteristics of lower current devices.

The 3-amp regulator is virtually blowout proof. Current limiting, power limiting, and thermal shutdown provide the same high level of reliability obtained with these techniques in the LM109 1 amp regulator.

No external components are required for operation of the LM 123. If the device is more than 4 inches from the filter capacitor. However, a 1 μF solid tantalum capacitor should be used on the input. A 0.1 μF or larger capacitor may be used on the output to reduce load transient spikes created by fast switching digital logic, or to swamp out stray load capacitance. An overall worst case specification for the combined effects of input voltage, load currents, ambient temperature, and power dissipation ensure that the LM 123 will perform satisfactorily as a system element.

Operation is guaranteed over the junction temperature range -55°C to 150°C. An electrically identical LM 223 operates from -25°C to +150°C and the LM 323 is specified from 0°C to 125°C junction termerature. A hermetic TO-3 package is used for high reliability and low thermal resistance. Pin connections and application hints are shown in Figures 12.13, 12.14 and 12.15.

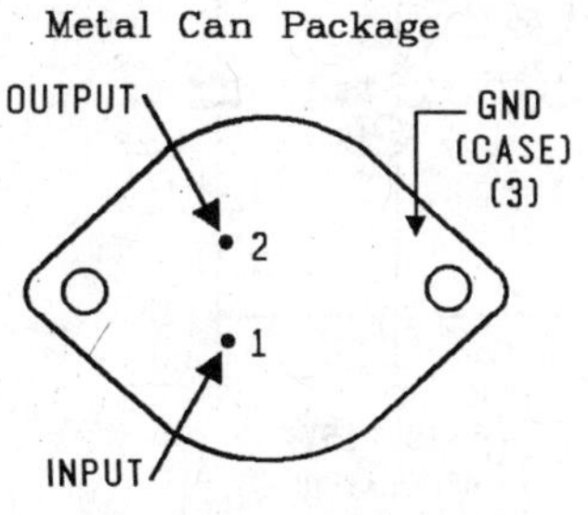

Figure 12.13
Pin connections of LM 123, LM 223 and LM 323 regulators.

Absolute Maximum Ratings

Input voltage	20 V		
Power dissipation	Internally limited		
Operating junction temperature range			
LM 123	-55°C	to	+150°C
LM 223	-25°C	to	+150°C
LM 323	0°C	to	+125°C
Storage temperature range	-65°C	to	+150°C
Lead temperature (soldering, 10 sec)	300°C		

Application Hints

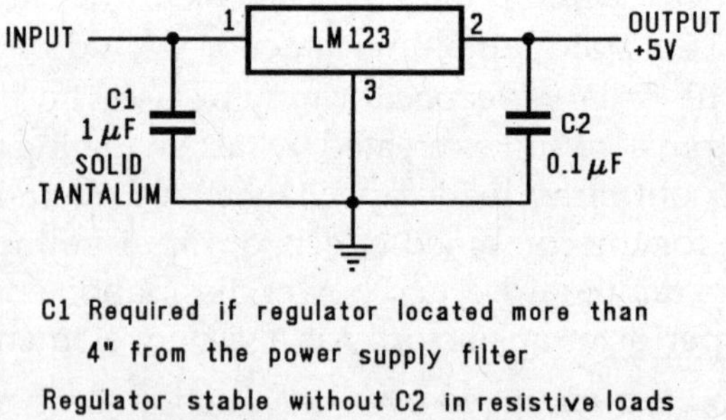

Figure 12.14: Basic 3-amp regulator.

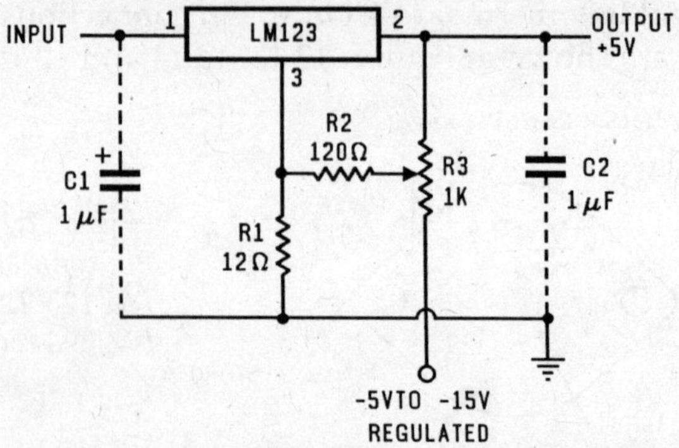

Figure 12.15: Trimming output to 5 volts.

Voltage Regulators
LM 125/LM 325/LM 325A
LM 126/LM 326

General Description

These are dual polarity tracking regulators designed to provide balanced positive and negative output voltages at currents up to 100 mA. The devices are set for ± 15 V, and ± 12 V outputs respectively. Input voltages up to ± 30 V can be used and there is provision for adjustable current limiting. These devices are available in three package types to accommodate a variety of power requirements and temperature ranges. Pin connections and application hints are shown in Figure 11.16 and 11.17 respectively.

Features

- ☐ ± 15 V and ± 12 V tracking outputs
- ☐ Output currents to 100 mA
- ☐ Output voltages balanced to within 1% (LM 125, LM 126, LM 126A)
- ☐ Line and load regulation of 0.06%
- ☐ Internal thermal overload protection
- ☐ Standby current drain of 3 mA
- ☐ Externally adjustable current limit
- ☐ Internal current limit

Absolute Maximum ratings

Input voltage	± 30 V
Forced V_O^+ (Min)	- 0.5 V
Forced V_O^- (Max)	+ 0.5 v
Power dissipation	Internally limited
Output Short Circuit Duration	Indefinite

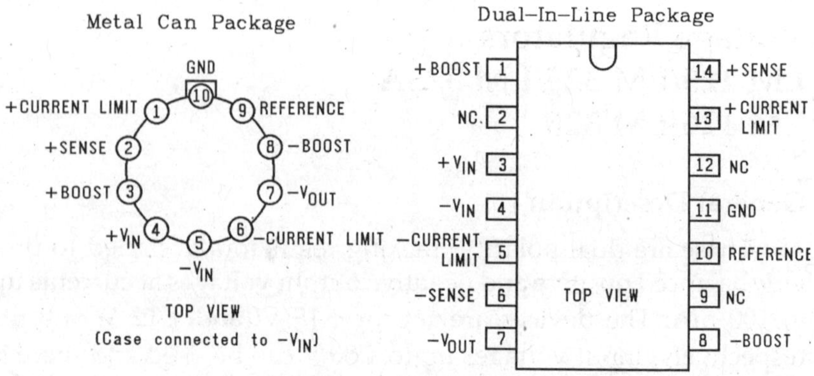

Figure 12.16: Pin connections for LM 125/LM 325/LM126/LM 326.

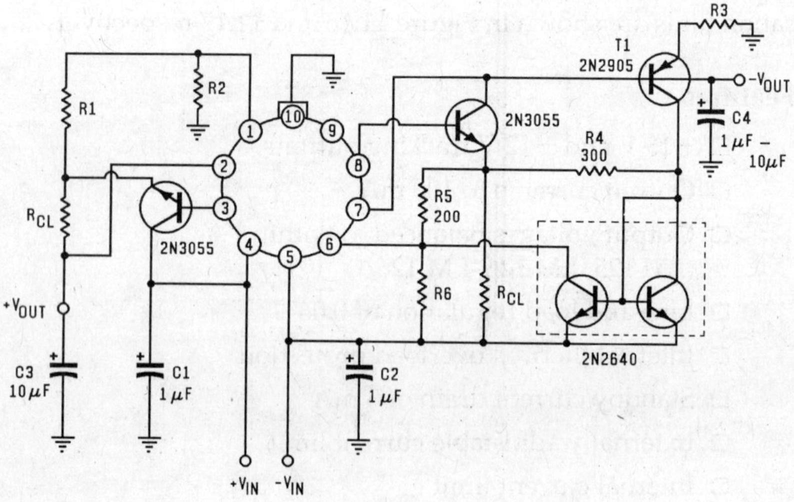

Positive Regulator

I_{MAX} = 2.0 A

I_{SC+} = 750 mA

@T_A = 25°C

$+V_{IN}$ = +25V

Negative Regulator

I_{MAX} = 2.0 A

I_{SC+} = 750 mA

@T_A = 25°C

$-V_{IN}$ = −25V

	LM125	LM126
R1	18	20
R2	310	180
R3	2.4k	1.35k
R6	300	290
R_{CL}	0.7	0.9

Figure 12.17: Boosted regulator with foldback current limiting.

3 Amp Adjustable Power Regulators LM150, LM350

General Description

The LM 150/LM 350 are adjustable 3-terminal positive voltage regulators capable of supplying in excess of 3 A over a 1.2 to 33 V output range. They are exceptionally easy to use and require only two external resistors to set the output voltage. Further, both line and load regulation are comparable to discrete designs. Also, the LM150 is packaged in standard transistor packages which are easily mounted and handled.

In addition to higher performance than fixed regulators, the LM 150 series offers full overload protection available only in IC's. Included on the chip are current limit, thermal overload protection and safe area protection. All overload protection circuitry remains full functional even if the adjustment terminal is accidentally disconnected.

Normally, no capacitors are needed unless the device is situated more than 6 inches from the input filter capacitors in which case an input bypass is needed. An optional output capacitor can be added to improve transient response. The adjustment terminal can be bypassed to achieve very high ripple rejection ratios which are difficult to achieve with standard 3-terminal regulators.

Besides replacing fixed regulators or discrete designs, the LM150 is useful in a wide variety of other applications. Since the regulator is "floating" and sees only the input-to-output differential voltage, supplies of several hundred volts can be regulated as long as the maximum input to output differential is not exceeded, i.e., avoid short-circuiting the output.

Also, it makes an especially simple adjustable switching regulator, a programmable output regulator, or by connecting a fixed resistor between the adjustment pin and output, the LM 150 can be used as a precision current regulator. Supplies with electronic shutdown can be achieved by clamping an adjustment terminal to ground which programs the output to 1.2 V

The LM 150K/LM 350K are packaged in standard steel TO-3 transistor packages. The LM 350T is packaged in TO-220 plastic package. The LM 150 is rated for operation from -55°C to + 150°C and the LM 350 from 0°C to 125°C

Pin connections for these regulators are shown in Figure 12.18 and application hints are shown in Figures 12.19 and 12.20.

- ☐ Adjustable output down to 1.2 V
- ☐ Guaranteed 3 A output current
- ☐ Line regulation typically 0.005%/V
- ☐ Load regulation typically 0.1%
- ☐ Guaranteed thermal regulation
- ☐ Output is short circuit protected
- ☐ Current limit constant with temperature
- ☐ 100% electrical burn-in in thermal limit
- ☐ 86 dB ripple rejection

Absolute Maximum Ratings

Power dissipation	Internally limited
Input-Output Voltage Differential	35 V
Operating Junction Temperature range	
LM 150	-55°C to +150°C
LM 350	0°C to +125°C
Storage Temperature	-65°C to +150°C
Lead Temperature (soldering, 10 s)	300°C
Plastic Package (soldering, 4 s)	260°C

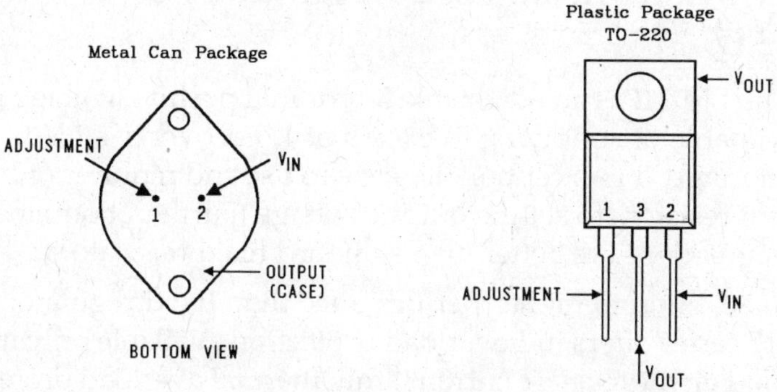

Figure 12.18: Pin connections for LM 150 and LM 350 packages.

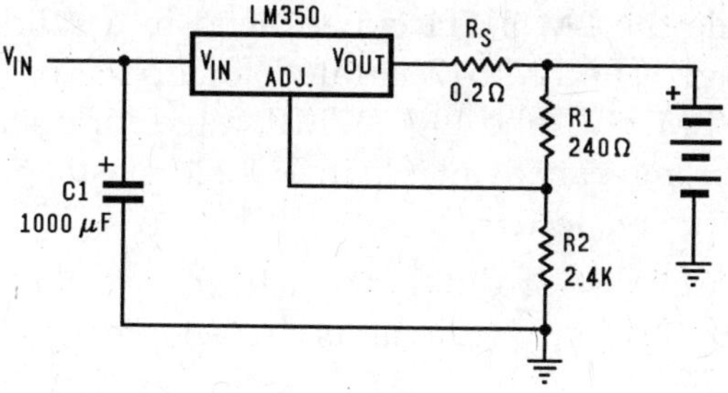

Figure 12.19: Simple 12-V battery charger.

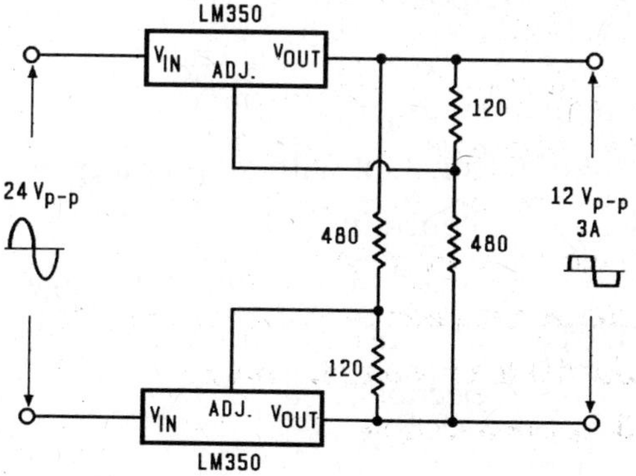

Figure 12.20: AC voltage regulator.

3-Terminal Adjustable Regulator LM 317

The LM 317 is an adjustable 3-terminal positive voltage regulator capable of supplying in excess of 1.5 A over a 1.2 V to 37 V output range. It is exceptionally easy to use and requires only two external resistors to set the output voltage. Further, both line and load regulations are better than standard fixed regulators.

In addition to higher performance than fixed regulators, the LM 317 series offers full overload protection available only in ICs. Included on the chip are current limit, thermal overload protection and safe area protection. All overload protection circuitry remains fully functional even if the adjustment terminal is disconnected.

The LM 317K is packaged in the standard TO-3 transistor package while the LM 317H is packaged in a solid Kover base TO-39 package. The LM 317 is rated for operation from 0°C to +125°C. The LM 317T and LM 317MP, rated for operation over a 0°C to 125°C range, are available in TO-220 plastic package and a TO-202 package respectively.

Pin connections for different packages are shown in Figure 12.21 and application circuit hints are shown in Figures 12.22, 12.23 and 12.24.

Features

- ☐ Adjustable output down to 1.2 V
- ☐ Guaranteed 1.5 A output current
- ☐ Line regulation typically 0.01%
- ☐ Current limit constant with temperature
- ☐ 100% electrical burn-in
- ☐ Eliminates the need to stock many voltages
- ☐ Standard 3-lead transistor package
- ☐ 80 dB ripple rejection

Absolute maximum ratings

Power dissipation	Internally limited
Input-Output voltage differential	40 V
Operating junction temperature range LM 317	0°C to +125°C
Storage temperature	-65°Cto +150°C
Lead temperature (soldering 10 s)	300°C

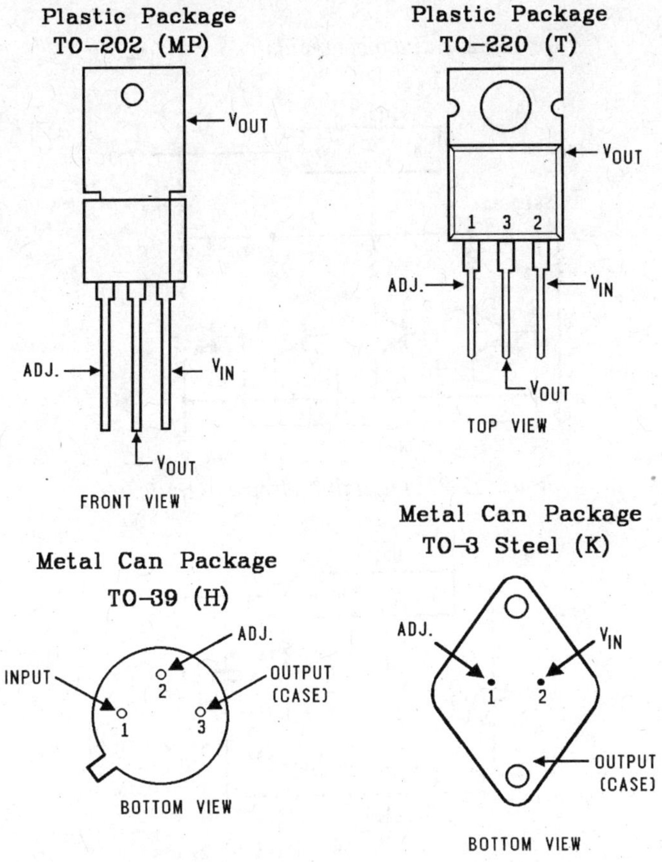

Figure 12.21: Pin connections for LM 317 regulator.

Application Hints

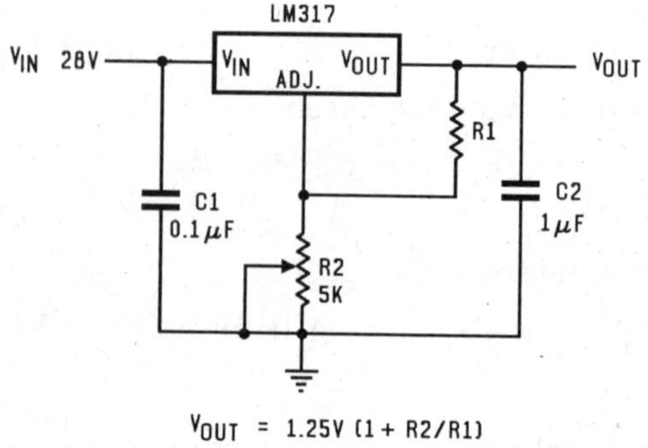

Figure 12.22: Adjustable regulator 1.2 V to 25 V.

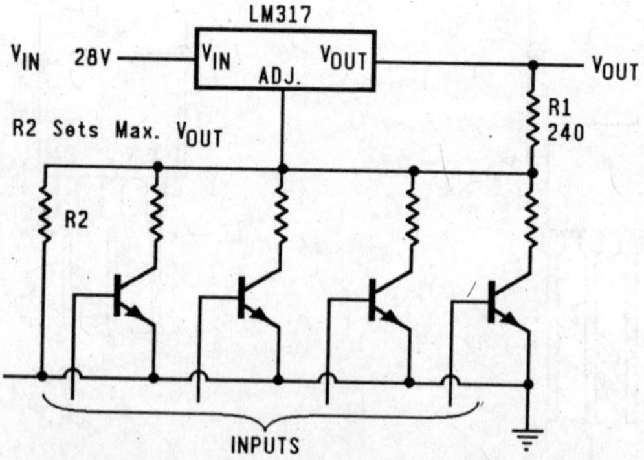

Figure 12.23: Digitally selected outputs.

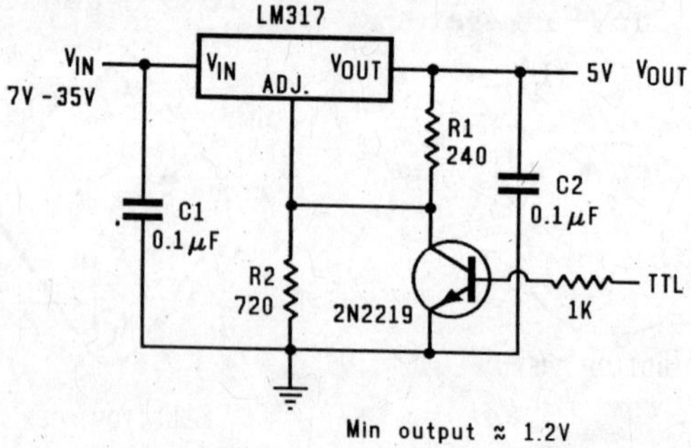

Figure 12.24: A 5V logic regulator with electronic shutdown.

3-Terminal Adjustable Negative Regulator LM 337

The LM 337 is an adjustable 3-terminal negative voltage regulator capable of supplying in excess of -1.5 A over an output voltage range of -1.2 V to -37 V. This regulator is exceptionally easy to apply, requiring only 2 external resistors to set the output voltage and 1 output capacitor for frequency compensation. The circuit design has been optimized for excellent regulation and low thermal transients. Further, the LM 337 series features internal current limiting, thermal shutdown and safe-area compensation, making them virtually blowout-proof against overloads.

The LM 337 serve a wide variety of applications including local on-card regulation, programmable-output voltage regulation or precision current regulation. The LM 337 is an ideal complement to the LM 317 adjustable positive regulator.

Pin connections for the LM 317 regulators are shown in Figure 12.25 and application hints are shown in Figures 12.26, 12.27, and 12.28.

Features

- ☐ Output voltage adjustable from -1.2 V to -37 V
- ☐ Line regulation typically 0.01%
- ☐ Load regulation typically 0.3%
- ☐ Excellent thermal regulation, 0.002%/W
- ☐ 77 dB ripple rejection
- ☐ Excellent rejection of thermal transients
- ☐ 50 ppm/°C temperature coefficient
- ☐ Temperature independent current limit
- ☐ Internal thermal overload protection
- ☐ 100% electrical burn-in
- ☐ Standard 3-lead transistor package

Absolute maximum ratings

Power dissipation	Internally limited
Input-output voltage differential	40 V
Operating junction temperature range	0°C to 125°C
Storage temperature	-65°C to 150°C
Lead temperature (soldering 10 s)	300 °C

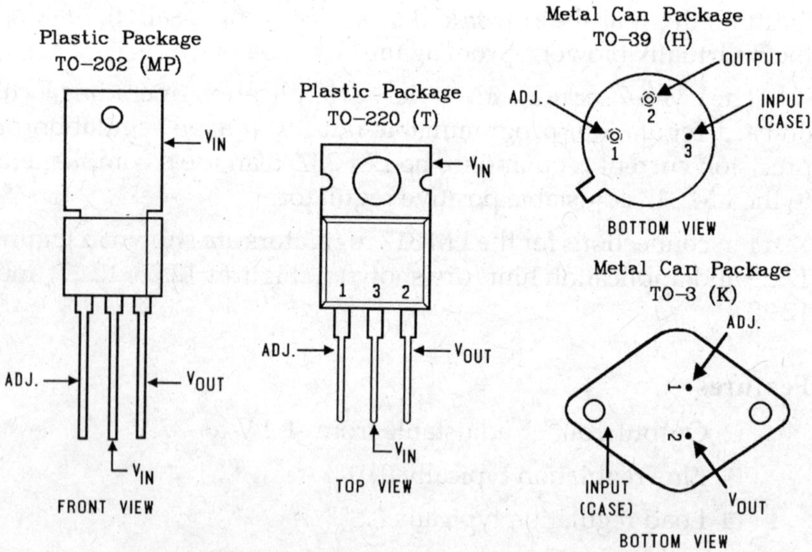

Figure 12.25: Pin connections for LM 337 regulator.

Application Hints

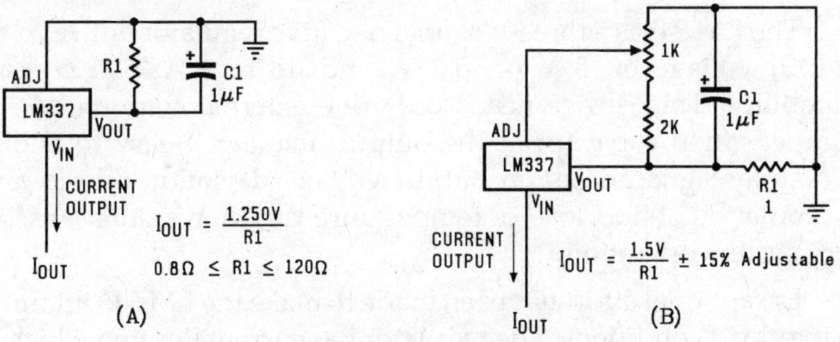

Figure 12.26: Current regulator (A) and adjustable current regulator (B)

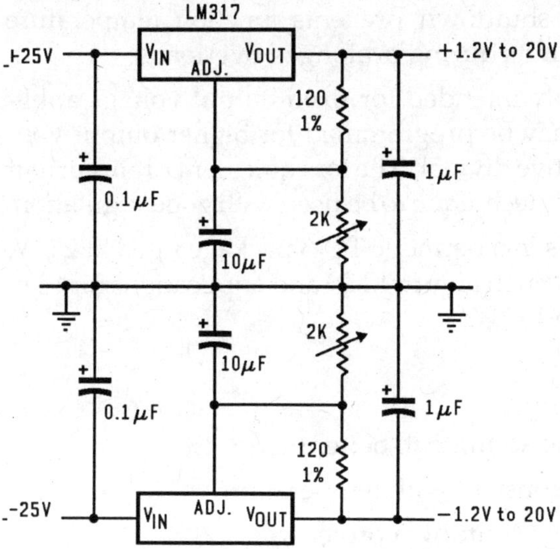

Figure 12.27: Adjustable lab voltage regulator.

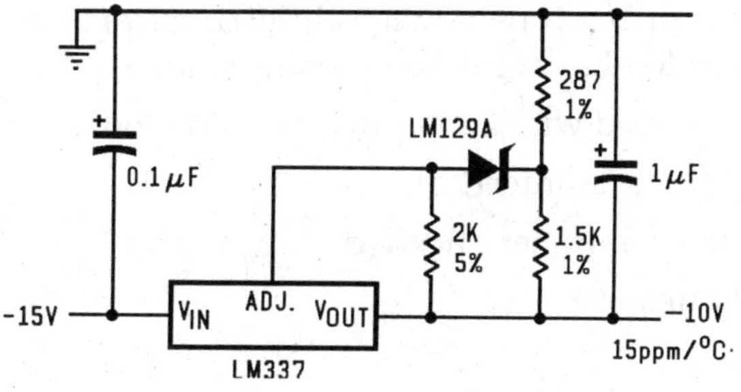

Figure 12.28: High reliability -10 V regulator.

Negative 3-Amp Regulator LM 345

The LM 345 is a three-terminal negative regulator with a fixed output voltage of -5 V or -5.2 V, and up to 3 A load current capability. This device needs only one external component – a compensation capacitor at the output, making it easy to apply. Worst case guarantees on output voltage deviation due to any combination of line, load or temperature variation assure satisfactory system operation.

Exceptional effort has been made to make the LM 345 immune to overload conditions. The regulator has current limiting which is independent of temperature, combined with thermal overload protection. Internal current limiting protects against momentary faults while thermal shutdown prevents junction temperature from exceeding safe limits during prolonged overloads.

Although primarily intended for fixed output voltage applications, the LM 345 may be programmed for higher output voltages with simple resistive divider. The low quiescent drain current of the device allows this technique to be used with good regulation.

The LM 345 comes in a hermetic TO-3 package rated at 25 W. Pin connections are given in Figure 12.29 and applications hint s in Figures 12.30, 12.31 and 12.32.

Features

- ☐ Output voltage accurate to better than ± 2%
- ☐ Current limit constant with temperature
- ☐ Internal thermal shutdown protection
- ☐ Operates with input-output voltage differential of 2.8 V at full rated load over full temperature range
- ☐ Regulation guaranteed with 25 W power dissipation
- ☐ 3 A output current guaranteed
- ☐ Only one external component needed
- ☐ 100% electrical burn-in

Absolute maximum ratings

Input voltage	20 V
Input-output differential	20 V
Power dissipation	Internally limited
Operating junction temperature range	0°C to 125°C
Storage temperature range	-65°C to +150°C
Lead temperature (soldering, 10 s)	300°C

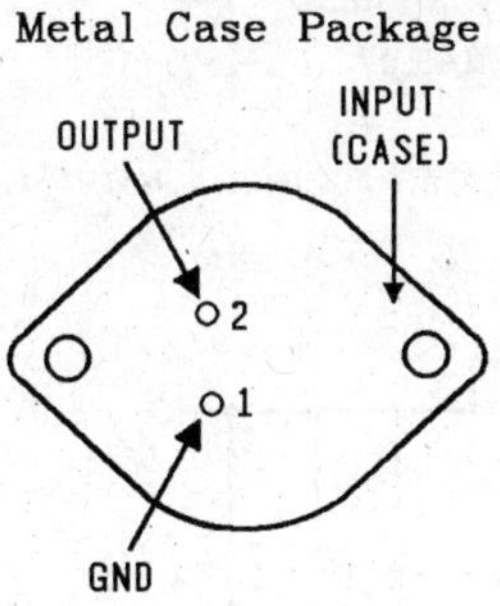

Figure 12.29: Pin connections for LM 345.

Application Hints

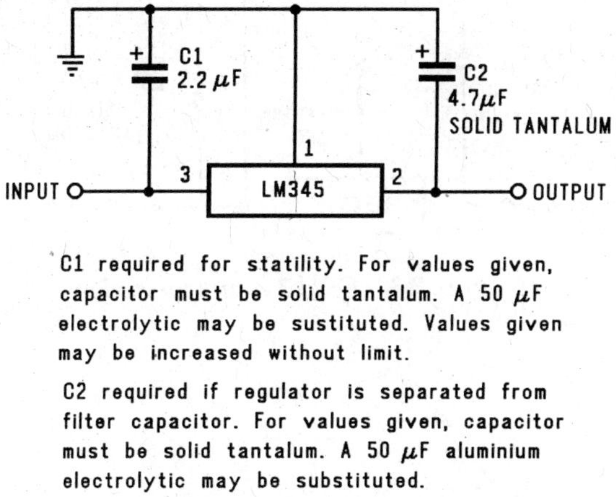

Figure 12.30: Fixed voltage regulator using LM 345.

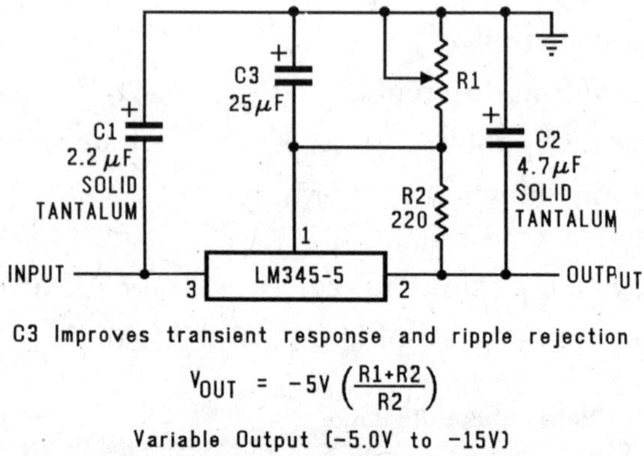

Figure 12.31: Variable output regulator using LM 345.

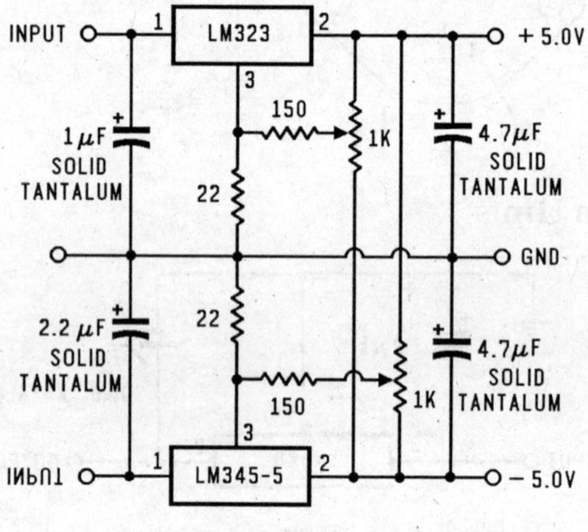

Figure 12.32: Dual 3-amp trimmed supply.

13

Adjustable Output Voltage Regulators

Three-terminal regulators discussed in the previous chapter are primarily meant for fixed voltage output applications. However, with a few external components, they can be made to deliver outputs that are different from their preset outputs. Special integrated circuits are available in 4-pin and 5-pin versions for delivering outputs that can be set to the desired voltages. These ICs are available in versions that can give positive or negative outputs.

Adjustable regulators like μA78G and μA79G series have an edge over μA78XX and μA79XX series regulators in the sense that you do not need to store the entire range of fixed voltage regulators to meet your regulator requirements.

For instance, you might have to store as many as ten types of fixed voltage regulators to cover a range of 5 V to 30 V. This range can be covered by μA78G or μA79G depending upon whether you want a positive or a negative output. In addition, the performance of these 4-terminal adjustable regulators are guaranteed for all regulated output voltages in the specified range. This does not hold good when a 3-terminal IC is used as an adjustable regulator.

In this chapter we discuss the characteristics and circuit configurations using these regulators.

As an illustration, presume that we are using μA7815 and a μA78G separately to get a regulated output of +15 V. If we go through the specifications sheets of the two types of regulators, we find that under identical load current and input voltage conditions both types have identical line and load regulations (typically 1 percent of V_{OUT}).

Let us see what happens when a μA7805 and a μA78G are wired to get a regulated voltage of +7.5 V, if μA7805 is also being used as an adjustable output regulator. Specifications sheets tell us that though μA78G maintains its line and load regulation specifications, that in case of μA7805 becomes worse by a factor of 2.

Other salient features of μA78G and μA79G regulators include:

- ☐ Internal thermal overload protection
- ☐ Internal short circuit current protection
- ☐ Output transistor safe area protection
- ☐ Adjustable output voltages from 5 V to 30 V (78G and 78MG) and -2.2 V to -30 V (79G and 79MG)

Four-Terminal Positive Regulators

μA78G and μA78MG Regulators

Four-terminal adjustable output voltage regulators like μA78G are capable of delivering continuous load currents up to 1 ampere without the use of external series pass transistors. Regulators like μA78MG are identical to μA78G with the difference that these regulators are capable of delivering a continuous total current of 500 mA without the use of external series pass transistors.

Regulators like μA78G are available in either TO-3 or power TAB (4-pin TO-202) packages and regulators like μA78MG are available in TO-39, power mini DIP and power TAB packages. Pin connections for these devices are shown in Figure 13.1.

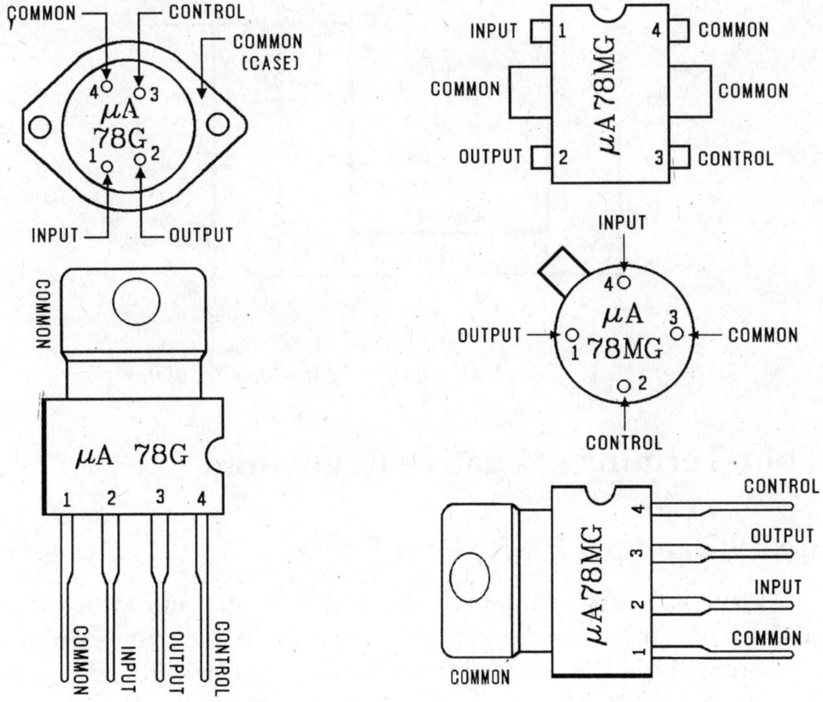

Figure 13.1: Pin connections of μA78G and μA78MG regulators.

Application Hints

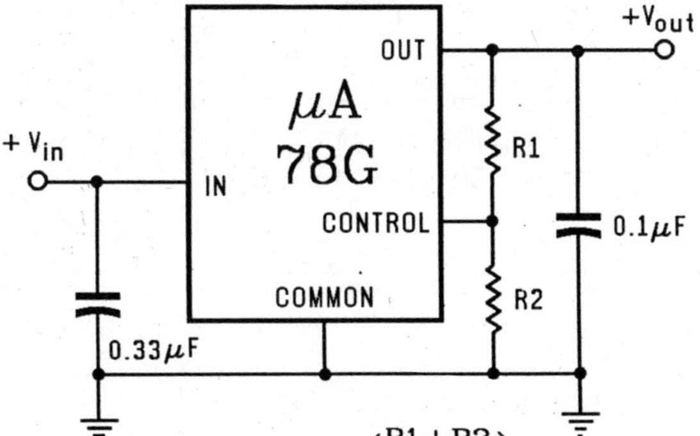

$$V_{OUT} = V_{CONT}\left(\frac{R1 + R2}{R2}\right)$$

Figure 13.2: Basic positive regulators.

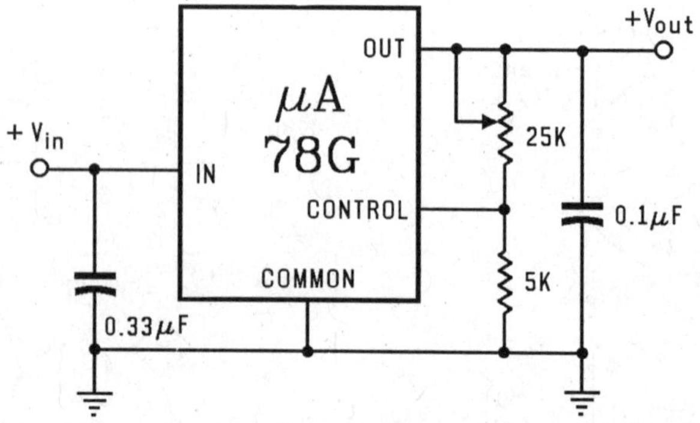

Figure 13.3: Positive 5 V to 30 V adjustable regulator.

Four-Terminal Negative Regulators

μA 79G and μA 79MG Regulators

Four-terminal adjustable output voltage regulators like μA79G are capable of delivering continuous load currents up to 1 ampere without the use of external series pass transistors. Regulators like μA79MG are identical to μA79G with the difference that these regulators are capable of delivering a continuous total current of 500 mA without the use of external series pass transistors.

Regulators like μA79G are available in either TO-3 or power TAB (4-pin TO-202) packages and regulators like μA79MG are available in TO-39, power mini DIP and power TAB packages. Pin connections for these devices are shown in Figure 13.4.

The salient features of these regulators have already been given above. Application hints are given in Figures 13.5 and 13.6.

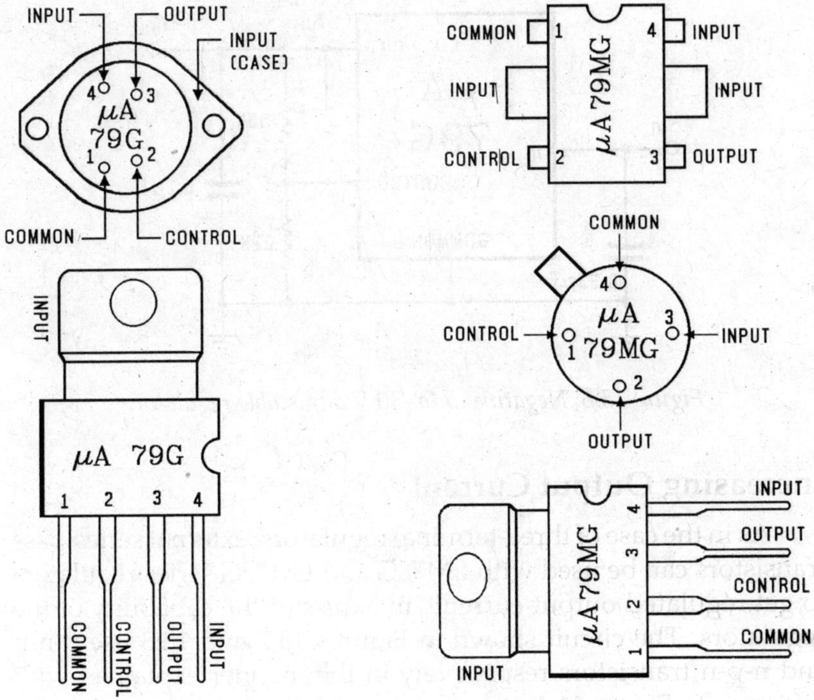

Figure 13.4: Pin connections for μA79G and μA79MG regulators.

Application Hints

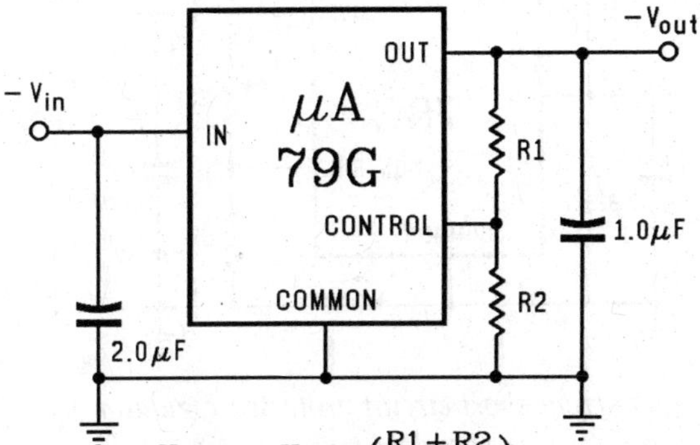

$$V_{OUT} = V_{CONT}\left(\frac{R1+R2}{R2}\right)$$

Figure 13.5: Basic negative regulators.

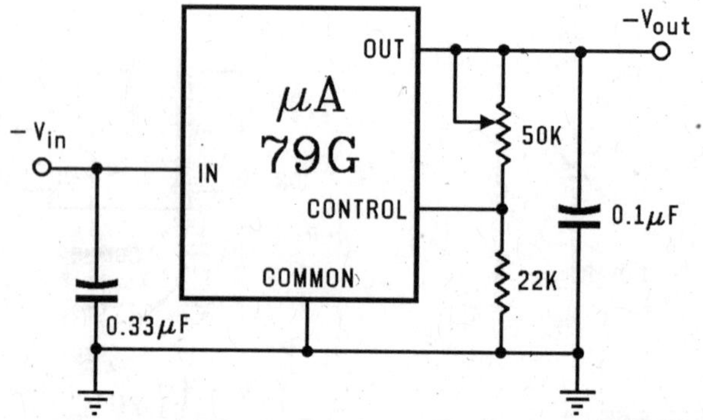

Figure 13.6: Negative -5 to -30 V adjustable regulator.

Increasing Output Current

As in the case of three-terminal regulators, external series-pass transistors can be used with LM78G and LM79G series regulators to get regulated output currents in excess of the capability of the regulators. The circuit shown in Figures 13.7 and 13.8 use p-n-p and n-p-n transistors respectively in the manner explained with reference to Figure 11.6.

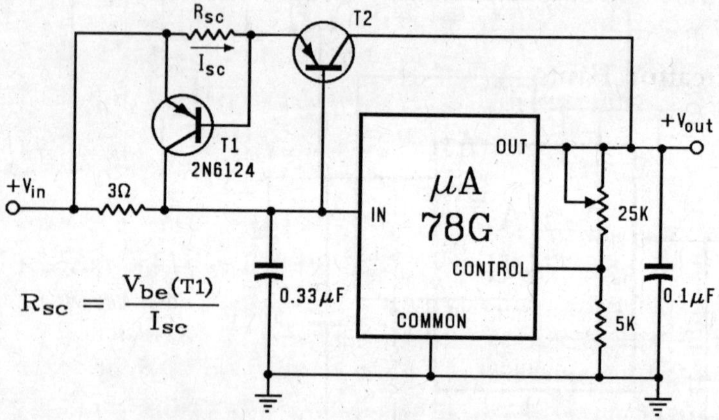

Figure 13.7: Positive high current short circuit protected regulator using external p-n-p type series pass transistor.

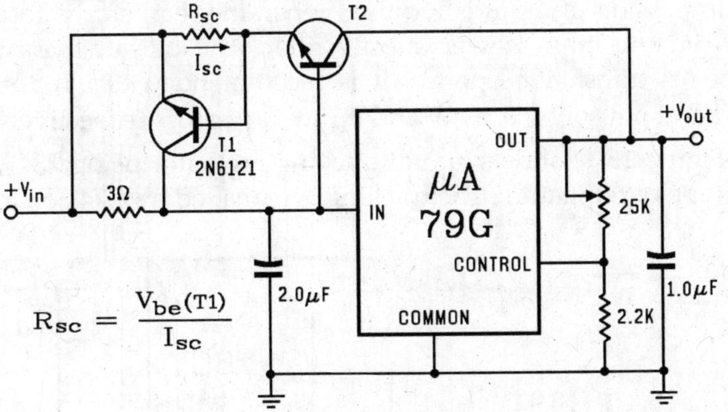

Figure 13.8: Negative high current short circuit protected regulator using external n-p-n type series pass transistor.

Figure 13.9 shows a dual tracking regulator using one 78MG and one 79MG type of regulators. Input and output capacitors may not be required in many applications.

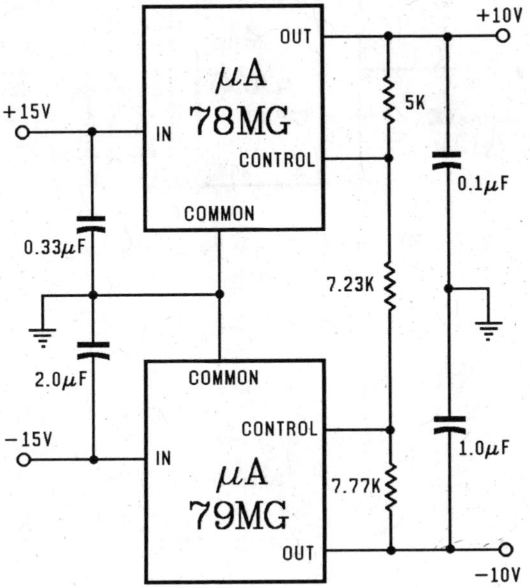

Figure 13.9: Dual tracking regulator

Bypassing of input is required when the regulator is located far from the main power supply filter. The output bypassing improves transient response. It is recommended to bypass the input and output by 0.33 μF and 0.1 μF capacitors respectively.

Figure 13.10 shows a dual tracking regulator using 78G and 79G type of regulators. The tracking is managed by a 741 op amp.

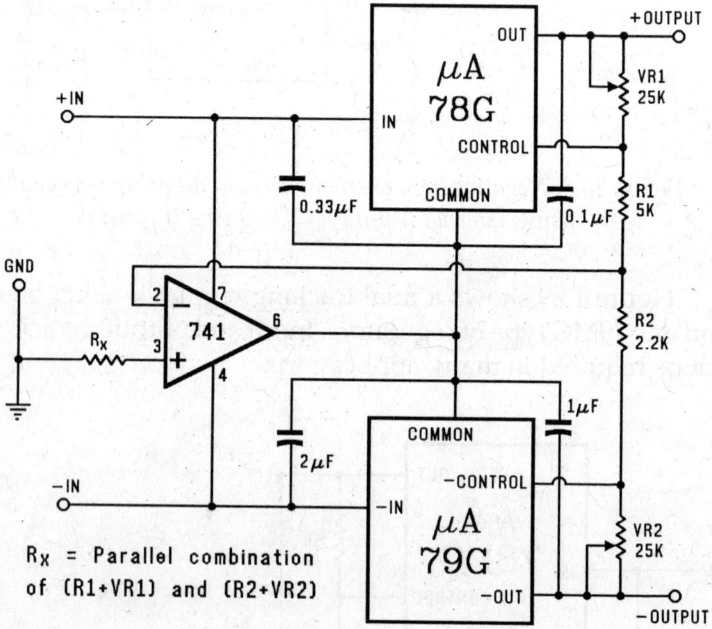

Figure 13.10: A dual tracking regulator using an op amp for managing the tracking of the two output voltages.

Five Terminal Voltage and Current Regulator L200

General Description

The L 200 is a silicon monolithic integrated circuit for voltage and current programmable regulation. It is available in Pentawatt package or 4-lead TO-3 metal case. Current limiting, power limiting, thermal shutdown and input over-voltage precision (up to 60 V) make the L 200 virtually blowout proof. The L 200 can be used to replace fixed voltage regulators when high output voltage protection is required and eliminates the need to stock a range of fixed voltage regulators.

Pin connections for the regulator IC packages are shown in Figure 13.11.

Absolute maximum ratings

V_i	DC input voltage	40 V
V_i	Peak input voltage (10 ms)	60 V
$\Delta V_{i\text{-}o}$	In-out operating differential voltage	32 V
I_o	Output current	Internally limited
P_{tot}	Power dissipation	Internally limited
T_{stg}	Storage temperature	-55°C to +150°C
T_{op}	Operating junction temperature for	
	L 200 C	-25°C to 150°C
	L 200	-55°C to +150°C

Thermal Data

				TO-3	Pentawatt®
R_{th}	j-case	Thermal resistance junction-case	max	4°C/W	3°C/W
R_{th}	j-amb	Thermal resistance junction-amb.	max	35°C/W	50°C/W

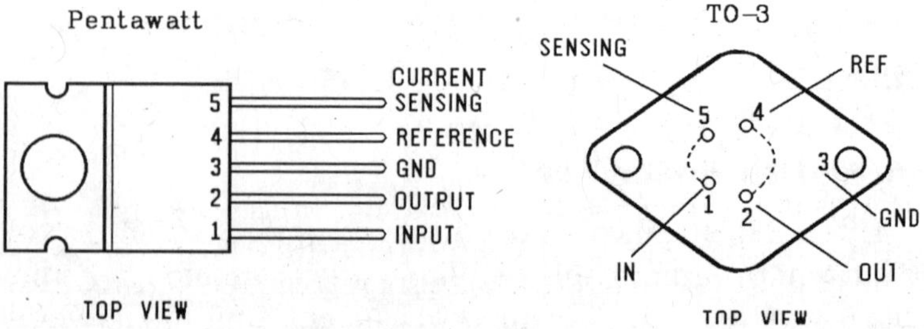

Type	Pentawatt	TO3
L200		L200T
L200C	L200CH, L200CV	L200CT

Figure 13.11: Pin connections of L200 regulator packages.

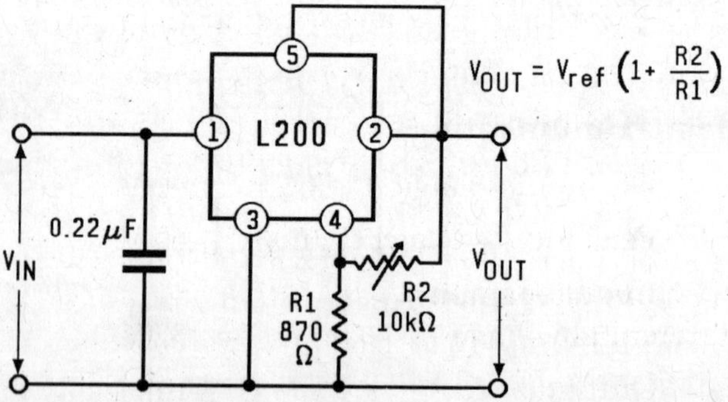

Figure 13.12: Programmable voltage regulator using L200.

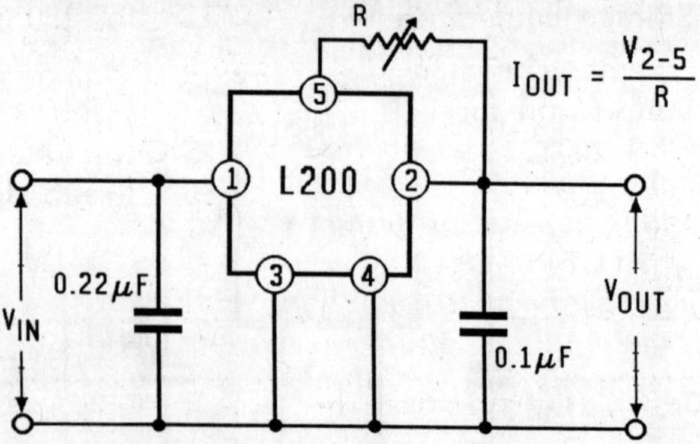

Figure 13.13: Programmable current regulator using L200.

14

Practical Circuits

Electronic circuits use a wide variety of power supplies to meet specific requirements. In previous chapters we have so far described the basic principles of operation of different types of circuits and other associated components that make regulated and unregulated power supplies. In this chapter we describe some practical circuits to illustrate the way electronic power supply circuits are used in practical applications. Knowing the principles of operation, it is easy to understand the working of these circuits.

For the experimenters, it is possible that some components like transistors or zener diodes are not available locally. However, after understanding the basic principles it is easy to estimate the characteristics of those components and substitute them with components available focally. In most cases, the components are not critical and substitutes are likely to work satisfactorily.

In small prints, there can be a confusion about the decimal point. For example, if a particular resistor is 33k or 3.3k. To avoid this type of confusion, many circuits show the designating letter in place of the decimal point. In this case, a 33k resistor will be shown as 33k, but a 3.3k resistor will be shown as 3k3. Similarly, 47 microfarad will be shown as 47 μF or simply as 47μ. but a 4.7 microfarad capacitor will be shown as 4μ7. Capacitances of small capacitors are shown in picofarads (pF) and nanofarads (nF). We know that 1000 picofarads make one nanofarad.

Experimenter's Current Limiter

Experimenters mostly use a battery (often in a battery box of the type shown in Figure 7.1) as power supply. It is quite likely that during experimentation the battery terminals get shorted, or there may be defect in the experimenter's circuit and it draws excessive current. Batteries are capable of supplying too much of current and damage the components in the circuit, or the battery itself can be damaged under overload/short-circuit conditions. To protect the battery as well as the circuit under test, the two-terminal current limiter shown in Figure 14.1 can be inserted in series with the battery, or in series with a battery eliminator or an a.c. powered power supply, when that is used in place of battery.

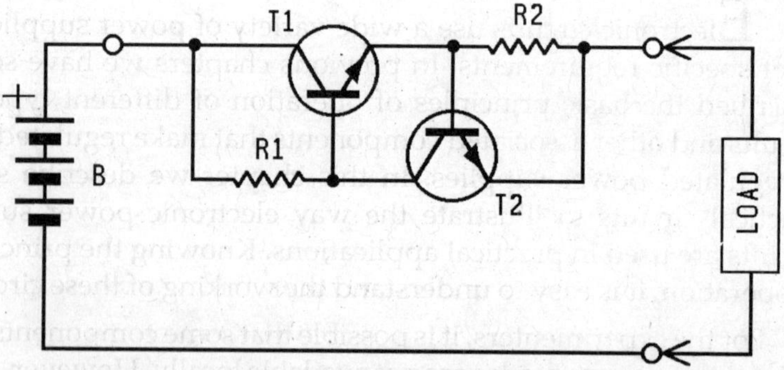

Figure 14.1. Simple current limiter for experimenters.

The circuit uses only two general-purpose transistors and two resistors to perform the current limiting function. Transistor T1 is the series element which allows current, up to a desired maximum, to flow to the load. Resistor R1 provides a suitable forward bias to T1 to permit such forward current flow. R2 is a current sensing resistor interposed between the series transistor and the load, and provides forward bias to T2. Normally, this bias is low enough to make T2 conduct. When excess current flows through R2 as a result of a circuit malfunction or a short-circuit, the voltage drop across R2 rises, biasing T2 into conduction. When T2 turns on, it bypasses the bias current of T1 directly to output and limits the current flow. The maximum load current can be set by changing the value of R2.

T1 can be a power transistor like SL100 and T2 can be a low-current transistor like BC148. A resistance of 47 ohms for R1 and 10 ohms for R2 will limit the current to about 60 mA.

Switchable AC Power Supply

With reference to Figure 3.4 we explained that the voltages induced in the secondary windings of a transformer can be connected in series to buck or boost a.c. voltage. If the two voltages are in phase, they will add up; and if they are in phase opposition, the output will be the algebraic difference of the two voltages. This principle is used in Figure 14.2 to get different a.c. voltages.

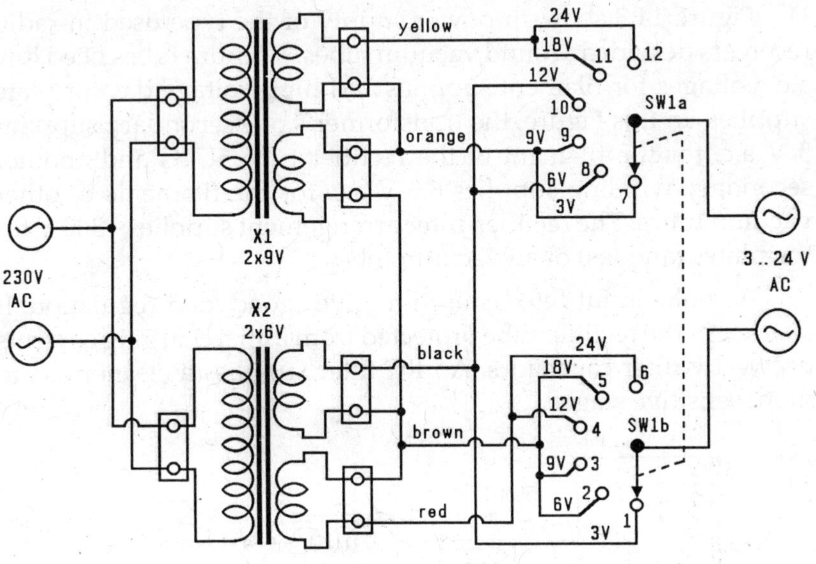

Figure 14.2: Switch-selectable a.c. supply.

Two transformers, X1 with two secondary windings of 9 V and X2 with two windings of 6 V can be connected in different ways by a two-pole six-way rotary switch SW1a and SW1b. This switch skips two possible combinations, 21 V and 30 V, while all other values, 3 V, 6 V, 9 V, 12 V, 18 V and 24 V are available at the output.

In case you do not get the voltages as shown in Figure 14.2, one of the secondary winding may be in reverse phase. This situation can be easily analyzed and corrected by a.c. voltmeter measurements. Remember that the open circuit voltage of secondary windings can be a little higher than the rated voltage. Also, ensure that the switch used is a break-before-make type. Otherwise short circuits will result during switching.

For experimenters, low power transformers will do. If you want to use bigger transformers, be sure to see that the switch contacts are also rated accordingly.

Vacuum Tube Power Supply

Figure 14.3 shows a power supply of the type used in radio receivers designed around vacuum tubes. Vacuum tubes need low a.c. voltages for filament supplies and high voltage d.c. for plate supplies. In this Figure, the transformer's one secondary supplies 5 V a.c. for the filament of the rectifier tube 5U4G and another secondary winding supplies 6.3 V a.c. for the filaments of other vacuum tubes. The rectifier-filter arrangement supplies +300 V for the plates (anodes) of all vacuum tubes.

A choke input filter is used to give good load regulation. It also keeps the rectifier tube protected from initial charging currents drawn by filter capacitors. An RC filter supplies a cleaner d.c. to more sensitive stages.

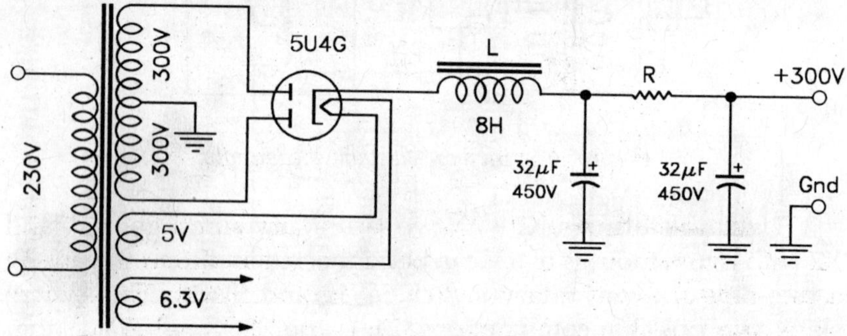

Figure 14.3: A vacuum tube power supply for vacuum tube receivers.

A Lightweight Transceiver Power Supply

The power supply shown in Figure 14.4 converts 110- or 220-volt a.c. power to provide suitable voltage and current values for operation of a transceiver (transmitter-receiver).

The power transformer X1 furnishes all voltages and currents for the power supply outputs. It has two primary windings that may be switched to parallel connection for 110-volt operation and to series connection for 220-volt operation. The change over is done by double-pole double-throw switch S1. This method of connection was explained with reference to Figure 3.4.

The transformer has four secondary windings. One winding furnishes the power for the high-voltage plate supply rectifiers. It is not tapped, and is connected to a voltage-doubling rectifier circuit, D1 through D4. The output of the voltage doubler is filtered by C1, C2, C3, and C4 (see Figure 7.11). Resistors R2, R3, R4, and R5 equalize voltages across the capacitors. This high voltage is needed for the plate supply of transmitter vacuum tubes.

Another secondary winding furnishes power for the +275-volt output for the plates of vacuum tubes in the receiver section. It is connected to a full-wave bridge rectifier consisting of rectifier diodes from D5 through D8. The output is filtered by a pi-section filter comprising C7A, C7B and filter choke L1.

A third winding furnishes voltage for the adjustable bias supply rectifier D9. This is a half-wave rectifier. Its output is filtered by C5, C6, R8 and part of R9. Resistor R9 is adjustable and, with R8 and R10, forms a voltage divider from which the adjustable bias voltage output is taken. All rectifiers in the supply are the small silicon type 1N4007 (or 1N1492 units). The rectifiers are stacked in series in each leg of the high-voltage doubler supply to limit the voltage across each unit to a value that is safely within the limits of tolerance of the rectifiers

In the low-voltage and bias supplies, a surge-suppressing resistor (R6 and R7) is inserted to limit the peak current through the rectifiers. In the high-voltage supply, the transformer winding itself offers enough resistance and no extra resistor is needed.

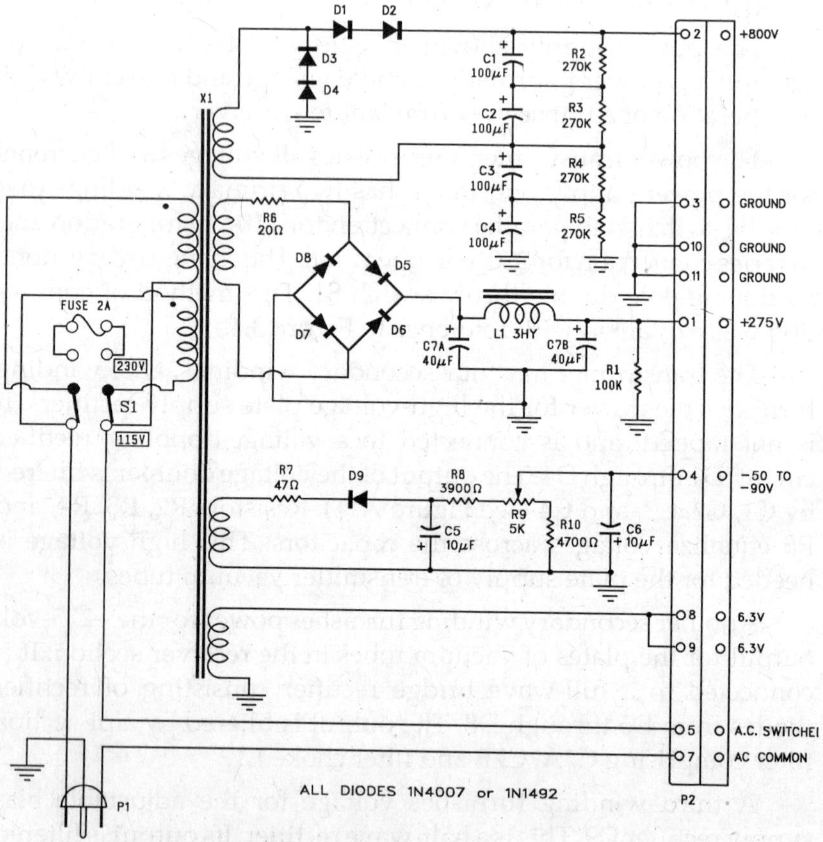

Figure 14.4: A lightweight transceiver power supply.

The fourth winding furnishes 6.3 V a.c. filament power for the vacuum tubes.

The outputs of the power supply and the primary switching leads are connected to pins of an 11-pin female cable connector P2. This connector mates with the power plug of the cable of the transceiver. This enables the power supply to be turned on and off from the front panel of the transceiver. A 3-pin plug provides the ground connection for safety.

Specifications

Input requirements	110 / 220 volts, a.c.; 4 / 2 amps
Output (nominal)	
Heater power	6.3V a.c. at 11 amperes
Low-voltage B+	+275 volts at 175 mA +260 volts at 210 mA
High-voltage B+	+700 volts at 230 mA (key down intermittent)
Bias voltage	-50 to -90 volts, no current requirement

Parts List

C1, C2, C3, C4	Capacitor, electrolytic, 100 μF, 350 V d.c.
C5, C6	Capacitor, electrolytic, 10 μF, 150 V d.c.
C7	Capacitor, electrolytic, two section each section 40 μF, 450 V d.c.
D1 thru D12	Rectifier diodes 1N4007 or 1N1492 or equivalent
F1	Fuse, 4 amp. 250 V for 115 V operation
F2	Fuse 2 amp. 250 V for 230 V operation
L1	Choke, 3 hy. at 210 mA, 50 ohms
R1	Resistor, 100 k ± 10%, 2W, bleeder
R2, R3, R4, R5,	Resistor, 270 k ± 10%, 1W
R6	Resistor, 20 ohms, 5W, wire wound
R7	Resistor, 47 ohms, ± 10%,1W
R8	Resistor, 3900 ohms ± 10%, 1W
R9	Potentiometer, 5000 ohms ± 10%,
R10	4700 ohms, ± 10%, 2W

A High Voltage Power Supply

The high voltage power supply shown in Figure 14.5 is a simple unregulated supply capable of delivering 3400 volts at a load current of 1 ampere. The circuit consists of a step-up transformer with primary rated for 230 V, 10 A and secondary rated for 2500 V at 1 A. Additional surge protection is affected by the use of the 10-Ω, 10-watt resistors in the secondary of power transformer.

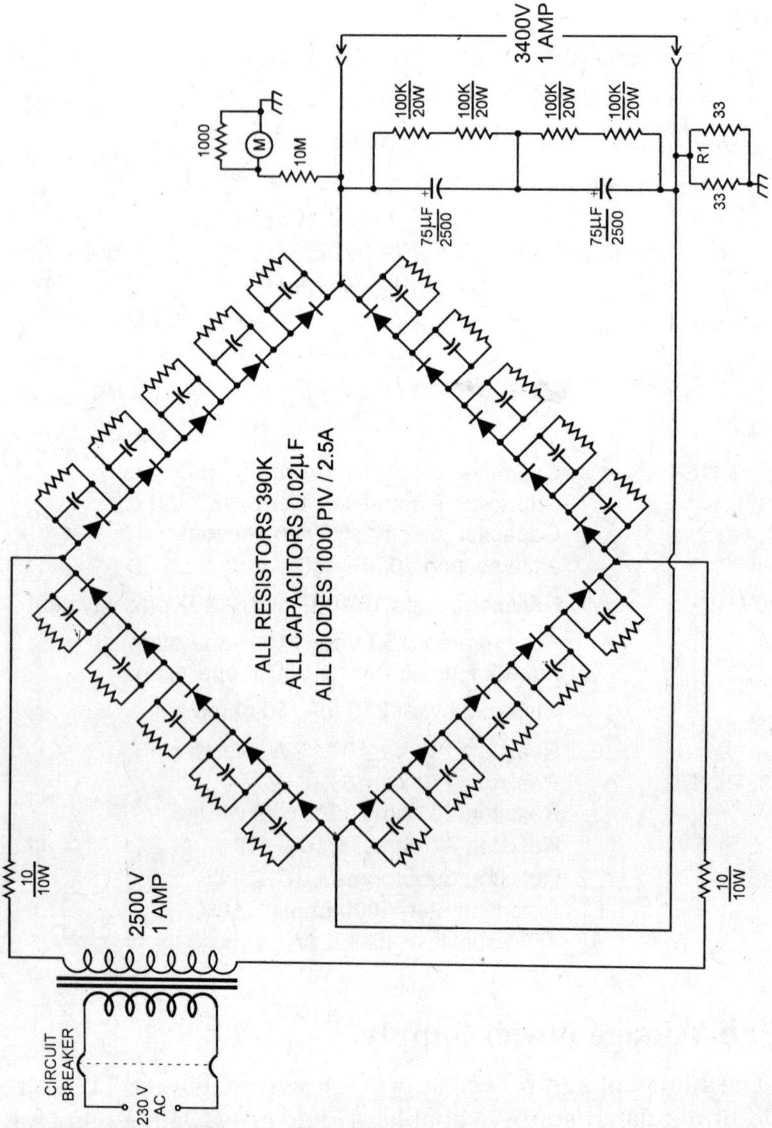

Figure 14.5: A High Voltage Supply.

The paralleled 33-W resistors (R1) are for current metering purposes using the negative lead. They can be omitted and the minus lead tied to ground if a different system of metering is used.

A 3200-V Power Supply

The very high-voltage supply shown in Figure 14.6 is rated at 3200 V at a continuous output current of 1.2 A, and an intermittent peak current of 2 A for supplying power to a 1-kW amplifier of a transmitter. The power supply also has a 12 V, 1 A output used to power the control circuitry in a transmitter.

A simple, effective step/start circuit is incorporated to ensure that the rectifier diodes are not over stressed by the heavy surge current into the output capacitor when the power supply is first turned on. A 5 kV meter is used to monitor high-voltage output.

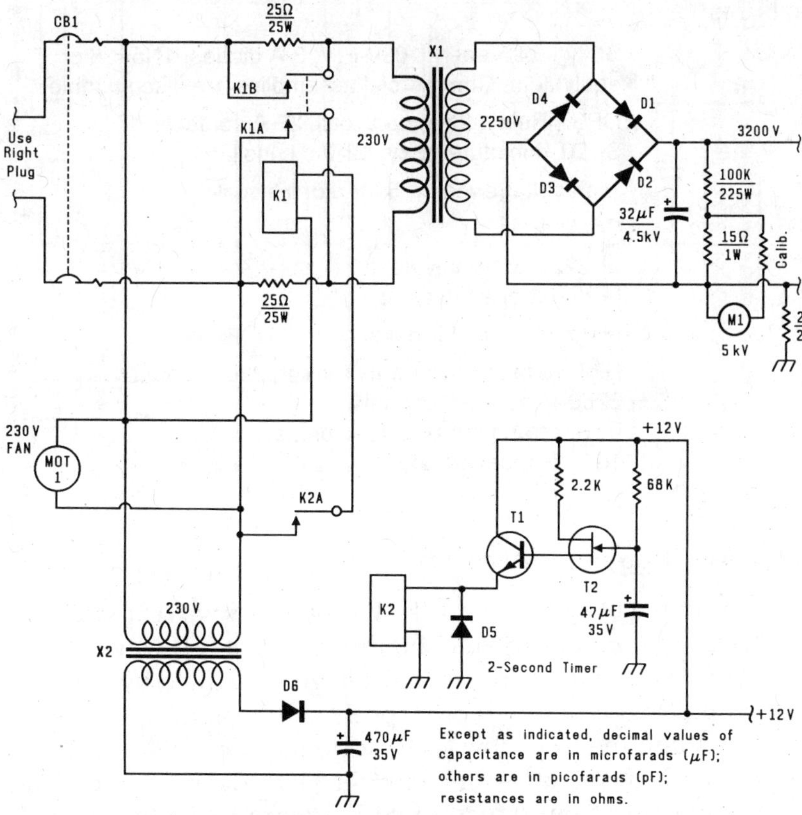

Figure 14.6: A 3200 V d.c. power supply.

Power Supply Operation

When the front panel magnetic circuit breaker is turned on, two 25-Ω resistors in the primary circuit limit inrush current as the filter capacitor charges. After two seconds, step-start relay K1 activates, shorting both resistors and allowing the full line voltage to be applied to the transformer.

Caution:-As with all high-voltage power supplies, be extremely careful! Before opening the cabinet, remove the a.c.-line plug from the receptacle, and confirm that the filter capacitors are discharged before working on the supply.

Parts List

D1 - D4	Strings of 5 each, 1000-PIV, 3-A diodes, 1N5408 or equivalent. Commercial assemblies are also suitable.
K1	DPST Relay, 230-V a.c. coil, 20-A contacts.
K2	SPDT miniature relay, 12-V d.c. coil.
M1	High voltage meter, 5 kV d.c. full scale
MOT1	Cooling fan
T1	2N2222A or equivalent
T2	MPF102 or equivalent
CB1	Magnetic circuit breaker
X1	High voltage power transformer, 230-V primary, 2250-V, 1.2-A secondary
X2	Power transformer, 240-V primary, 16-V, 1-A secondary

Fullwave Phase Controller

Controlling of a.c. power by SCRs and triacs was explained in Chapter 4. The circuit shown in Figure 14.7a is a full-wave power controller. The available a.c. mains voltage is rectified by a bridge rectifier and the pulsating d.c. is applied to the anode of an SCR. Power through the load flows only as long as the SCR conducts. The start of the firing of SCR is controlled by a delay circuit which can delay the start from zero to about 175 degrees, and thus control the total power flowing into the load.

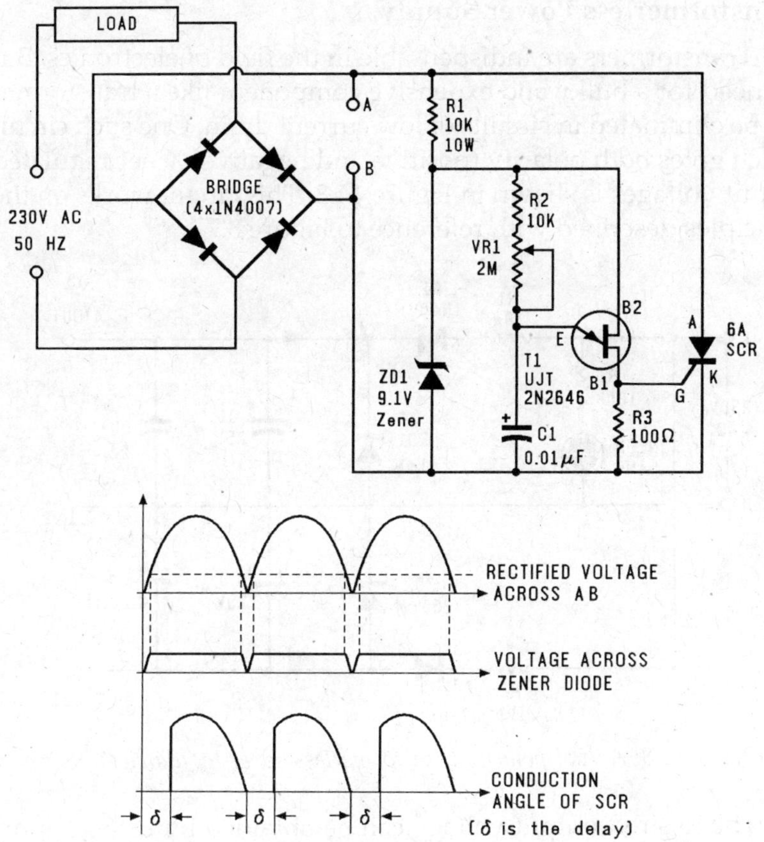

Figure 14.7: A full-wave phase controller.

Resistor R1 and zener diode ZD1 form a 9V d.c. supply. R2+VR1 with C1 form the time delay network. Everytime the a.c. voltage starts rising, capacitor C1 starts charging through R2+VR1. When the voltage across the capacitor reaches the firing voltage of the unijunction transistor, T1 conducts heavily. This discharges C1 and the voltage drop across R3 fires the SCR. The cycle is reset every time the a.c. voltage becomes zero in its a.c. cycle.

The working of the circuit is explained by waveforms across different points. Terminals A and B can be used as test points to measure the rectified output voltage from the bridge rectifier.

Transformerless Power Supply

Transformers are indispensable in the field of electronics. But the need for a bulky and expensive component like a transformer can be eliminated in circuits of low current drain. One such circuit which gives both polarity (positive and negative) zener regulated output voltages is shown in Figure 14.8. This circuit works on the principles described with reference to Figure 8.7.

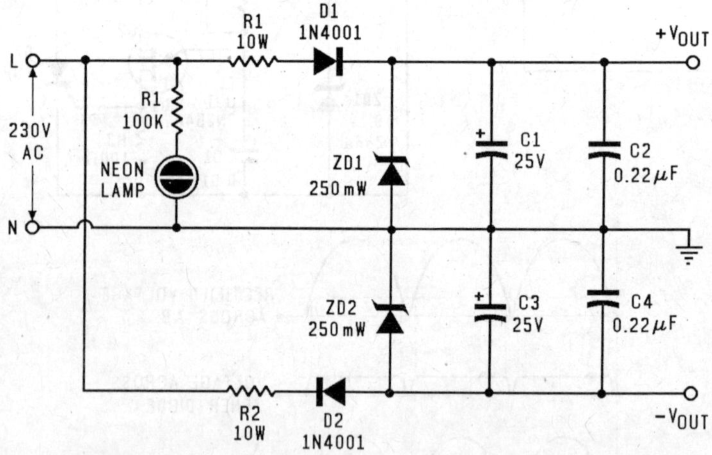

Figure 14.8: A dual-polarity transformerless zener regulated supply.

The desired output voltage can be obtained by using appropriate zener diode. If only one kind of voltage is needed, the other half need not be used.

The circuit is capable of providing up to 25 mA of load current. The output voltage is stable even if the line voltage varies from 195 to 270 volts. However, the power dissipation in the series dropping resistance increases with the rise in line voltage and increase in load current. It is better to use such circuits in applications of low current drains only.

Warning: An important point to note in this circuit is that the supply is not isolated from the mains supply and hence, extreme care should be taken while using the supply to minimize the chances of an electric shock.

The following table gives the values of components to be used for different voltages:

VOUT	R2	D2	C1
3V	15 k	3 V	220 μF
6V	12 k	6.2 V	220 μF
9V	10 k	9 V	220 μF
12V	10 k	12 V	470 μF
15V	10 k	15 V	470 μF
18V	10 k	18 V	1000 μF
24V	8.2 k	24 V	1000 μF

Mains Powered Low Current Power Supply

Many circuits can be powered directly from the mains with the aid of a series capacitor (C1) instead of a series resistor for dropping the voltage. The disadvantage of this approach is that usually only one half-cycle of the mains waveform can be used to produce a d.c. voltage. An obvious solution is to use a bridge rectifier to perform full-wave rectification, which increases the amount of current that can be supplied and allows the filter capacitor to be smaller. The circuit shown in Figure 14.9 in fact does this, but in a clever manner that uses fewer components. Here we take advantage of the fact that a zener diode is also a normal diode that conducts current in the forward direction. During one half-wave of the a.c. cycle, the current flows via D1 through the load and back via D4, while during the other half wave it flows via D2 and D3.

An uncharged capacitor draws a lot of charging current initially. Resistor R1 limits this current to a safe value.

Bear in mind that with this circuit (and with the bridge rectifier version), the zero voltage reference of the d.c. voltage is not directly connected to the neutral line of the 230 V circuit. This means that it is usually not possible to use this sort of supply to drive a triac, which normally needs such a connection. However, circuits that employ relays can benefit from full-wave rectification.

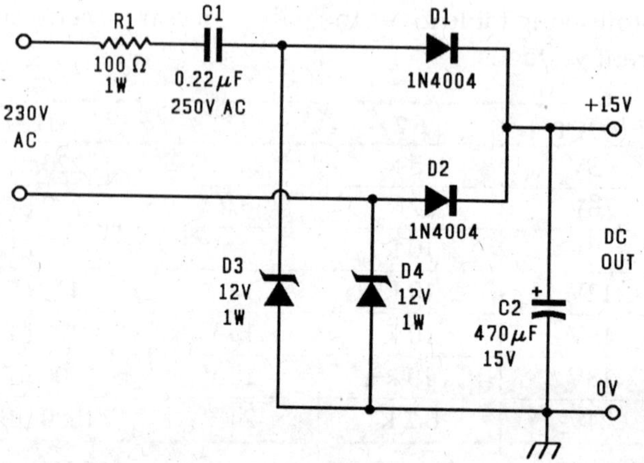

Figure 14.9: Mains powered low current d.c. supply.

The value of the supply voltage depends on the specifications of the zener diodes that are used, which can be freely chosen. C2 must be able to handle at least this voltage. The amount of current that can be delivered depends on the capacitance of C1. With the given value of 0.22 μF (220 nF), the current is approximately 15 mA

Warning: This circuit is directly connected to mains voltage, which can be lethal. You must never come in contact with this circuit when it is connected to the mains! It is essential to house this circuit safely in a suitable enclosure.

Battery Eliminator

Many hobbyists use an unregulated battery eliminator for their projects, or design a separate regulated power supply for each project. The circuit shown in Figure 14.10 is useful for those who need a variable and economical battery eliminator.

The circuit uses a standard adjustable voltage regulator type LM317T whose output can be selected by rotary switch S1. Resistors R1 to R6 are selected to give outputs of 1.5, 3, 4.5, 6, 7.5 and 9 volts. The output rises as the value of resistor between pin 2 (ADJ.) of the IC and ground increases.

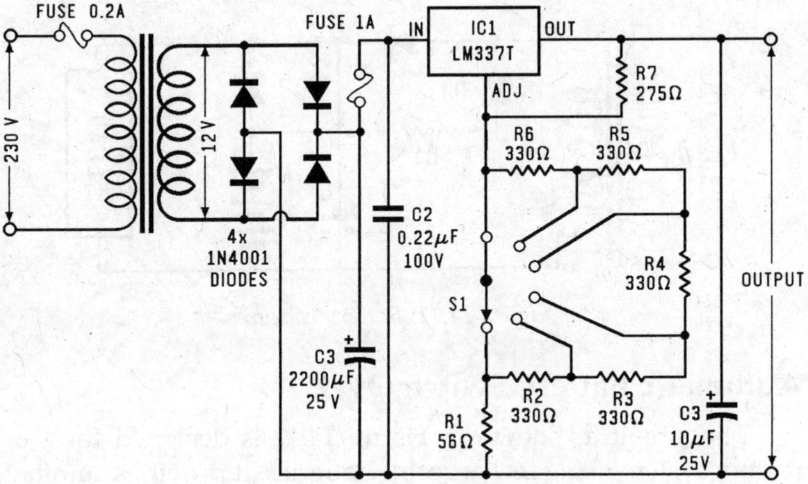

Figure 14.10: A simple battery eliminator.

The accuracy of output voltages with the values given in the circuit is good enough for most practical purposes. Those desiring more accurate voltages can use a 470 ohm preset resistor in place of R1.

Battery Backup

In many applications a battery backup is required which, in the case of mains supply failure, should take over instantaneously, without any interruption. The circuit arrangement for this is simple application is shown in Figure 14.11.

The a.c. voltage at the secondary of transformer is rectified by D2 and filtered by C1. So long as this d.c. output is higher than the battery voltage, diode D3 remains reverse biased and the battery remains cut off. When there is a failure of the mains supply, or the mains voltage falls too low to give the required d.c. output, the battery starts supplying load current immediately via D3.

Diode D1 and resistor R1 form a charging circuit. They can be omitted if charging is not needed. The values of the secondary voltage and R1 and C1 are determined by the application.

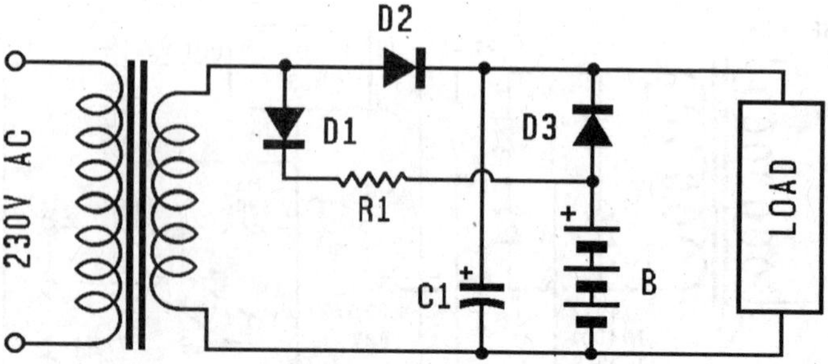

Figure 14.11: Battery backup.

Automatic Battery Switch Over

The circuit as shown in Figure 14.12 is designed for a 6 V battery system using rechargeable batteries. It provides automatic switch over to a charged reserve battery when the voltage of the main battery approaches the fully discharged level. The clever thing about this circuit is that the actual switch over of the battery is voltage-dependent and is adjustable, even though this circuit essentially consists of only two components. The active element is a small SCR (type BRX49), which is connected between the load and the reserve battery. For calibrating the switch over point, an incandescent lamp is connected as a load. In actual use, however, the load can be anything having a maximum current consumption of 300 mA at about 5 V.

The cathode of the SCR is connected to the load, with the anode connected to the positive terminal of the reserve battery. The gate of the SCR is also connected to the positive terminal of the reserve battery via a 47-kΩ trimpot (P1).

This already clearly suggests how the automatic switch over works. The voltage across the SCR is the difference between the two battery voltages. Initially, with both batteries fully charged, the voltages of the two batteries are nearly the same. The SCR is thus cut off, since there is not any significant voltage across it and no gate current flows via P1. As the main battery becomes increasingly discharged, the voltage across the load decreases, which means that the voltage at the cathode of the SCR decreases. The anode becomes positive with respect to the cathode and the gate. A gate current starts to flow via P1, with the magnitude depending on the voltage and the value of P1. As soon as the gate current exceeds the triggering threshold of the SCR, the SCR fires (which means that it suddenly starts to conduct) causing the reserve to be connected to the load. Since the voltage of the reserve battery is higher than that of the discharged main battery, diode D1 is cut off, preventing any current from flowing from the reserve battery into the main battery.

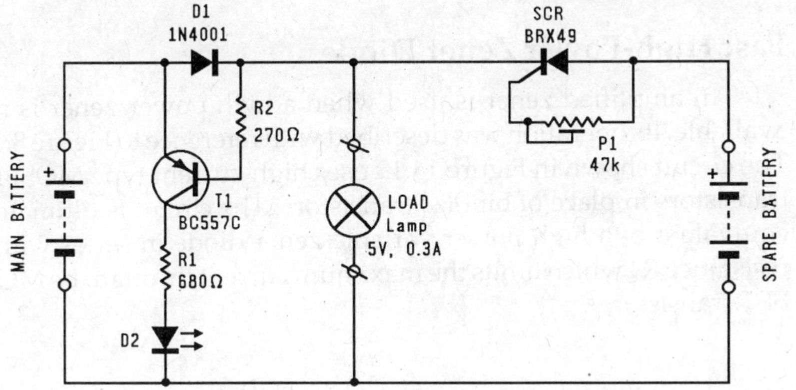

Figure 14.12: Automatic battery change over.

Diode D1 not only prevents a reverse current flow from the reserve battery into the main battery, it also forms the part of an indicator circuit in combination with transistor T1 and LED D2. As long as a current flows from the main battery through D1, a base current also flows through the base-emitter junction of the transistor, which, together with base resistor R1 is connected in parallel with D1. This causes the transistor to conduct, so LED D2 is illuminated to indicate that the main battery is in use. When the circuit switches over to the reserve battery, T1 is also cut off, and the LED is immediately extinguished.

To calibrate the circuit, connect only the reserve battery to the circuit (a non-rechargeable battery can also be used), and replace the mains battery by an adjustable power supply whose output voltage can be varied over (at least) the range of 5 V to 8 V. After first setting P1 to maximum resistance, adjust the power supply voltage to approximately 5.3 V, and then slowly rotate the trimpot until the switch over occurs which can be recognized by the LED going out. Next, increase the power supply voltage and then reduce it again, in order to verify the operation of the circuit.

If necessary, the setting of the trimpot should be readjusted until switch over occurs reliably when the desired voltage is present at the main battery terminal.

Fast High-Power Zener Diode

An amplified zener is used when a high power zener is not available. Its operation was described with reference to Figure 8.11. The circuit shown in Figure 14.13 uses high-current type MOSFET transistors in place of bipolar transistors. The circuit is ultimately equivalent of a high power variable zener diode in series with a resistance R4 which limits the maximum current through the MOSFET transistor.

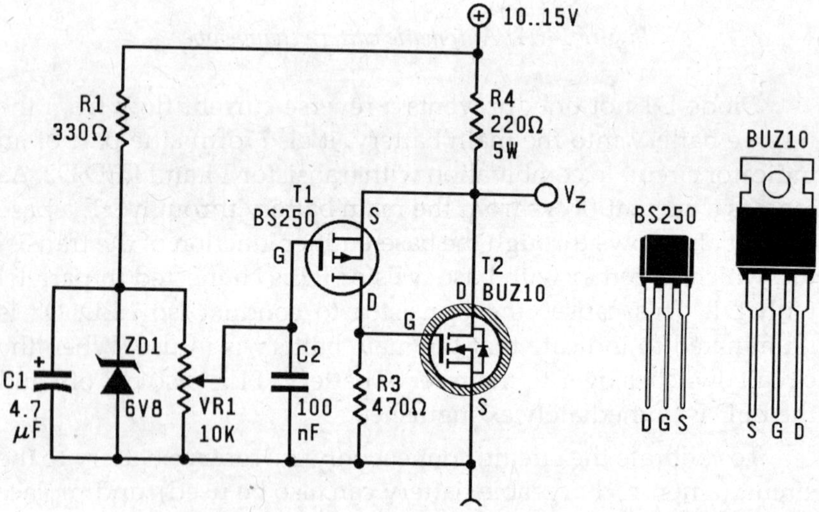

Figure 14.13: Fast high-power zener diode.

From an external power source, whose voltage must be about 2 V higher than the wanted stabilized potential, zener diode ZD1 in combination with R1, C1 and VR1 derive a stable voltage in the range 0 — 6.8V. This voltage switches on MOSFET T1, which in turn switches on another MOSFET T2. If the output voltage V_Z drops slightly, the potential at the gate of T2 also drops, whereupon T2 raises its on-resistance, so that the output voltage rises. The control speed is determined by R4.

The circuit shown in Figure 14.13 delivers output voltages in the range 3.9 — 9.6 V at a supply voltage of 15V.

3V Regulated Output Power Supply

The three-volt output power supply shown in Figure 14.14 is designed using discrete components. It can be used as a substitute for 3 V batteries in many types of battery operated electronic gadgets like walkman or small radios. It will not operate the electric motor but will supply enough power for other operations.

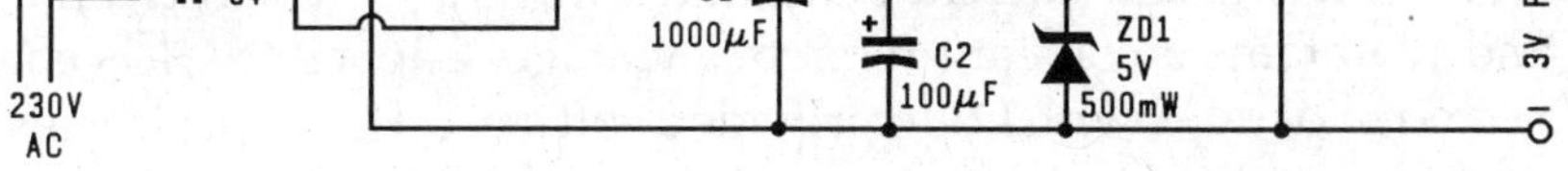

Figure 14.14 : A 3V regulated power supply.

The circuit is based on the principle explained with reference to Figure 9.1. The two direct-coupled transistors behave as a single transistor of high current gain, but the output voltage in this case is the voltage of the zener diode ZD1 minus the base-emitter voltage drops in the two transistors T1 and T2.

A full-wave rectifier using a bridge is used so that a 1000 microfarad filter capacitor gives sufficient filtering of the ripple.

Discrete Voltage Regulator

The circuit shown in Figure 14.15 is very simple. It uses three semiconductors, three resistors, a capacitor and a diode only. The maximum output current depends on the output transistor. With BD680, as used here, a current of 4 A can be delivered at a collector-emitter voltage of 10 volts with adequate cooling. The peak current is 6 A, which is not easily matched with an integrated voltage regulator.

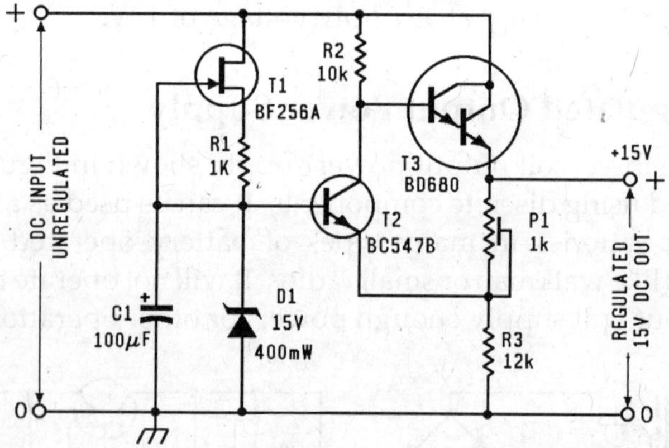

Figure 14.15: Discrete voltage regulator.

The working of the circuit is very simple. Zener diode D1 receives a constant current of around 1 mA by T1 which is a JFET and maintains a constant reference voltage. Capacitor C1 is connected in parallel with D1 to provide a soft start. This capacitor also provides additional buffering and decouples noise and other disturbances. The startup time is around three seconds.

The only additional item needed for the voltage regulator is an output buffer for the reference potential. This takes the form of a sort of super-darlington consisting of T2 and T3. This works very well but has the disadvantage that the output voltage is one diode drop lower than the zener voltage. Preset P1 may be added to correct this at the cost of a reduced regulation of the circuit. If the voltage difference is not important, it is better to replace P1 with a wire bridge. The maximum output voltage is limited to 30 V.

Simple Variable Power Supply

The low cost power supply shown in Figure 14.16 has an output voltage range of 1.5 V to 15 V at a maximum current of 500 mA. Regulation is better than 2% for output currents not exceeding 350 mA. Voltage adjustment is effected by a potentiometer and an acoustic overload indication is provided.

Diode D1 establishes a reference voltage of 15 V. Transistor T4 compares the voltage at the wiper of P1 with the output voltage. When this is 0.65 V higher than the set voltage, transistor T4, and hence transistor T2 is switched on. This removes the base current from the darlington power stage T3-T5 and bypasses it directly to ground. In this manner the output voltage of the supply is held at 0.65 V higher than the reference potential at the base of T4., which is derived from a 15 V zener diode D1.

The a.c. voltage at the 18 V, 1A winding of the transformer is rectified by a bridge rectifier and smoothed by capacitor C1. All the load current passes through the current sensing resistor R1. With R1 equal to 1.5 ohms, transistor T1 starts to conduct when the load currents exceeds about 500 mA.. When this happens, a simple acoustic overload alarm (Bz1) is actuated. Note that the exact level of actuation depends on the electrical specifications of the buzzer, which should be a 24 V, self oscillating type.

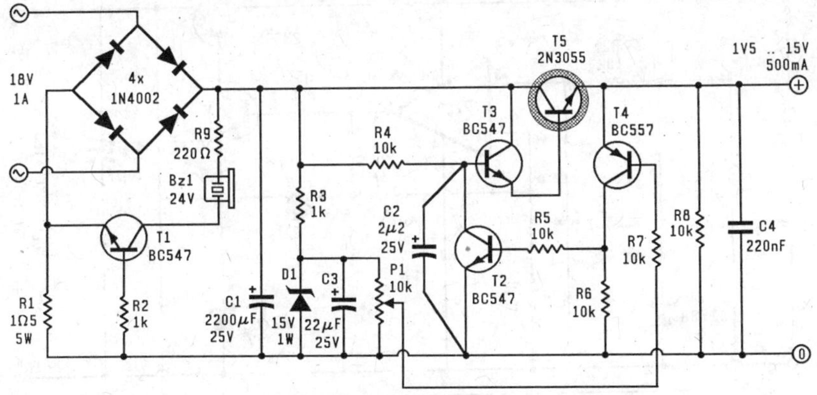

Figure 14.16: A simple variable power supply.

This power supply is, in principle, not short-circuit protected although the use of a generous heat sink for T5 will enable the transistor to withstand the maximum dissipation of about 30 W for the few seconds that lapse before the supply is switched off.

Regulated Supply Using an Op Amp

A practical circuit using an operational amplifier (see Figures 9.2 and 9.3) for voltage comparison and regulation is shown in Figure 14.17. In this case, the current for the zener diode is taken from the regulated supply rather than the unregulated supply. This minimizes noise and increases voltage stability. Zener current is decided by resistor R1.

A popular commonly available type of operational amplifier type 741 is used for comparison of voltages. The full zener voltage is applied to the non-inverting input of the op amp 741, while a fraction of the output voltage, as decided by the voltage divider R4 and R3 is applied to the inverting input of the op amp. If input to the (+) input is higher than that at the (-) input terminal, the positive going output of the op amp makes T1 conduct more and that increases the output voltage. The reverse happens if the voltage at (+) input is lower than that at the (-) input. The high gain of the operational maintains close regulation.

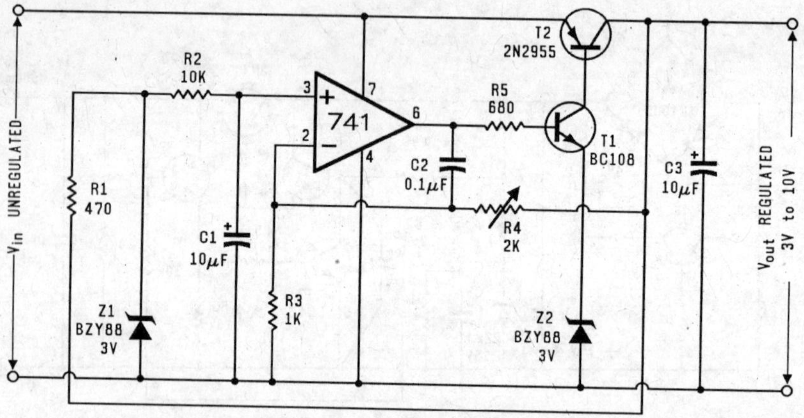

Figure 14.17: Voltage regulator using an op amp and discrete components.

Compact Power Supply

The design of the regulated power supply shown in Figure 14.18 is based on IC1, a 5-pin voltage regulator Type L200 with integral, presettable current limiting as well as thermal overload protection. Pin connections and characteristics of this IC have been described with reference to Figure 13.11.

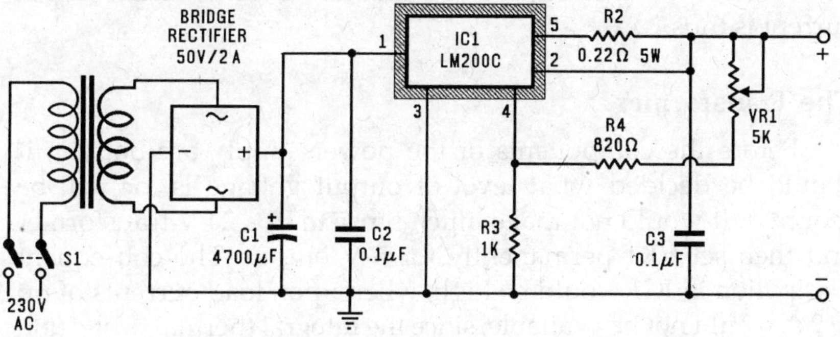

Figure 14.18: A compact regulated power supply.

To ensure correct regulation of the output voltage, the device has an internal 2.75 V reference voltage source. The voltage at pin 4 is compared with the fraction of the output voltage as decided by the voltage divider comprising R3, R4 and VR1. With values shown in Figure 14.18, the output can be set between 5 V and 26 V with VR1.

It should be borne in mind that the maximum input voltage to IC1 is 40 V and the maximum difference between the output and input voltages is 32 V. That means that the secondary voltage of the mains transformer should not exceed 24 V a.c.. This gives an unregulated voltage input into IC1 of 32 V, resulting in a maximum output voltage of 24 V. To set VR1 for an output voltage greater than 24 V is not a good idea, because the regulation at fairly high currents would then no longer function correctly, owing to the inadequate voltage reserve. The result would be an unacceptable 100 Hz ripple on the output because the differential voltage between input and output of the IC will be lower than the minimum.

To enable correct functioning of the integral current limiting of the IC, the voltage drop across R2, connected between pins 2 and 5 is monitored. When this voltage drop reaches 450 mV, current limiting begins.

The maximum output current I_O is therefore given by:

$$I_O = 450/R2 \text{ (mA)}$$

With R2 = 0.22 ohms, as in the diagram, the maximum output current is thus 2 A.

The Transformer

Before the components of the power supply are bought, it should be decided what level of output voltage is, or will be required. It would not make much sense to fit a 24 V transformer and then set VR1 permanently for 5 V output. The consequent dissipation in IC1 would be high, whereupon load currents of up to 2 A would not be available, since the integral thermal protection circuit would limit the output current. Component values differing from those shown in the diagram are, therefore given in the table below for the most usually required voltages.

Output Volts	X1 Secondary	C1
5V and 6V	8 V	16 V
9 V	9 V	25 V
12 V	12 V	25 V
15 V	15 V	25 V
18 V	18 V	40 V

It should be borne in mind when ordering the transformer that the alternating secondary current should be about 1.4 times the desired direct output current. At the same time, the value of R2, the rating of C1 and the current rating of rectifier bridge should also be considered.

Power Supply with Sense Lines

There are applications in which it is important for the supply voltage to be largely independent for the level of the output current, which is, of course, particularly so in the case of variable loads.

When the load is linked to the power supply by relatively short wires, a good variable power supply maintains the output voltage at a virtually constant level. Unfortunately, in practice, these wires can be fairly long, and since they have resistance, there is a voltage drop across them. This interferes with a good regulation; the only way of avoiding this problem is to link the control part of the power supply to the load via separate lines (see Figure 9.8).

Unfortunately, this cannot be done readily in every power supply without some tedious work, but as the diagram in Figure 14.19 shows, in the case of L200 it presents no problems.

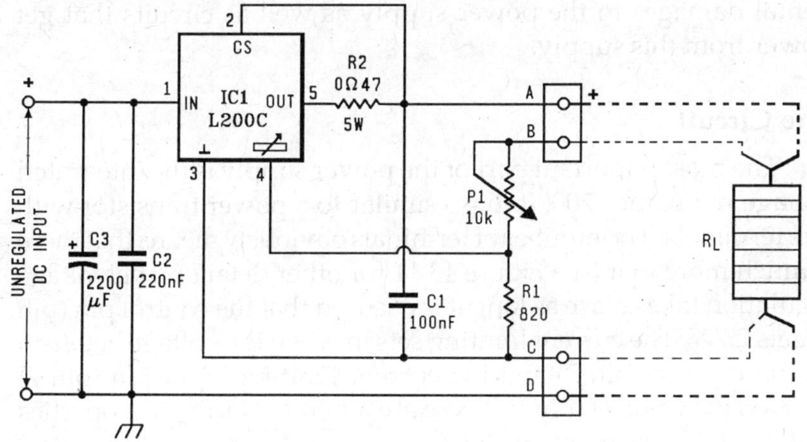

Figure 14.19: Power supply regulator with sense lines.

In Figure 14.19, A and D are the usual output terminals, while B and C are the sense input terminals. The output voltage is given by:

$$V_O = 2.77(1 + R_{P1}/R1)$$

where V_O is in volts and R_{P1} is the effective resistance of P1.

Resistor R2 in series with terminal A provides current limiting. The peak level of the output current Io in amperes is

$$Io = 0.45/R2$$

The maximum input voltage to the regulator is 40 V, and the peak output current is 2 A.

The regulator has on-board thermal protection, but this does not mean, of course, that is should not be mounted on a heat sink when dissipation is high.

Universal Power Supply

The power supply shown in Figure 14.20 has two output ranges; one from 2.7 V to 12 V, and the other from 10 V to 24 V. The output is short circuit protected. Current limiting is adjustable. These two features are mainly incorporated for preventing accidental damages to the power supply as well as circuits that get power from this supply.

The Circuit

The most important part of the power supply is the integrated voltage regulator L200. It looks similar to a power transistor with five terminals. The number of terminals obviously means that there is much more in it (see Figure 13.11 for other details). The voltage regulation takes place as long it is ensured that the control pin (pin 4) gets 2.7 V. The current limiting sets in when the voltage between the output pin (pin 5) and the current limiting input pin (pin 2) exceeds the value of 0.46 V. These are inherent electrical properties of the L200 and cannot be altered.

The output voltage has been divided in two ranges, to avoid heavy losses and overheating at low voltages.

If suppose the input voltage is 28 V and we need an output voltage of 5 V, then somehow the regulator must "destroy" these extra 23 volts. This will result in loss of power as well as overheating of the regulator IC. If we have an input of 14 V and the output is 5 V, then the regulator has to reduce the input only by 9 V. By dividing the output into two different ranges the efficiency has been increased and heating reduced.

The switching over from one range to another is done by switch S2. In the lower range diodes D1 and D3 form a full-wave rectifier with the center-tapped transformer. In the higher range the bridge rectifier action doubles up the end-to-end voltage of the secondary winding. This doubles the d.c. voltage input to the regulator circuit.

Potentiometer P1 and resistors R4 ... R9 set the output voltage. Potentiometer P2, transistor T3 and associated circuits set the current limit. Transistors T1, T2 and the red LED D5 give the indication of the current limiting action.

Potentiometer P1 and resistors R4 to R9 form a voltage divider network. The junction of R4 and R8 or R8 and R6 (depending on the range) is connected to the control input. The regulator adjusts the output voltage in such a way that voltage at the control input remains constant at 2.7 V. The output voltage thus depends on the setting of the potentiometer P1.

The setting of current limiting is somewhat difficult to understand. It becomes effective when the voltage difference between pins 2 and 5 of the regulator increases beyond 0.45 V.

The voltage present here is equal to the voltage drop over R12 plus the voltage at the sliding contact of potentiometer V2. The constant current source formed using T3 provides for the fact that the value of output voltage does not affect the current limiting function. As the voltage drop across R12 depends upon the current drawn by the load and the voltage at pin 2 is derived from a constant current source, the setting of current limiting is independent of the voltage at the output. The green LED D8 has a dual function. It severs for switching control as well as for control of the constant current source T3.

Transistors T1 and T2 and the red LED D5 are used for fault indication, when output current reaches the limit value that is set for it. In normal condition, the voltage on pin 4 is 2.7 V, which is greater than the threshold voltage of D6, D7 and the base-emitter

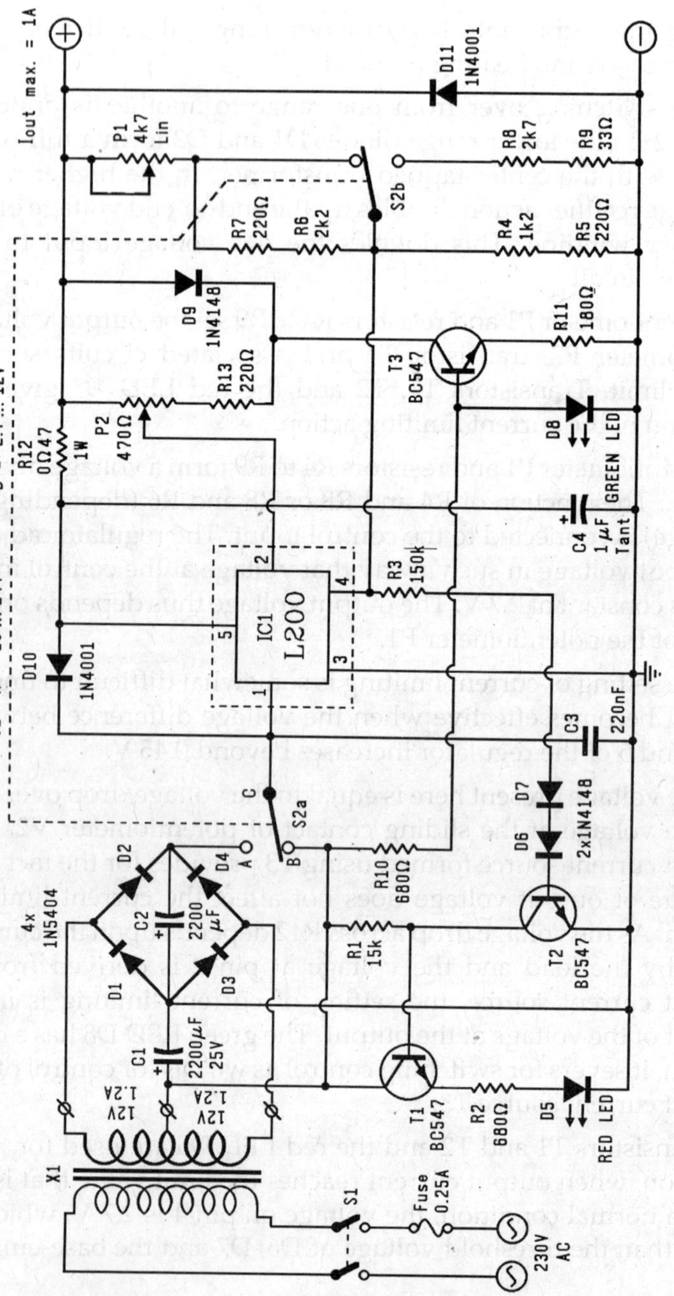

Figure 14.20: A universal power supply.

junction of T2. Thus, T2 is always conducting. This causes the base of T1 to remain at zero volts and it is always off. LED D5 also remains off. But as soon as the voltage on pin 4 reaches about 2 V, which is less than the required voltage to keep T2 on, T2 is turned off and T1 then starts conducting. Red LED D5 glows, indicating fault condition.

Diodes D10 and D11 serve as the protection for the circuit.

Testing the Power Supply

If the power supply is assembled correctly, the circuit must function correctly after it is switched on. Both the voltage ranges can be confirmed with a multimeter.

Whenever the regulator switches off due to overload, the power supply from mains must be switched off for some time and then restarted.

If the circuit does not function as expected, it can be tested stage-by-stage with a multimeter. First of all, the interconnections must be checked. The diode and capacitor polarities must be correct. Pin orientation of the IC and transistors must be carefully checked.

It must also be ensured that the cooling fin of the IC is not short circuited with any other pin or with any other component. It must be connected only to the minus pole of the power supply.

Before starting further measurements on the circuit, point C must be disconnected from the pole of switch S2a. By doing this, we have isolated the regulator part of the circuit from the rectifier. Both LEDs will now glow when the power is switched on.

The following values should be measured:

- Voltage at each transformer winding : 12V a.c.
- Voltage over both windings together : 24 V a.c.
- Voltage over C1 (Point B to earth) : 18 V d.c.
- Voltage over C2 (Point A to earth) : 36 V d.c.
- Voltage over both LEDs : 2 V

If all these are correct, it means that the transformer and rectifier part of the circuit is correct. As the next step, connection between P2 and R13 is desoldered. Now if a multimeter set to 100 mA is inserted between point B and collector of T3, it should indicate a current of 13 mA approximately.

The desoldered connection can now be restored and the following voltages should be measured:

- Pin 4 of IC1 and the earth : 2.7 V d.c. independent of the potentiometer setting. Reduces only when current limiting sets in.
- Voltage between D6 and D7 : 1.2 V d.c.
- Voltage between base of T3 and earth : 3 V d.c.
- Voltage between emitter of T3 and earth : 2.5 V d.c. independent of potentiometer setting.
- Voltage over D9 : Approximately 0.8 V d.c.

Any wrong indication during the test procedure indicates that the particular component is either wrongly connected or it is defective.

Parts List for universal power supply

R1	=	15 kΩ	C1	=	2200 μF/25 V
R2	=	680 Ω	C2	=	2200 μF/40 V
R3	=	150 kΩ	C3	=	220 nF
R4	=	1.2 kΩ	C4	=	1 μF/35 V, Tant.
R5	=	220 Ω	T1, T2, T3	=	BC547B
R6	=	2.2 kΩ	D1, D2, D3, D4	=	1N5404
R7	=	220 Ω	D5	=	LED Red
R8	=	2.7 kΩ	D6, D7, D9	=	1N4148
R9	=	33 Ω			or 1N4002
R10	=	680 Ω	D8	=	LED, Green
R11	=	180 Ω	IC1	=	L200
R12	=	0.47 Ω, 1 W	X1	=	12-0-12 V/1.2 A
R13	=	220 Ω	S1	=	DPST main switch
P1	=	4.7 kΩ, Linear	D2	=	SPDT Switch, 2A rating
P2	=	470 Ω, Linear	F1	=	250 mA fuse with holder

Balanced Power Supply

The design shown in Figure 14.21 is useful in those cases where balanced power lines providing a relatively small output current are needed. The diagram shows a ± 15 V supply that can provide a continuous output current of 25 mA, or 100 mA peak. With other transformers and/or voltage regulators, the supply can be adapted for output voltages of ± 5 V, ± 9 V, ± 12 V, ± 18 V, and ± 24 V. For the latter two voltages, however, the negative voltage regulator may be hard to obtain. This supply, because of its small size, can be easily incorporated into an existing equipment.

A disadvantage of small (low-VA) mains transformers as used for this supply is that they often supply relatively high no-load secondary voltages. Under no-load conditions the specified transformer, for example, supplies as much as 32 V to the regulator inputs (measured at a mains voltage of 230 V). In some cases, the no-load secondary voltage may exceed the maximum permissible input voltage of the low-power voltage regulator. Typically, this will be 30 V for 5-V regulators, 35 V for 12-V and 15-V types, and 40 V for 18-V and 24-V types. When the no-load voltage is likely to approach the absolute maximum level specified for the voltage regulator, shunt resistors (bleeders) should be used across the

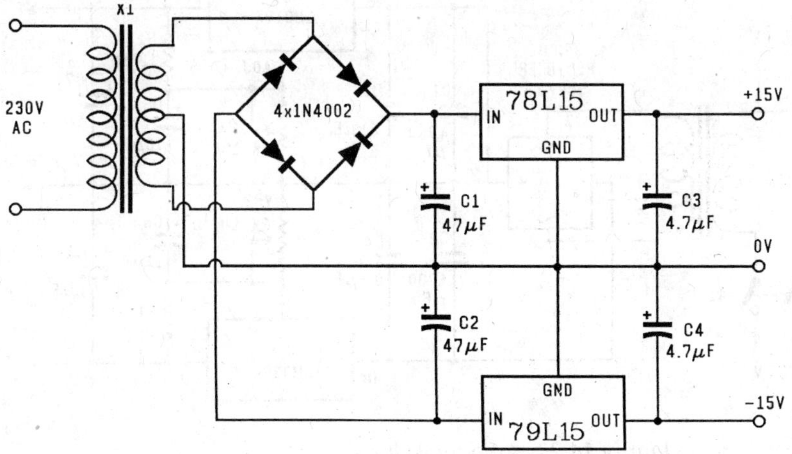

Figure 14.21: A balanced power supply.

transformer secondaries. The value of these resistors should be as high as possible to avoid unnecessary dissipation. In most cases, a bleeder current of a few mA is sufficient to drop the regulator input voltage to a safe level.

Adjustable Bipolar Voltage Regulator

A simple circuit which gives ± 1.2 V to ± 20 V output is shown in Figure 14.22. Two ICs type LM317K and LM337K are used here as positive and negative voltage regulators respectively. These ICs have built-in current limit and power dissipation limit features. The maximum dissipation is 15 watts for the TO-220 package and 20 watts for the TO-3 package. They should be mounted on heat sinks of adequate heat-dissipation capability.

The outputs are varied independently by VR1 and VR2. Capacitors C5 and C6 are used to improve a.c. ripple voltage rejection. However, if a short-circuit occurs across the regulator output, C5 will adjust the currents in the terminal. The output can be calculated by the formula:

$$V_{out} = 1.25\text{ V}\left(1 + \frac{VR1}{R1}\right)$$

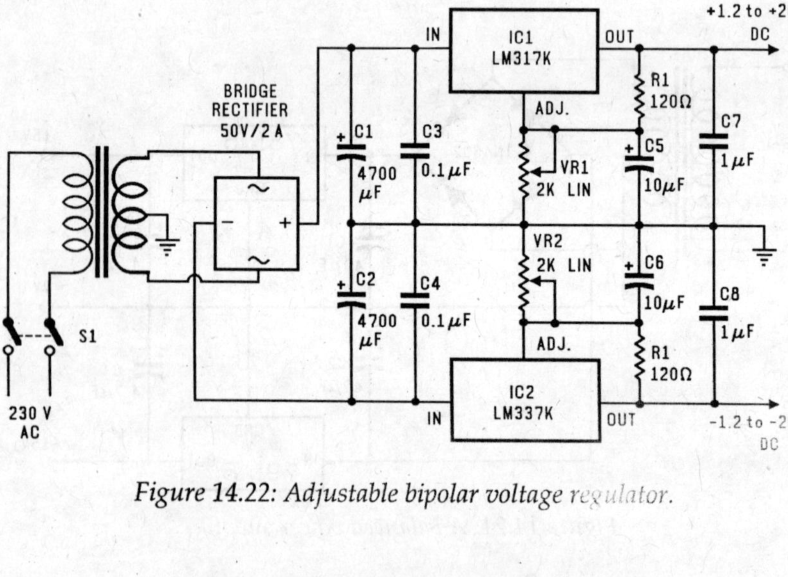

Figure 14.22: Adjustable bipolar voltage regulator.

3A, 5V Power Supply

A standard voltage regulator type 7805 is inexpensive and easily available, but its current of 1 A can at times prove a handicap. However, this current may be increased by adding a power transistor T3 mounted on a heat sink as shown in Figure 14.23. A p-n-p transistor is used as explained with reference to Figure 11.6.

When the current drain is small, the 7805 continues to function as before and transistor T2 remains inactive because the voltage drop across R4 is insufficient to forward bias T3. When the current rises above 15 mA, however, the potential drop across R4 is large enough to switch on T3. This transistor is protected against short circuits by T2. When the current through MJ2955 rises above 3 A, the voltage drop across R3 is large enough to switch on and reduce the forward bias on T3 to a low value. The output current then cannot rise further

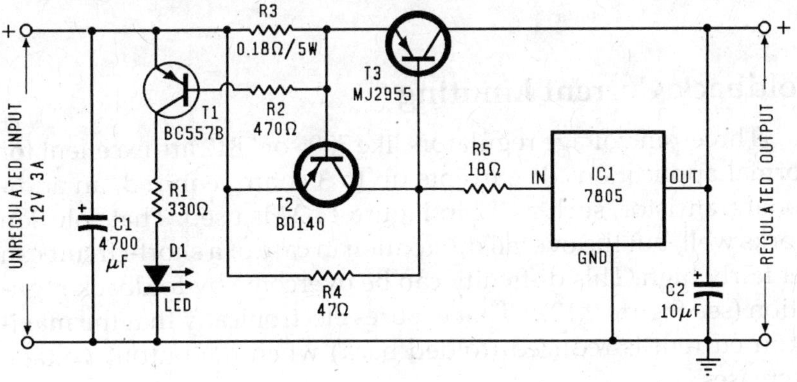

Figure 14.23: A 3-V, 5-A power supply.

In parallel with T2 is a transistor T1, that switches on an LED as soon as current limiting occurs. Resistor R5 has been added to limit the current through the regulator as soon as the current limiting operates, since R4 is than short circuited by T2. In the absence of R5, the full current will flow through the 7805.

Of course, there is a price to be paid for the higher output current: The input voltage must be 10 V for an output current of 3A instead of 8.5 V for currents up to 1 A.

The current limiting comes in relative gradually: When the output is short-circuited, a current of up to 6 A may flow for a short period. Obviously, that situation should not be allowed to last for long.

In the construction, take care that T2 and T3 are insulated from the heat sink. The 7805 does not really need a heat sink, but it does no harm to fit it also to the heat sink.

Parts List for Figure 5V, 3A Power Supply

R1	330 Ω	C2	10 μF, 16 V
R2	470 Ω	D1	LED, Red
R3	0Ω18, 5W	T1	BC557
R4	47 Ω	T2	BD140
R5	18 Ω	T3	MJ2955
C1	4700 μF, 16 V	Heat Sink	2-3 K/W

Foldback Current Limiting

Three-pin voltage regulators like 7805 or 7812 are excellent for normal applications. If currents up to 3 A are required, an additional transistor, such as T2 in Figure 14.24 is used. That solution works well, but the overall dissipation in case of a short-circuit can get fairly high. This difficulty can be overcome by foldback regulation (see Figure 9.12). . That ensures electronically that the maximum current is reduced (folded back) when the output voltage decreases.

Only a few additional components are needed for foldback regulation. In Figure 14.24, transistor T1 provides current limiting. As soon as the voltage drop across R2+R3 becomes greater than 0.6-0.7 V, the transistor is switched on, which reduces the base current of T2 to virtually zero. The voltage regulator is more or less on its own, but it has very good thermal protection and limits its output current well before any harm is done. The voltage at which the protection circuits come into operation is the sum of the potentials across R2 and R3.

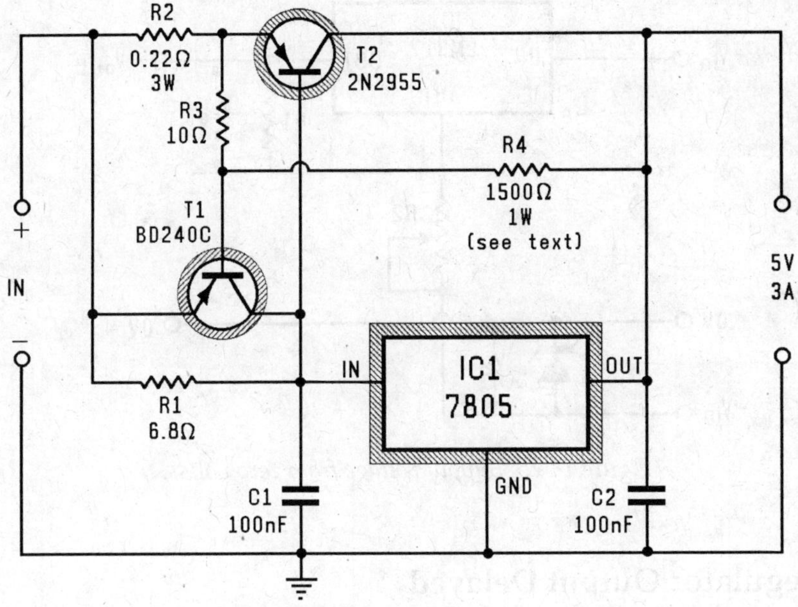

Figure 14.24: Foldback voltage regulator.

Resistors R3 and R4 form a voltage divider for the potential across T2. The dissipation in T2 is directly proportional to the collector-emitter voltage, which is thus used here to control the current. In this way, the regulation characteristic is a function of the level of the input voltage.

Power Supply From 0 Volt

The special characteristic of the regulator shown in Figure 14.25 is that the output voltage may be adjusted down to zero volt.

The regulation is provided by an integrated regulator Type LM317. As is normal in supplies that can be adjusted to 0 V, this IC is used in conjunction with a zener diode. This zener diode provides a reference voltage that is equal but of opposite sign to the reference voltage of the regulator (V_{ref}) across R1 as shown in Figure 14.25. Potential divider R1-R2 enables adjustment of the output voltage

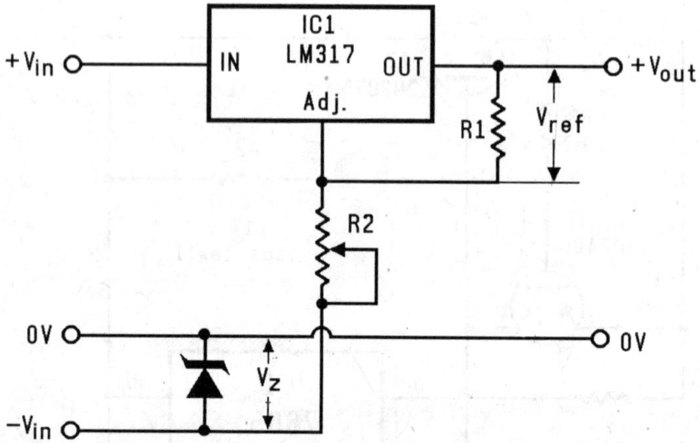

Figure 14.25: Supply voltage from zero volts.

Regulator Output Delayed

The circuit shown in Figure 14.26 uses a Type LM317 voltage regulator but supplies the regulated output rises slowly and the final value of 15 V is reached only after a time delay of 5 seconds. If the output voltage is set at 7.5 V, that value will be reached after 2.5 seconds. In other words, the time delay is directly proportional to the set output voltage.

Diode D1 protects the regulator if the output voltage were (accidentally) to drop faster than the output voltage.

Diode D2 protects the base-emitter junction of T1 against too high a reverse bias. Without the diode, if the output of the regulator were short-circuited, charged capacitor C3 could cause a V_{BE} that exceeds the permissible maximum 6 V.

The circuit draws a current of only a few mA, although the peak current may amount to 1.5 A. An adequate heat sink should be provided for the IC, and the constant output current should not exceed 500 mA.

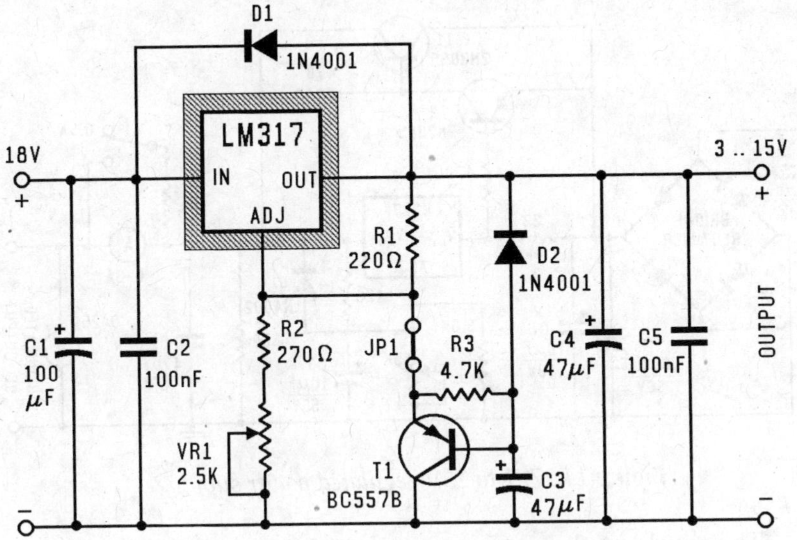

Figure 14.26: Regulator output delayed.

1.3 to 25 V Power Supply

For most amateur work the voltages needed from transistorized power supplies fall into two general ranges; 5 volts for digital circuits and 12 volts. On occasion there is a need for other values and this power supply is capable of providing voltages from 1.3 V to 25 V at a load current of up to 3 amperes.

The heart of the supply is the National LM317K. The LM317K, available in three case styles, is a completely self-contained adjustable, regulated supply requires no more than a d.c. source, the regulator and two resistors. In the basic configuration the regulator is capable of supplying up to 1.5 A output current. To provide higher output current an external pass transistor has been added in Figure 14.27. C1 and C2 are the usual capacitors to improve transient response and reduce the noise on the output voltage. C1 may be eliminated if the regulator is located physically close to the filter capacitor. The 5.6K resistor is added in parallel with the potentiometer to limit the output voltage to approximately 25 V.

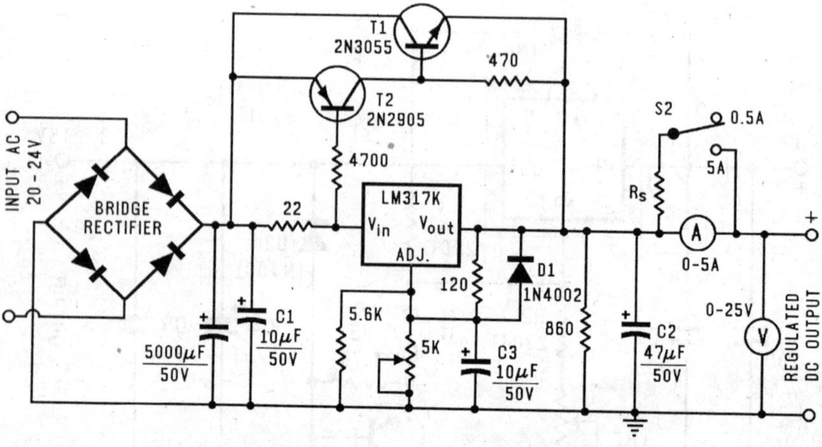

Figure 14.27: 1 to 25V regulated power supply.

The ripple on the output voltage is decreased by the addition of C3 to the adjustment lead. If the regulator is going to be used over 25 volts, a diode D1 should be connected for protection in case the input or output is shorted.

Construction hints

The transformer secondary voltage should be in the range of 20 to 24 volts a.c. when drawing slightly more than the desired load current. In addition to the load current, the transformer must supply a small amount of power to the regulator. The rectifier can be four individual diodes or an encapsulated bridge module. If individual diodes are used their rating should be at least 50 PRV and 1.5 amperes (3 A or more if pass transistors are used.

The 2N3055 and the regulator will both require a heat sink which can be home-made. A suitable commercially available heat sink can also be added.

One point that should be noted is the proper wire size for the interconnecting wiring. The load regulation is a function of the resistance of the wiring. A value as small as 0.05 ohm can decrease the regulation by as much as a factor of ten.

The value of the 5.6k resistor may be changed to suit the builder. Lowering its value will decrease the maximum output voltage available. No other adjustment is necessary for operation of this power supply.

Depending upon the builder's preferences, meters may be added to the power supply. A shunt resistor was added to the meter reading the output current. The method of calculating the value of shunt is discussed in Author's book *Fast Servicing With Modern Multimeters,* published by BPB Publications

Linearly Variable Power Supply

In variable voltage output power supplies, the output voltage variation with the rotation of the shaft is not linear. This results in making the setting of correct voltages a little tricky. If the relation were linear, a simple linear potentiometer would suffice. One such power supply is shown in Figure 14.28.

The only difference between the present and a traditional power supply is that the wiper and the earthy terminal of the relevant potentiometer are inter-linked. This simple fact allows a number of equivalent circuits to be improved. It is, of course, necessary to use a linear potentiometer.

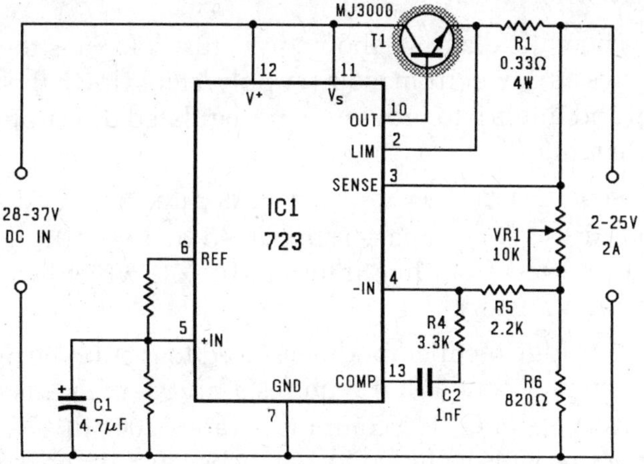

Figure 14.28: Linearly variable power supply.

28V, High Current Power Supply

Many modern high power transistors used in RF power amplifiers require 28-V d.c. collector supplies, rather than the traditional 12-V supply. By going to 28 V (or even 50 V), designers significantly reduce the current required for an amplifier in the 100-W or higher output class. The power supply shown in Figure 14.29a through 14.29c is conservatively rated for 28 V at 10 A (enough for a 150-W output amplifier)—continuous duty! It is designed with simplicity.

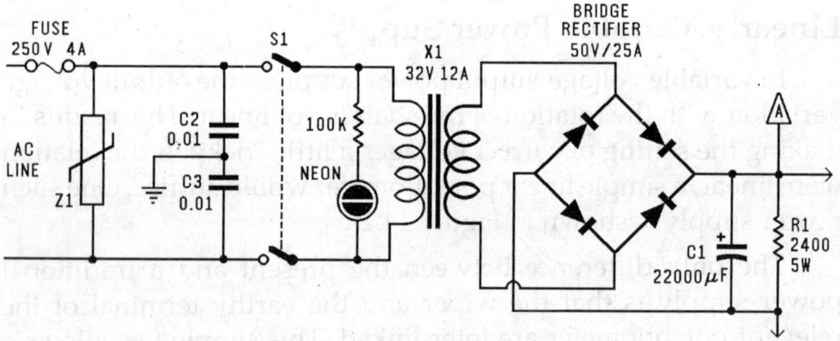

Figure 14.29a: Unregulated supply for the 28 V 10 A supply.

Circuit Details

The schematic diagram shown in Figure 14.29a is the section which draws its power from mains supply, steps down the voltage, rectifies it and filters it to provide an unregulated d.c. supply to the regulator circuit.

The secondary of the transformer is rated for 32 V at 15 A, continuous duty. The primary is bypassed by two 0.01 μF capacitors and protected from line transient by Z1, which is an MOV (Metal Oxide Varistor).

A 25-A bridge rectifier module is used to rectify the output of the transformer secondary. It requires a heat sink in this application. Filter capacitor C1 is a computer grade 32000 μF electrolytic. Bleeder resistor R1 is included for safety because of the high value of C1; bleeder current is about 12 mA.

There is a trade off between the transformer secondary voltage and the filter capacitor value. To maintain regulation, the minimum supply voltage to the regulator circuitry must remain above approximately 31 V. Ripple voltage must be taken into account. If the voltage on the bus drops below 31 V in ripple valleys, regulation may be lost.

The regulator circuitry shown in Figure 14.29b takes the advantage of commonly available parts. The heart of the circuit is IC2,

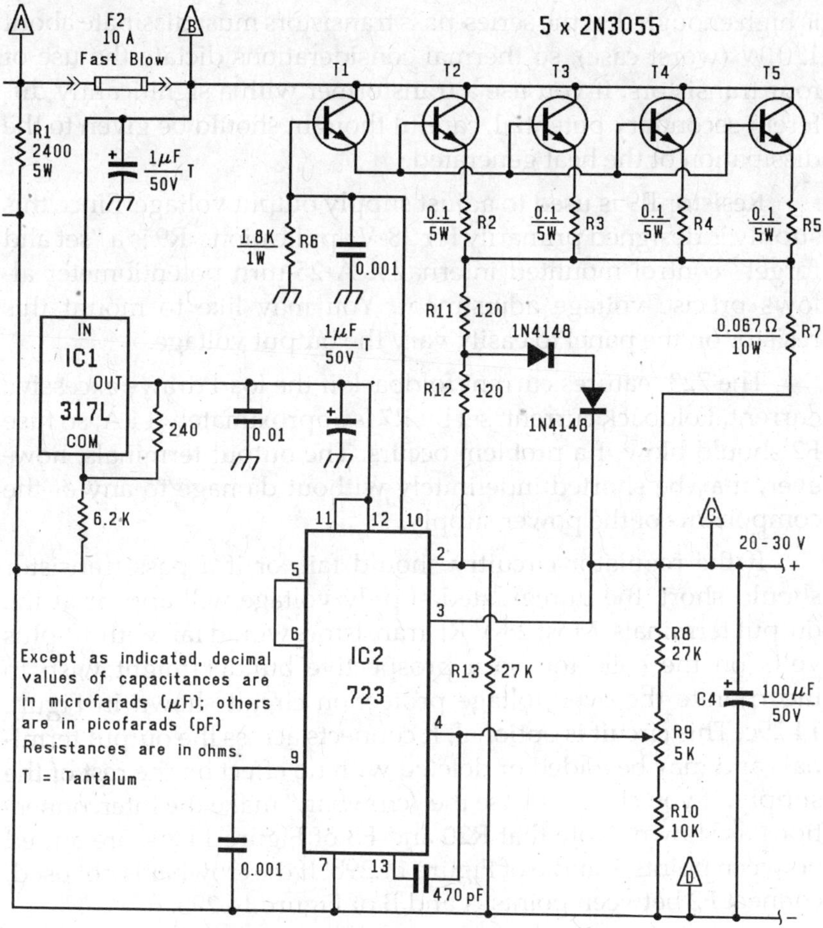

Figure 14.29b: 28V, high current power supply.

a 723 voltage regulator IC. The values of R8, R9 and R10 are chosen to allow the output voltage be varied from 20 V to 30 V. The 723 has a maximum input voltage rating of 40 V, somewhat lower than the filtered bus voltage. IC1 is an adjustable 3-terminal regulator; it is set to provide approximately 35 V to power IC2. Integrated circuit IC2 drives the base of T1 which in turn drives series pass transistors T2 to T5 in parallel. This arrangement takes the advantage of commonly available components. At first glance, the number of pass transistors seems high for a 10-A supply. Input voltage is high enough that the series-pass transistors must dissipate about 120 W (worst case), so thermal considerations dictate the use of four transistors. If you use a transformer with a significantly different secondary potential, careful thought should be given to the dissipation of the heat generated.

Resistor R9 is used to adjust supply output voltage. Since this supply is designed primarily for 28-V applications, R9 is a "set and forget" control mounted internally. A 25-turn potentiometer allows precise voltage adjustment. You may like to mount this resistor on the panel to easily vary the output voltage.

The 723 features current foldback if the load draws excessive current. Foldback current, set by R7 is approximately 14 A, so fuse F2 should blow if a problem occurs. The output terminals, however, may be shorted indefinitely without damage to any of the components of the power supply.

If the regulator circuitry should fail, or if a pass transistor should short, the unregulated supply voltage will appear at the output terminals. Most 28-V RF transistors would fail with 40-plus volts on the collector, so a prospective builder might wish to incorporate the over-voltage protection circuit shown in Figure 14.29c. This circuit is optional. It connects across the output terminals and may be added or deleted with no effect on the rest of the supply. If you choose to use the "crowbar," make the interconnections as shown. Note that R20 and F3 of Figure 14.29c are added between points A and B of Figure 14.29b. If the crowbar is not used, connect F2 between points A and B of Figure 14.29b.

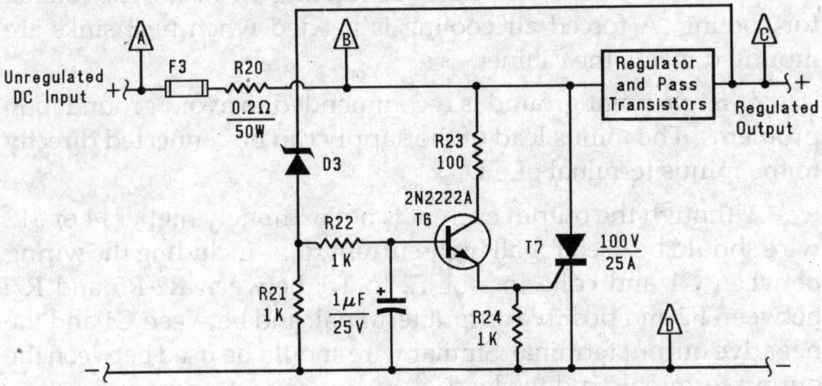

Figure 14.29c: Crow bar for protection of supply shown in Figure 14.29b.

The crowbar circuit functions as follows: The zener holdoff diode (D3) blocks the positive regulated voltage from appearing at the base of T6 until its avalanche voltage is exceeded. In the case of the device used in this circuit, this voltage level is 33 V, which allows for small overshoots that might occur with sudden removal of the output load (switching of a load, for instance).

In the event the output voltage exceeds 33 V, diode D3 will conduct, and forward bias transistor T6 through R22. The 1 μF capacitor at the base eliminates short duration transients and noise. When T6 is biased on, trigger current flows through R23 and T6 into the gate of SCR T7, turning it on and shorting the raw d.c. source, forcing fuse F3 to blow. Since some SCRs have a tendency at high temperature to turn themselves on, resistor R24 shunts any internal leakage current.

Construction Hints

The transistors should be mounted to the heat sink with insulating washers and thermal heat-sink compound to aid heat transfer. The heat sink surface under the transistors must be absolutely smooth. Carefully deburr all holes after drilling and lightly sand the edges with fine emery cloth.

Resistor R7 can be made with two 0.1 ohm, 5-W resistors in parallel. These resistors get warm under sustained operation.

A five inch fan is helpful in circulating air inside the cabinet for cooling. A forced air cooling is needed when heat sinks are mounted inside the cabinet.

A single point ground is recommended to avoid ground loop problems. The minus lead of the supply can be connected directly to the minus terminal of C1.

Although the output current is not extremely high, #14 or #12 wire should be used for all high-current runs, including the wiring between C1 and collectors of T2 to T5; between R2-R5 and R7; between F2 and positive output terminal; and between C1 and the negative output terminal. Similar wire should be used between the output terminals and the load.

Testing

First connect the transformer, bridge rectifier and C1 and verify that the no-load voltage is approximately 44 V d.c. Then, connect unregulated voltage to the regulator circuitry. Leave the base lead of T6 disconnected to disable the crow bar circuit. You should be able to adjust the output voltage between approximately 20 V and 30 V. Set the output to 28 V.

Next, short the output terminals to verify that the current foldback is working. Voltage should return to 28 V when the shorting wire is disconnected. This completes testing and setup.

Parts List

C1	Electrolytic capacitor, 22,000 μF, 50 V
C2, C3	AC rated bypass capacitors
C4	Electrolytic capacitor, 100 μF, 50 V
IC1	Three-terminal adjustable voltage regulator, 100 mA (LM-317L or equivalent)
IC2	723-type adjustable voltage regulator IC, 14-pin DIP package
Neon	Neon indicator lamp, 230 V a.c.
T1 - T5	NPN power transistor , 2N3055 or equivalent
R2 - R5	Power resistor 0.1 Ω, 5 W (or greater), 5%

R7	Power resistor, 0.067 Ω, 10 W (or greater), made from three 0.2-Ω, 5-W resistors in parallel
Rectifier	Bridge rectifier, 50 PIV, 25 A
Transformer	Power transformer, primary 230 -V a.c.; secondary, 32 V, 15 A
ZD1	250-V MOV (Metal Oxide Varistor)

A 4.5-to 25-V, 2.5-A Power Supply

For home-laboratory requirements, a series-regulator supply is simpler and less expensive than a switching power supply. Series-regulated supplies are also free of electrical switching noise, which is a problem with some switching supplies. During tests of sensitive low-level circuitry, the power supply output should be pure d.c.; it should not contribute to problems in the circuit under test.

Features

The supply shown in Figure 14.30 is designed to meet the following objectives:

- ☐ Continuously variable output voltage from 4.5 V to 25 V.
- ☐ Tight load voltage regulation.
- ☐ Excellent response to load fluctuations and transients (low output impedance).
- ☐ Very low a.c. ripple, less than 2 μV RMS with 2.0-A load.
- ☐ Very low random noise, less than 2 μV from 0.1 Hz to 500 kHz.
- ☐ Load currents up to 2.5 A, continuous duty.
- ☐ Switch selectable current limiting at 0.1, 0.5 or 2.5 A to protect delicate circuits under test.

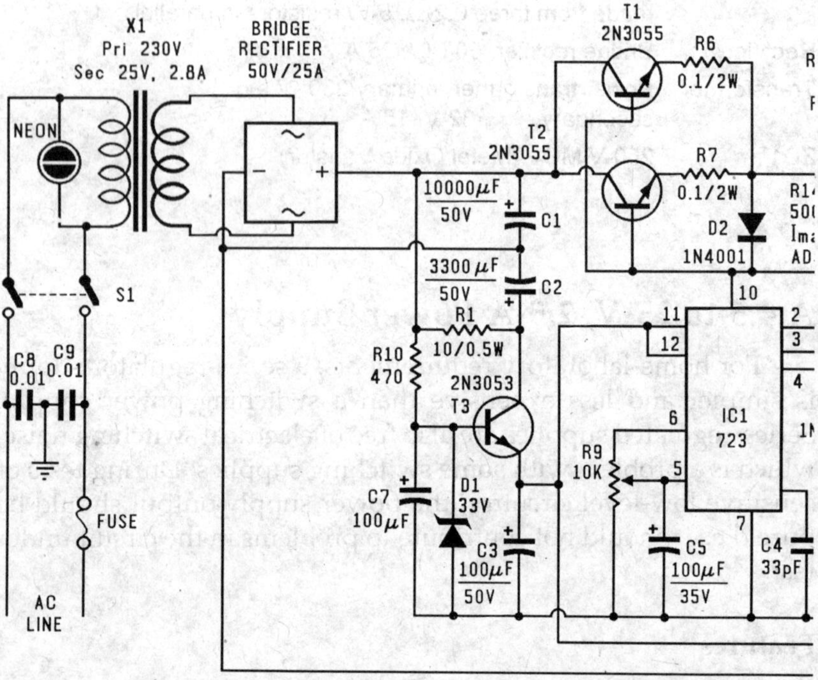

Figure 14.30a: A 4.5- to-25-V, 2.5-A power supply.

Design and Working of Circuit

An LM723 regulator is used because it is simple, and its reference voltage is available (pin 6) for filtering. Reference noise (typical of zener diodes) is reduced to a very low level via C5. The current limiting circuitry is also accessible (at pins 2 and 3). It senses the voltage drop across R2 and R3.

The combination of R2 and R3 sets the current limit given by 0.62/(R2+R3) amperes. Resistor R3 (R3a and R3b together) forms an 0.11-Ω resistor, which acts as a shunt for the digital meter that measures load current. R8 provides meter adjustment without affecting R3 significantly. S3 selects from three values of R2 to set current limits at approximately 0.1, 0.5 and 2.5 A.

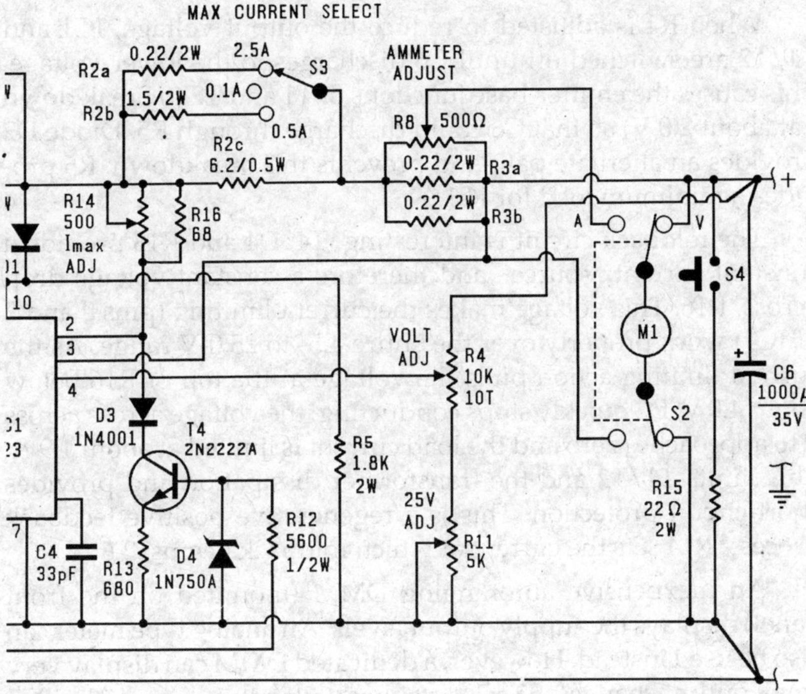

Figure 14.30b: A 4.5- to-25-V, 2.5-A power supply.

Capacitor C1 has a large value to reduce output ripple voltage. Smaller values would increase ripple and require a higher transformer voltage to prevent regulator drop out. Other voltage drops between the T1/T2 emitters and the output terminal are minimized to assure that a standard 25-V transformer can do the job. The R1-C2 combination reduces a.c. ripple at the 723 IC by a factor of 25; this eliminates the need for an extremely large value of C1.

The circuitry of T3, R10, C7 and D1 prevents the voltage on pins 11 and 12 of IC1 from exceeding the 40-V maximum rating of the 723 (especially with light loads and high line voltage). C7 eliminates a very small a.c. ripple at the d.c. output. As load current increases, the voltage at C1 decreases and T3 is saturated.

When R4 is adjusted to reduce the output voltage, IC1 and T1/T2 are switched off until C6 discharges to the lower voltage. This causes the emitter-base junctions of T1 and T2 to break down (at about 2.0 V) so that C6 could discharge through R5. Diode D2 provides an alternate path and prevents this breakdown. R5 provides a minimum load for IC1.

The foldback circuit is interesting. T4, D4 and R13 provide a constant current source, and therefore a constant voltage drop across, R16. This voltage makes the current limiting (pins 2 and 3 of IC1) work properly over the entire 4.5- to 25.0-V range. As the current limiting action pulls the voltage at the top of R16 below about 4.0 V, D3 quickly stops conducting, the voltage across across R16 approaches zero and the load current is limited at about 1.9 A. This limits T1/T2 and the transformer dissipation and provides short-circuit protection. This is a regenerative positive feedback process. R14 sets the current at which foldback begins, 2.6 A.

An inexpensive autoranging DMM (mounted on the front panel) displays the supply output levels. An analog type meter can also be used instead. However, a dedicated DMM can display very small output changes. S2 selects either voltage or current (divided by 10) for display. (The DMM is left in its voltage range for both measurements.) The voltage across R3 is 0.11 times the load current. R8 provides the required ammeter adjustment without significantly affecting R3. The meter shows 0.2 for 2 A. Ordinary wire-wound resistors are adequate for R3 up to 2.5 A.

Resistor R9 sets the reference voltage on pins 5 of IC1 at 4.5 V to establish the minimum output voltage. R11 sets the 25.0-V upper limit. A three-wire line cord assures that the supply chassis is always tied to the a.c.-line ground, for safety reasons. The d.c. output is not referenced to chassis ground, and performance is independent of the a.c.-ground connection. If the load current is very small, it may take a long time for C6 to discharge when the power is switched off; a press of S4 quickly discharges the capacitors (through R15) after turnoff. The mechanical construction can be done in a way to provide enough natural cooling and a cooling fan may not be necessary.

Construction Hints

Several areas are critical to good performance. Pay particular attention to these:

1. Wire C1 with short, heavy leads to present a minimum impedance to a.c. ripple. Also, connect C1 directly to the negative binding post with a heavy lead. This provides a low inductance path for load-current fluctuations.
2. Connect C2 directly to the negative lead of C1.
3. Connect C6 directly across the binding posts.
4. Connect R4 to R11 with low-impedance lead.
5. Use heavy-duty binding posts to reduce a small but significant voltage drop.

Adjustments

Several adjustments are needed: the dedicated AMMETER ADJUST (R8), the foldback limit R14, (MAX ADJ), the lower (R9, 4.5 VOLT ADJ), and upper (R11, 25-V ADJ) voltage limits. First obtain an accurate external ammeter and a test load that can draw at least 3 A at 12 V or so (automobile sealed beam headlights work well). Series connect the load and external meter to the supply binding posts. Set S3 to the 2.5 A range; set R4 for minimum output; set S2 to A and switch on the power. Advance R4 until the ammeter reads 2.0 A, and adjust R14 so that the current begins to decrease (foldback). Switch off the supply and remove the external meter and load.

Adjust the voltage limits with no load connected and S2 set to V. Switch on the supply, set R4 for minimum voltage and adjust R9 until the meter reads 4.5. Increase R4 to maximum voltage and adjust R11 so that the meter reads 25. The supply is ready for use.

Parts List

Bridge	25 A, 50 V
C1	1000 µF, 50 V
C2	3300 µF, 50 V
C3, C7	100 µF, 50 V
C4	33 pF, 50 V
C5	100 µF, 35 V
C6	1000 µF, 35 V
C8, C9	0.01 µF, ac-rated r
D1	1N5257A 33V Zener
D2, D3	1N4001
D4	1N750A
Neon	Neon indicator lamp
F1	1.5 A slow-blow
Heat Sink	To fit T1 and T2
IC1	LM723 regulator IC
M1	DMM
T1, T3	2N3055
T3	2N3053
T4	MPS2222A or equivalent
R1	10 Ω, 0.5 W
R2a	0.22 Ω, 2 W
R2b	1.5 Ω, 2 W
R2c	6.2 Ω, 5%, 0.5 W
R3	0.15 Ω, 2 W
R4	10 kΩ, 10-turn potentiomete
R5	1.8 kΩ, 2 W
R6, R7	0.1 Ω, 2 W
R8, R14	500 Ω potentiometer
R9	10 kΩ potentiometer
R10	470 Ω, 0.25 W
R11	5 kΩ potentiometer
R12	5.6 kΩ, 0.5 W
R13	680 Ω, 0.25 W
R15	22Ω, 2W
X1	Transformer, primary 230-V Secondary, 25V, 2.8 A

(For more info on dmm, see Author's book on multimeters)

Power Supply For Ham Radio Transceivers

The power supply shown in Figure 14.31 is based on the popular LM338 3-pin voltage regulator. The LM338 is capable of supplying 5 A over an output voltage range of 1.2 - 32 V with all standard protections and overload, thermal shutdown, overcurrent, internal limit, etc., built in. In this power supply, some extra protection have been added to make it particularly suitable for use with low to medium-power portable and mobile VHF/UHF (ham) transceivers.

Diodes D4 and DD5 provide discharge paths for capacitors C1 and C2. Diode D8 protects the supply against reverse polarity being applied to the output terminals. Capacitor C1 assists in RF decoupling and also increases the ripple rejection from 60 dB to about 80 dB. When junction R1 - R2 is not grounded by switch S1A,

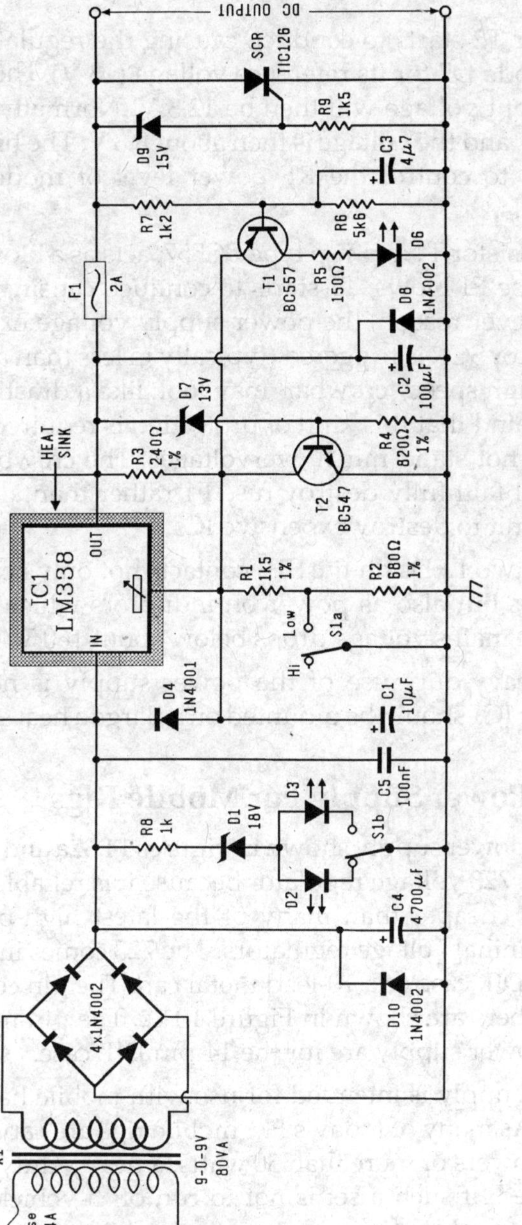

Figure 14.31: A power supply for ham transceivers.

transistor T2 starts to conduct, causing the regulator to switch to zener diode D7 for its reference voltage (13 V). The power supply unit output voltage will then be 12.3 V. Normally T2 will be off, however, and the voltage is then about 8.8 V. The high/low switch is useful to control the RF power level of modern VHF/UHF handhelds.

Transistor T1, a p-n-p type BC557, acts as a blown fuse sensor. When fuse F1 blows, T1 starts to conduct, causing D6 to light. If, for whatever reason, the power supply voltage exceeds about 15 V, thyristor SCR is triggered (typically in less than a microsecond). Such a high speed 'crowbar' may look like a drastic measure, but bear in mind that this kind of protection is required by digital ICs that will not stand much overvoltages. The crowbar, when actuated, will faithfully destroy fuse F1 rather than allow the power supply unit to destroy expensive ICs.

The two LEDs on the S1B contacts not only act as 'high/low' indicators but also as power on indicators which are turned off when the mains voltage drops below about 160 V.

If heavy-duty use of the power supply is needed, voltage regulator IC1 should be mounted on as large a heat sink as possible.

13.8 V Power Supply For Mobile Rigs

The power supply shown in Figures 14.32a and 14.32b is based on a type 723 voltage regulator because it is reliable, widely available and cheaper than many of the latest high-power (>1.5 A) three-terminal voltage regulators. The 723 comes in two packages; a 14-pin DIL case or a 10-lead metal can. The pin connections and pin numbers are shown in Figure 10.12. The pin numbers shown in this power supply are for the 14-pin DIL case.

The supply is intended for use with mobile ham radio transceivers. As many of today's FM mobile rigs are capable of supplying r.f. powers of more than 50 watts, a heavy-duty power supply is required, if such a set is not to run off a vehicle battery. This supply is capable of providing output currents of up to 12 amperes at 13.8 V.

The regulator, IC1, is conventionally wired, drives a power transistor array and monitors the output current by measuring the voltage drop across series resistor Rsc. The normal supply output voltage of 13.8 V is set with preset P1. The input voltage for the supply is obtained from a 22-25 V, 15 A transformer, a 25-A bridge rectifier and a 10,000 μF smoothing capacitor. This section is shown in Figure 14.30a.

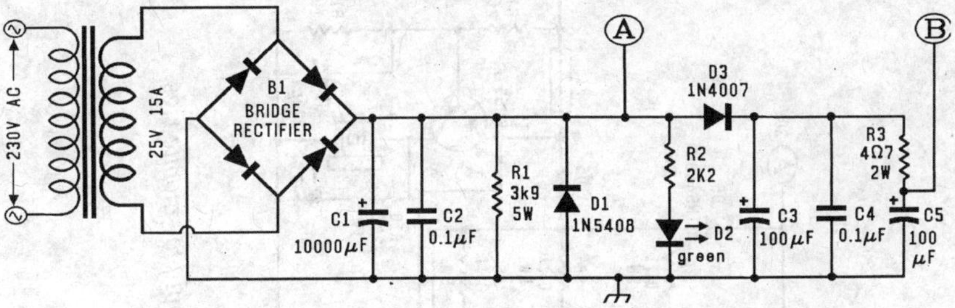

Figure 14.32a: Unregulated supply for feeding the regulator shown in Figure 14.30b.

The current booster consists of a darlington transistor T1 and three parallel-connected power transistors T2 - T4. Resistors R_A, R_B and R_C are emitter current distribution elements as explained with reference to Figure 9.6.

The current sensing resistor for the short circuit protection, Rsc, has a value of 0.05 Ω which results in a protection onset level of about 0.6/0.05 = 12 A. This resistor is either made from resistance wire or from two parallel connected 0.1 Ω, 5 W resistors.

A 12 A or 16 A fuse in the positive supply rail provides additional protection against output short circuits. If this fuse blows, transistor T5 briefly actuates an active buzzer and a flashing LED. It does so by draining the charge built up in C9.

The 15 V overvoltage protection of the output of the supply is a crowbar circuit. If the supply is set to an output voltage other than 13.8 V, the zener diode, thyristor and associated resistor have to be omitted or disconnected by breaking the wire link indicated in the circuit diagram. If used, the thyristor should have a current rating of about 25 A; a type 2N6506 is suggested.

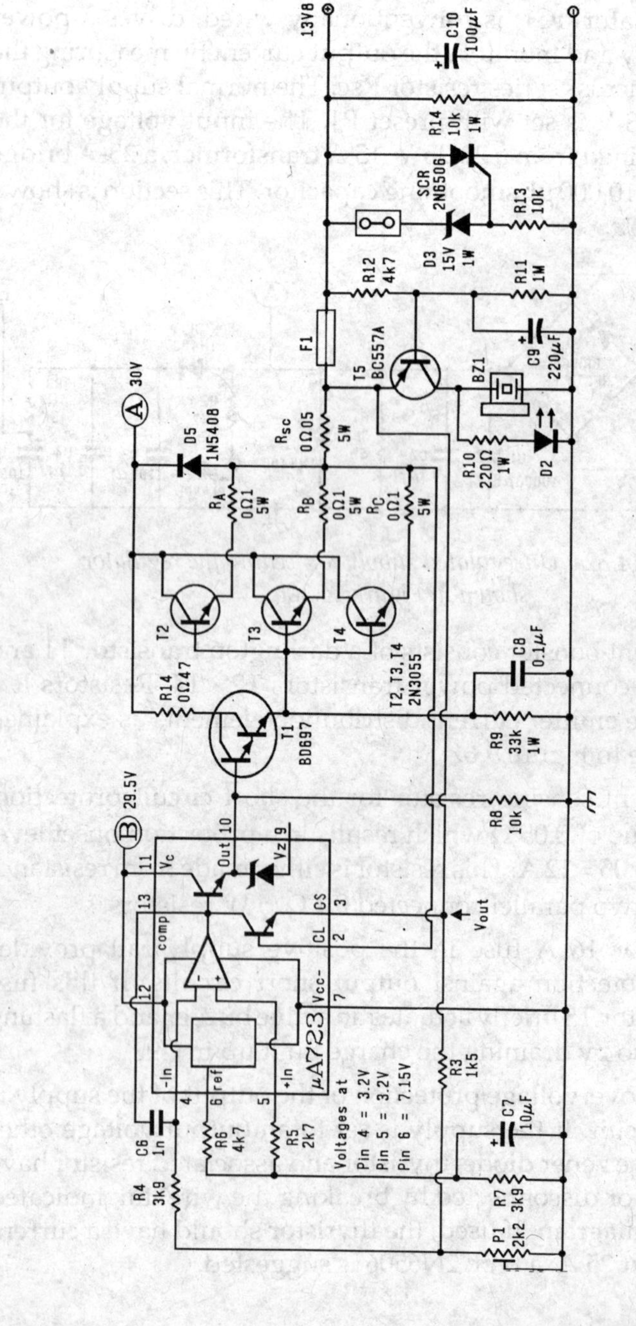

Figure 14.32: A 13.8b volt power supply for mobile rigs.

The darlington and power transistors have to be mounted on a large heat-sink with appropriate insulating washers. A small fan may be provided in necessary.

As the choice of components used for the supply is not critical, the circuitş easily modified for smaller output currents and/or different output voltages. The use of a 10 A transformer with two power transistors is fine for an 8 A supply. Similarly, a 5 A transformer and one power transistor are sufficient for a 4 A version of the power supply.

Simple Crowbar Circuit

Crowbar circuits are used to protect the circuit from, damages in the event of a breakdown of power supply. Its principle of working of a crowbar circuit was explained with reference to Figure 6.5. A practical circuit is shown in Figure 14.33.

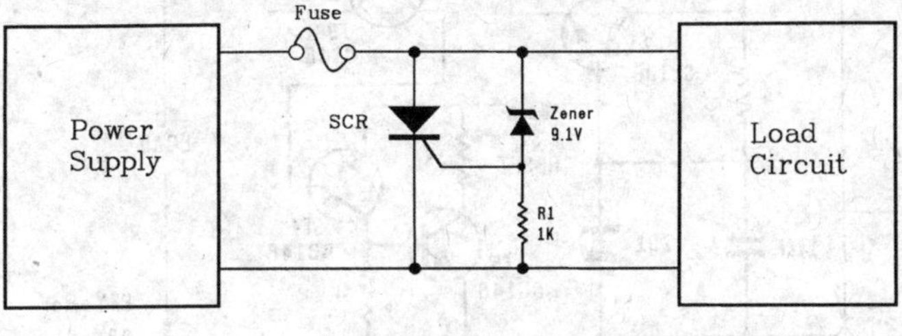

Figure 14.33: A simple crow bar circuit.

So long as the output voltage of the power supply is within limits, the zener diode is ineffective and the circuits work normally. If, for any reason, the output voltage builds up to a value higher than the zener voltage (9.1V in this case), the diode conducts and the voltage drop across R1 fires the SCR. A conducting SCR blows off the fuse quickly and protects the circuit from any damage. The circuit will work normally only when the normal voltage is restored. Otherwise,the fuse will blow again.

Crow bar circuits used in previous circuits also work in the same manner.

Electronic Fuse I

The electronic fuse shown in Figure 14.34 protects the power supply from excessive load currents and protects the load circuit from damages that can be caused by excessive currents that may be drawn in case of malfunction in the load circuit.

Transistors T1 and T2 and zener diode ZD1 form a simple zener regulated power supply of the type shown in Figure 9.1. The combination of T1 and T2 acts like a single high beta transistor.

The combination of p-n-p and n-p-n transistors T4 and T5 form an electronic latch circuit. When the circuit is turned on, this latch is non-conductive. The latch can be triggered by a small current applied to the base of T4. Once the circuit is triggered and turned on, it can be reset only by the removal of applied d.c.

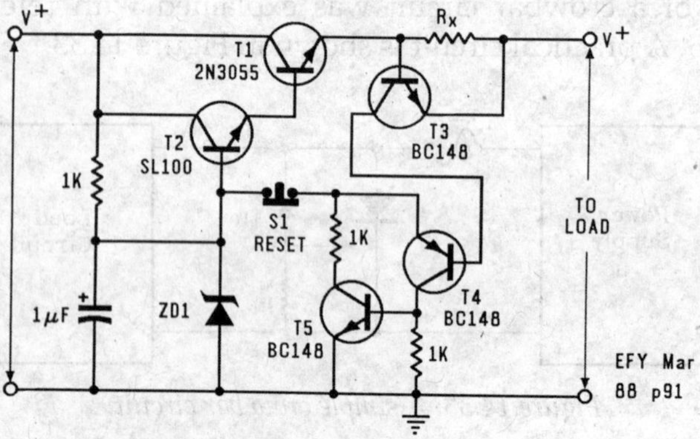

Figure 14.34: Electronic Fuse 1.

Transistor T3 is the current sensing transistor. In the normal state of working, the voltage drop across resistor R_x is lower than the V_{be} of T3 and transistor T3 therefore remains non-conducting. If the load current exceeds a certain value, the increased voltage applied to the base of T3 makes it conduct. This, in turn, triggers the latch. A conducting latches short circuits the zener diode and the output voltage, and hence the load current drops to zero.

After removing the overload, the circuit can be reset by a momentary push on the RESET switch S1.

Electronic Fuse II

The circuit shown in Figure 14.35a is the electronic version of the glass fuse, which melts when it carries a current exceeding its amperes specification. The blown fuse has to be replaced by a new one. This is annoying as well as expensive when fault-finding in equipment, or experimenting with circuit whose maximum current drain is not accurately known. A fuse for such applications should be resettable.

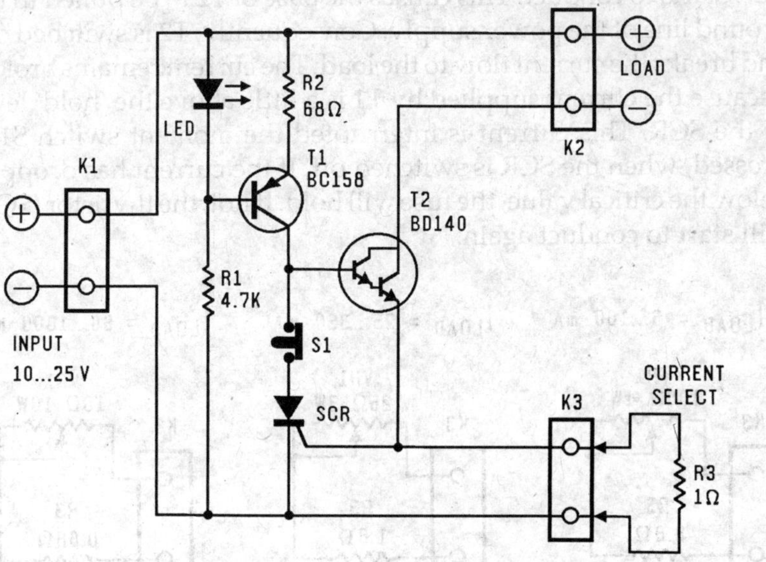

Figure 14.35a: Electronic Fuse II.

As shown in Figure 14.35a, the circuit consists of little more than two transistors and one thyristor (SCR). The electronic fuse is inserted into the supply line of the equipment to be protected. The current at which the circuit is actuated is defined with a single resistor, which is connected as an external part to connector K3.

The operation of the circuit is simple. Darlington transistor T2 serves as a current-controlled switch. Its collector current carries the load current connected to connector K2. Since transistor T1 feeds a current into the base of T2, the switch is normally closed, and the load powered. T1 functions as a current source, with the base of the transistor held at a constant voltage by LED. consequently, the emitter resistor of R2 will supply a constant voltage, which, in turn, results in a constant emitter current (and collector current), the magnitude of which is determined by the value of R2. In this case, it is just over 10 mA, ample to keep T2 conducting.

The load current also flows through R3. As soon as the current rises to a value that causes a voltage drop of about 0.6 V across R3, SCR starts to conduct. This causes the base of T2 to be pulled to the ground line of the power supply. Consequently, T2 is switched off, and breaks the current flow to the load. The current remains broken because the current supplied by T1 is a little above the 'hold' level of the SCR. This current is interrupted the moment switch S1 is pressed, when the SCR is switched off. If the current has dropped below the critical value, the fuse will hold. If not, the thyristor (SCR) will start to conduct again.

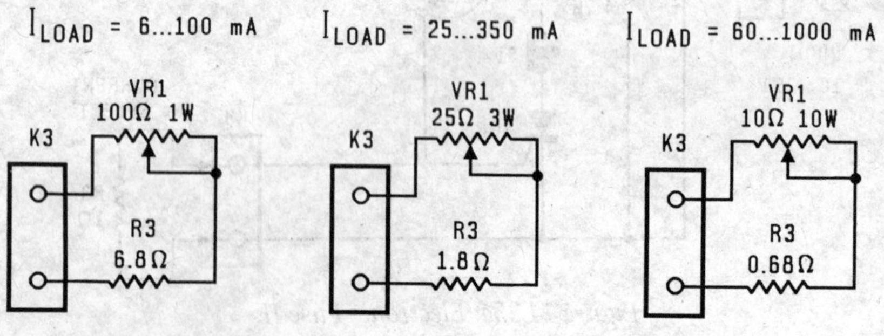

Figure 14.35b: Varying the fuse actuation current.

The current level at which the electronic fuse is actuated is simple to calculate using Ohm's law. As already mentioned , the fuse is actuated at a voltage of 0.6 V across R3. Therefore, the fuse comes into action at a current of 600 mA (0.6 A) with R3 = 1 Ω. Similarly, the actuation current is 125 mA with R3 = 4.7 Ω, or 50 mA with R3 = 12 Ω.

Looking at the most commonly used low-power fuse values, the value of R3 will usually lie between 1.2 Ω and 12 Ω. The minimum value that may be used is 0.68 Ω., which results in a 'trip' value of 1 A. That corresponds to the maximum current this circuit can handle safely.

Assuming that this electronic fuse is built into a power supply for use in your workshop, it is, of course, useful to make the current limiting action switchable or continuously variable. The former is achieved by fitting a rotary switch and a number of resistors. A continuously variable fuse actuation level is accomplished by connecting a potentiometer in series with R3. Note, however, that most ordinary pots can withstand a dissipation of about one watt only. Relatively high current therefore call for a wire-wound potentiometer. The resistor/potentiometer connections are shown in Figure 14.35b.

DC-to-DC Converter

The converter in Figure 14.36, based on the CMOS version of the 555 timer (TLC555), changes a positive potential of 9 V into a negative one of the same level and is ideal for use where a single battery is to power a circuit requiring a symmetrical supply.

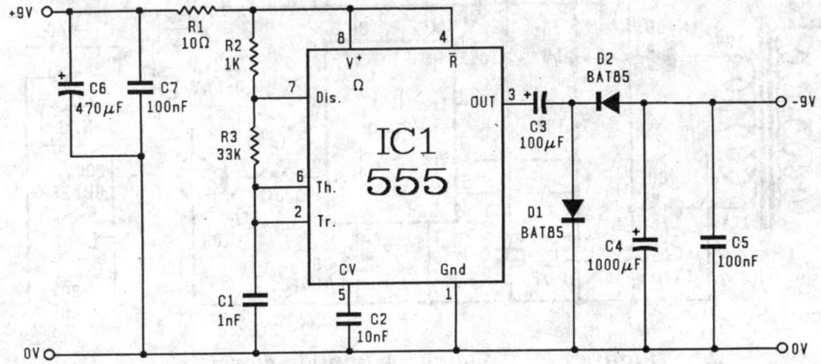

Figure 14.36: DC-to-DC converter.

If a TLC555 is not available, a Type 7555 may be used instead. The device is arranged as an astable multivibrator (AMV) with R2, R3 and C1 determining the operating frequency., which is about 20 kHz in this case.

The square wave produced by the AMV is fed to cascode rectifier C3-D1-D2-C4. Schottky diodes are better because of their lower forward voltage drop of about 0.4 V compared with 0.7 V in the case of silicon diodes and because of the higher frequency of oscillations. The rectified voltage is smoothed by C4, while C5 serves to bypass high frequency noise.

The input supply voltage to IC1 is decoupled by R1, C6 and C7. The current drawn by the converter depends largely on the load connected to the -9 V output. The output current may rise to about 10 mA before the output voltage collapses.

Mains Resumption Alarm

The audio annunciator shown in Figure 14.37 produces an alarm for predetermined period when power resumes after failure.

The circuit is built around 555 timer in monostable mode. The surge when the power resumes triggers the monostable which switches transistor T1 on. The relay energizes and the bell/buzzer sounds. After a time delay preset by VR1 and C1, the 555 is turned off and the alarm stops. Capacitor C2 should have low leakage.

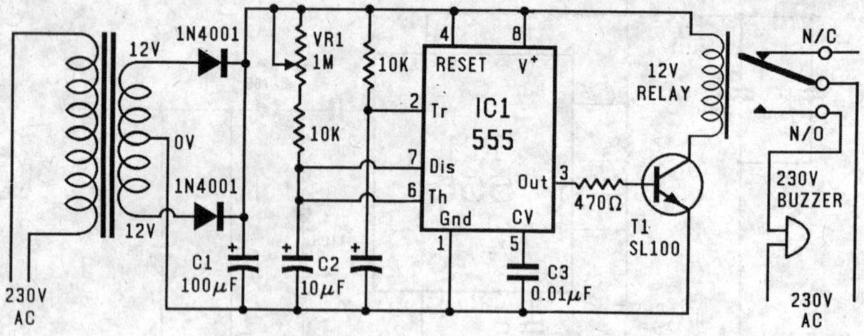

Figure 14.37: Mains resumption alarm.

15

Inverters and Converters

In those days when all radio and electronic equipment was designed around vacuum tubes, there was a need to design equipment like radio receivers for automobiles, audio amplifiers and transceivers which could be operated on 6V/12V storage batteries. The plates of vacuum tubes needed a high voltage supply of the order of 300 volts. Since transformers cannot step up d.c., a method was evolved to first convert d.c. into a.c. by chopping, then get a high voltage by using a setup-up transformer, and rectify and filter the output of the transformer to get high voltage d.c. These devices are now referred to as *DC-to-DC converters*. Since the end voltage is d.c., the frequency of chopping is immaterial. The choice of frequency depends on technology and materials available. This method can be used to get d.c. voltages at several different levels by using a transformer with more than one secondary winding.

In early days, electro-mechanical vibrators were used to chop the d.c. voltage from the battery and feed that square wave a.c. (chopped d.c.) to the primary of a transformer as shown in Figure 15.1. The electromechanical vibrator contains a reed which vibrates between two contacts. When battery is connected, a current flows through the lower half of the primary winding and the coil of the vibrator. This attracts the armature and closing of the lower contacts shorts the coil of the vibrator. The reed therefore goes up and connection is made with the upper contact which makes current from the battery to flow through the upper half of the primary winding. But opening of contacts energizes the coil of the vibrator again and the reed continues to vibrate at a frequency decided by the mechanical characteristics of the vibrator. It was about 50/60 times per second. The rectified output of the secondary is pulsating d.c. which is filtered and used as plate supply for vacuum tubes.

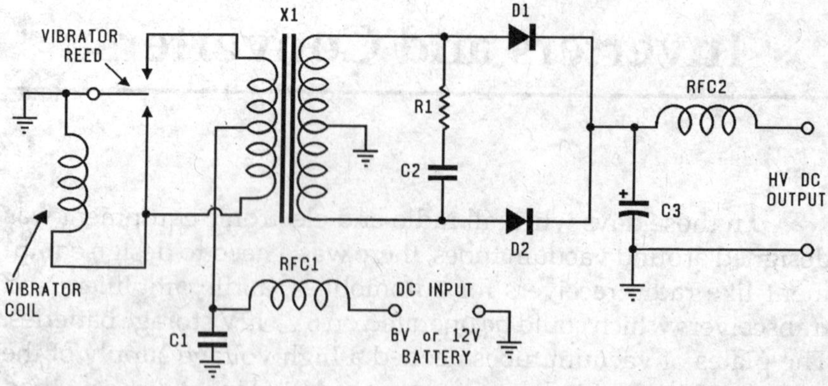

Figure 15.1: A vibrator type power supply using a mechanical vibrator.

The electro-mechanical vibrator has a limited life because of contact erosion. At the same time, sparking at the vibrator contacts generated r.f. interference or "hash". To minimize this, r.f. filters are incorporated, consisting of RFC1 with C1 in the battery circuit and RFC2 with C3 in the d.c. output circuit.

Semiconductors as Switches

With the availability of power transistors that could handle large currents, the vibrator was replaced by a free-running multivibrator driving power transistors. In these designs, the frequency of chopping is higher, and that reduces the bulk and size of the transformer, as explained in details in the book *Designing Coils and Transformers* by author and published by BPB Publications. But before we describe the use of semiconductor devices as switches, let us review the characteristics of switches.

Switch Characteristics

Whenever an electric current flows through a component or a device, it dissipates electrical power equal to V×I or I^2R where V is the voltage drop across the device, I is the current and R is the resistance. As we know already, V, I and R are related to each other by Ohm's law V = I×R. This electrical energy gets converted into heat and the temperature of the component or device starts to rise. If no action is taken to remove the heat generated, the temperature will rise to a point where the component or the device gets burnt off. This is not desirable and efforts are made to see that temperatures in all sections of an electronic circuit remain within safe limits.

Current through a load can be turned on or off by a switch as shown in Figure 15.2. The circuit shows the symbol for a switch. In practice, the switch can be either a mechanical switch or a semiconductor device like a diode or a transistor.

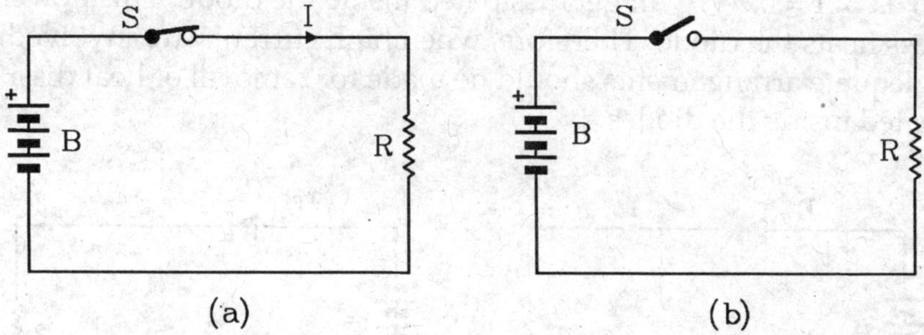

Figure 15.2: Using a switch to turn current on or off.

In Figure 15.2a, the switch S is closed and a current I flows through the load resistance R. An ideal switch has zero resistance when closed. Hence the power dissipated in the switch (I^2R) is zero. In practice, the contact may not be perfect and may have some resistance (r). In that case, power equal to I^2r will be dissipated in the switch and will not be available to the load. The power dissipated within the switch heats the contacts and the switch starts getting warm. The amount of heat generated within the switch contacts increases with increase in contact resistance.

In Figure 15.2b, the switch is open. An ideal switch, when open, has infinite resistance and hence, the current is zero. Again, no electrical power gets dissipated either within the switch or in the load. A practical switch can have some leakage resistance and that will allow a tiny current to flow in the load even if the switch is physically open. In that case, heat is generated within the switch.

Figure 15.3 shows the same circuits as in Figure 15.2, but the switch is replaced by a diode. We know that a forward biased diode conducts and acts like a closed switch that allows a current to flow through it, and a reverse biased diode acts like an open switch and does not allow current to flow. In Figure 15.3a the battery applies a forward bias to the diode. The diode conducts and a current I flows through the load. Though the diode allows the current to flow, there is a voltage drop across the semiconductor diode (barrier voltage) which is about 0.3 V for germanium diodes and about 0.7 V for silicon diodes. If the current is 1 A, electrical power equal to $0.7 \times 1 = 0.7$ W will get dissipated inside the diode. This power loss heats the diode. Therefore, when high currents are involved, adequate arrangements should be made for removal of heat dissipated inside the diodes.

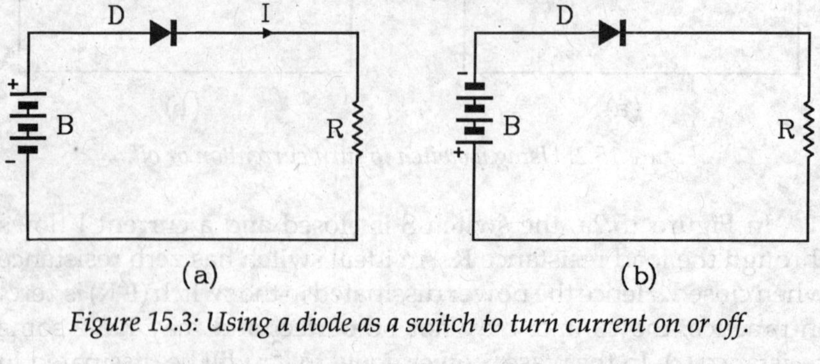

Figure 15.3: Using a diode as a switch to turn current on or off.

In Figure 15.3b, the battery is reversed. The diode does not conduct but a leakage current can still flow through the diode. The leakage currents is very small in silicon diodes and can be neglected for most practical purposes. The leakage current is significant in germanium diodes and it increases with increase in temperature..

Switching Speed

The vibrator type of power supply shown in Figure 15.1 is an example of DC-to-DC conversion. When the end product is d.c., the frequency of switching and the waveform of the a.c. is not important. It is advantageous to use higher frequencies because the size of the transformer goes on becoming smaller as the frequency of switching is increased. With modern core materials, it is possible to use switching speeds of several hundred kHz. However, as the switching speed is increased, other problems crop up because of limitations in the switching speeds of semiconductor devices.

Figure 15.4 shows a full-wave rectifier using two diodes and one transformer with a center-tap on the secondary. When the upper end of the secondary is positive as shown in Figure 15.4a, diode D1 conducts and a current I flows through the load. On the next half-cycle of a.c., the upper end of the secondary goes negative. Diode D1 stops conducting and diode D2 starts conducting. Again a current I flows through the load in the same direction.

Semiconductor devices like diodes and transistors do not start conducting or stop conducting instantaneously. They take some definite time. The time taken by the device to switch off (stop conduction from its fully conducting state) is important. Imagine that in Figure 15.4b, if diode D2 starts conducting *before* D1 has stopped conducting, there will be a moment when both diodes are conducting. This means a dead short across the secondary and a high current will flow through the diodes, though for a short time.

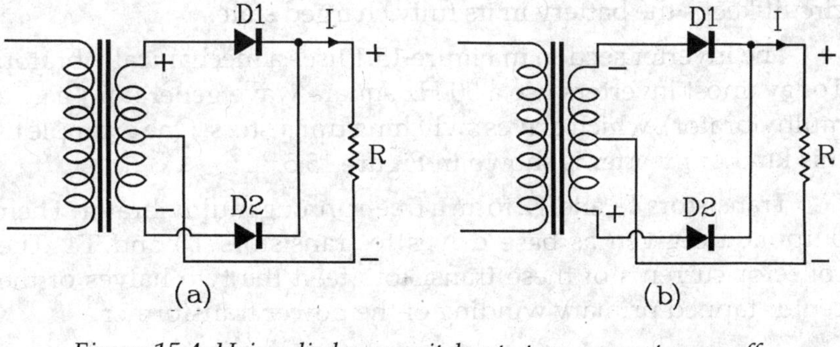

Figure 15.4: Using diodes as switches to turn current on or off.

In practice, the turn-on and turn-off times are very small and hardly have any effect at power line frequencies such as 50 or 60 Hz. But, when the a.c. frequency becomes 10,000 Hz or higher, they become important. It is therefore wise to use only those diodes or transistors in switching circuits which are designed for switching applications. Manufacturers of semiconductor devices therefore quote turn-on and turn-off times for their devices such as Schottky diodes, that are meant for switching applications .

Inverters

There are many applications where a 230 V a.c. at 50 Hz is needed to operate an appliance in the absence of mains supply. Devices to supply this kind of power are known as *inverters*. They draw electrical energy from a battery (a rechargeable battery in most cases), chop the d.c. as done in Figure 15.1, and then step up the a.c. voltage by a step-up transformer. The chopping frequency in most cases, is about 50 Hz. Frequency accuracy and stability is not important so long the device is not used to power a frequency sensitive appliance such as a synchronous motor of a clock.

Medium size inverters are used to operate things such as emergency lights, TV receivers and computers. High capacity units are used to power lights and fans in a house during failure of the mains supply. These inverters are designed to start up automatically and immediately as soon as the failure of mains supply occurs, and switch off themselves when the mains supply is restored. When the mains supply is available, a built-in charger circuit keeps the battery in its fully charged state.

The inverter section in Figure 15.1 uses a mechanical vibrator. Today, most inverters use a 50 Hz square wave generator (e.g., a multivibrator), which drives switching transistors. One example of this kind of inverter is shown in Figure 15.5.

Transistors T1 and T2 form a free-running multivibrator. Their outputs are given as base drives to transistors T3 and T4. The collector currents of these transistors feed the two halves of the center-tapped primary winding of the power transformer.

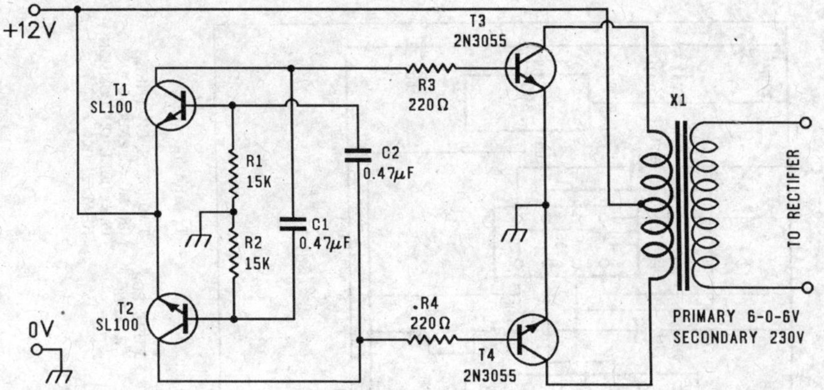

Figure 15.5: A simple inverter using a symmetrical multivibrator.

The block diagram in Figure 15.6 shows another method of generating symmetrical square waves to drive the switching circuits. An RC oscillator generates pulses at some multiple of the final frequency required. The pulses generated by the oscillator are divided by a suitable factor to get 50 Hz. In most cases, an RC oscillator is sufficient. If accuracy of frequency is important, The oscillator can be a crystal controlled oscillator.

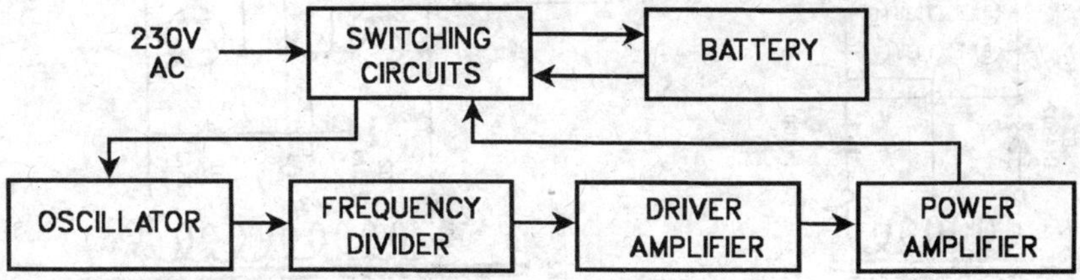

Figure 15.6: Block diagram of a driven inverter.

A practical circuit is shown in Figure 15.7. The circuit uses two transformers. Transformer X1 and rectifier diodes D3 and D4 provide d.c. power to the two relays and transformer X2 is for up/down stepping of voltage. When the mains supply is available, the relays are energized and indicator lamp N1 is lighted. In Figure 15.7, the relays are shown in their un-energized state.

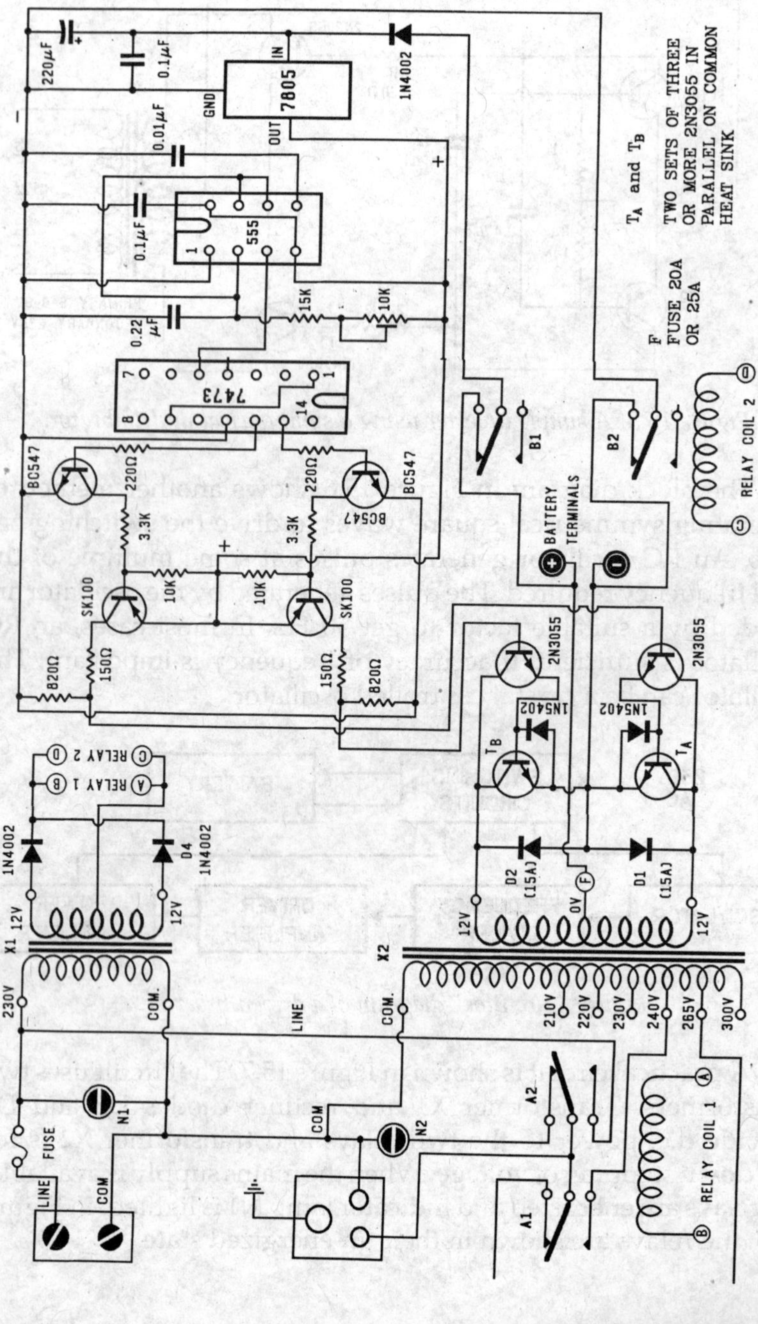

Figure 15.7: A power inverter

One terminal of the output socket is common with the mains supply line and is connected to it directly. The other line is connected to the other terminal of the output socket via a set of contacts A2 of relay 1 when that relay is energized. The other set of contacts of this relay connects this line to the 240 V tap on the transformer X2. Thus, mains power is supplied to the primary of the transformer X2 and the stepped down voltage in the secondary is rectified by diodes D1 and D2 to provide charging current to the battery via fuse F.

When mains supply is available, relay 2 is also energized. The two contacts of this relay disconnect both ends of the battery from the oscillator board. The output transistors are therefore inactive because they do not get their driving signal.

On failure of the mains supply, both relays are de-energized. One contact of relay 1 connects the 285 V tap on X2 to the output terminal, while the other contact connects this terminal to indicator lamp N2. Relay 2, when de-energized, connects battery power to the oscillator board and starts the oscillator. A diode on the board ensures protection against reverse connections of the battery. The oscillator board accommodates a 555 timer, a 7473 flip-flop, driver transistors and a 7805 voltage regulator that supplies +5 V regulated voltage for the 555 Timer and the 7473 TTL flip-flop.

The 555 generates pulses at 200 Hz. The frequency can be set to the correct value by a 10 k potentiometer. The output pulses of 555 are divided by four by the two flip-flops in the 7473 IC to give symmetrical square waves at 50 Hz. The two out-of-phase outputs of the flip-flop are applied as input to two sets of driver transistors. The chopped d.c. voltage in the 12 V winding of the power transformer X2 is stepped up and the high voltage a.c. is made available at the output socket. Indicator N1 goes off and N2 lights up to indicate that the inverter is working.

16

Overview of SMPS

Conceptually, three basic approaches exist for obtaining regulated d.c. voltages from an a.c. power source. These are:

1. Shunt regulation
2. Series linear regulation
3. Series switch mode regulation (mostly referred to as switch mode power supplies, or as SMPS)

All these require a.c. power line rectification. Briefly stated, if all three types of regulation can perform the same function, following are some of the key parameters to be addressed:

From an economical point of view, cost of the system is paramount

From an operations point of view, weight of the system is critical

From a design criteria, system efficiency is the first order of business

The series and shunt regulators operate on the same principle of sensing the d.c. output voltage, comparing to an internal reference level and varying a resistor (active device such as a transistor) to maintain the output levels within pre-specified limits. They are called *linear regulators*. We have studied them in previous chapters.

Switch Mode Power Supplies (SMPS)

Switch mode power supplies (SMPSes) have gained much popularity in recent years because of the benefits they offer. They are used now on a large scale in desk calculators, computers, as instrumentation supplies etc. The advantages of SMPSes are low weight, small size, high efficiency, wide a.c. input voltage range and low cost.

- Low weight and small size are possible because operation occurs at a frequency beyond the audible range. The inductive elements, which usually carry the bulk of weight and size, are therefore small.
- High efficiency because, for output regulation the power transistor is switched rapidly between saturation and cut-off and therefore has little dissipation. This eases heatsink requirements which also contributes to weight and volume reduction. Conventional linear regulator supplies may have efficiencies as low as 50%, or less, but efficiencies of 80% are readily available (see Figure 16.1) with SMPSes.
- Wide a.c. input voltage range because of the flexibility of varying the switching frequency in addition to the change in transistor duty cycle makes voltage adaptation unnecessary. Hence, no a.c. voltage stabilizers are needed.
- Low overall cost due to the reduced volume and dissipation means that less material is required and smaller semi-conductor devices suffice.

Switch mode power supplies also have slight disadvantages in comparison with linear regulators, viz. somewhat greater circuit complexity, tendency to r.f. radiation, slower response to rapid load changes and less ability to remove output ripple.

Switch mode power supplies (SMPS) are basically DC-to-DC converters operating at frequencies in the 20 kHz and higher region. Basically the SMPS is a power source which utilizes the energy stored in an inductor during one portion of its operating cycle to supply power to the output during the remaining segment of its operating cycle.

Linear regulators, both shunt and series, suffer when required to supply large currents with resultant high dissipation across the regulating device. Efficiency suffers tremendously. Efficiencies less than 40% are typical.

Switch mode power supplies operate at much higher levels of efficiency (generally in the order of 75% to 80%) thereby reducing significantly the energy wasted in the regulated supply. The SMPS does, however, suffer significantly in the ripple regulation it is able to maintain as opposed to a much higher degree of regulation available in series or shunt linear regulators.

The linear regulators obtain improved regulation by virtue of the series pass elements always conducting, as opposed to SMPS devices having their active devices operative only during a portion of the overall operating period. A comparison of the two types is shown in Figure 16.1.

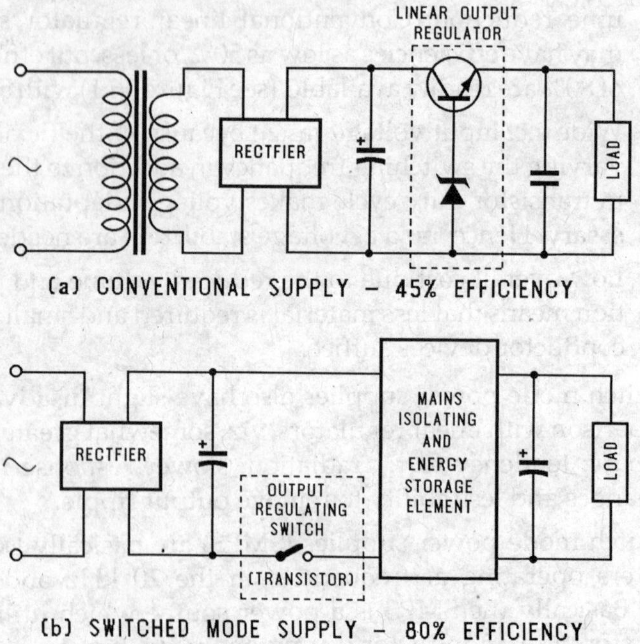

Figure 16.1: Efficiency comparison of the two types of supplies.

Some definitions and comparisons between linear regulators and switched mode power supplies are given below for reference:

Regulation

Line Regulation

Line regulation, sometimes also referred to as *static regulation*, refers to the changes in the output voltage (as a percent of nominal or actual value) as the input a.c. is varied slowly from its rated minimum value to its rated maximum value (e.g., from 210 V_{rms} to 250 V_{rms}.

Load Regulation

Load regulation, sometimes referred to as *dynamic regulation*, refers to changes in output (as a percent of nominal or actual value) when the load conditions are suddenly changed (e.g., minimum load to full load.

Thermal Regulation

Thermal regulation is referred to as changes due to ambient variations or thermal drift.

Transient Response

Transient response is the ability of the regulator to respond to rapid changes in either line variations, load variations, or intermittent transient input conditions. This parameter is often referred to as "recovery time".

AC Parameters

Current Limiting

Often referred to as "fold-back" where the amplifier segment of the regulator folds back the output current of the device when safe operating limits are exceeded.

Voltage Limiting

The regulator's ability to "shut down" in the event that the internal control elements fail to function properly.

Thermal Shutdown

The regulator's ability to shut itself down when the maximum die temperature is exceeded.

General Parameters

Power Dissipation

The maximum power the regulator can tolerate and still maintain operation within the safe operating area of its active devices.

Efficiency

The ratio in percent of the usable versus total power being dissipated (a.c. as well as d.c.) in a regulated supply.

EMI/RF

Generation of radio frequency interference signals and magnetic field disturbance especially in SMPS devices.

How SMPSes Operate?

The switch mode power supply is a modern version of its forerunner, the electro-mechanical vibrator (see Figure 15.1) used in the past to supply car radios. But the new concept is much more reliable because of the far greater life-time of the transistor switch. Figure 16.2 shows the principle of the a.c. fed SMPS. In this system the a.c. voltage is rectified, smoothed and supplied to the electronic chopper, which operates at a frequency above the audible range to prevent audible noise. The chopped a.c. voltage is applied to the primary of a transformer and the secondary voltage is rectified and smoothed to give the required d.c. output. The transformer is necessary to isolate the output from the input. Output voltage is sensed by a control circuit, which adjusts the duty cycle of the switching transistor, via the drive circuit to keep the output voltage constant irrespective of the load and line voltage changes. Without the input rectifier, this system can be operated from a battery or other d.c. source.

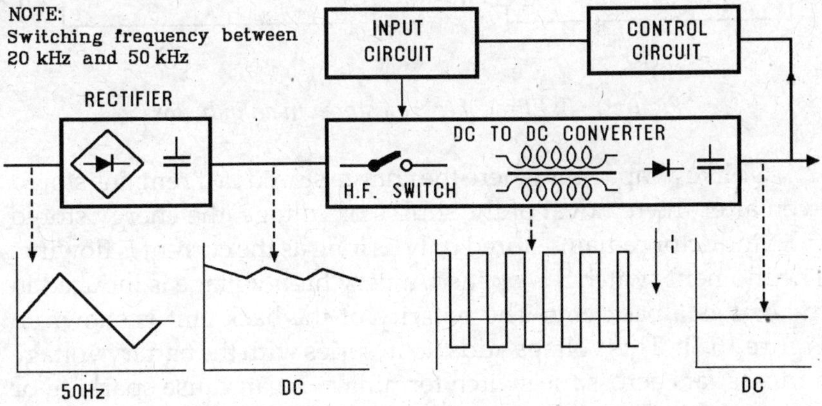

Figure 16.2: Block diagram of switched-mode power supply.

Depending on the requirements of the application, the d.c.-to-d.c. converter can be one of the three basic types: flyback converter, forward converter, or push-pull balanced converter.

Energy Storage in Inductors

Figure 16.3 shows a circuit in which a battery B is connected to an inductor of inductance L. When the switch is turned on as shown in Figure 16.3a, the current through the inductor rises slowly because of the self inductance L. The energy that goes into the inductor during the current build-up is stored in the inductor as magnetic energy. When the current reaches its final value I, the energy stored in the inductor is given by the equation $E = 0.5LI^2$ where L is the self inductance and I is the current. After this, there is no more storage of any energy and the value of current is determined by the d.c. resistance of the inductor and the energy is dissipated as heat in the inductor.

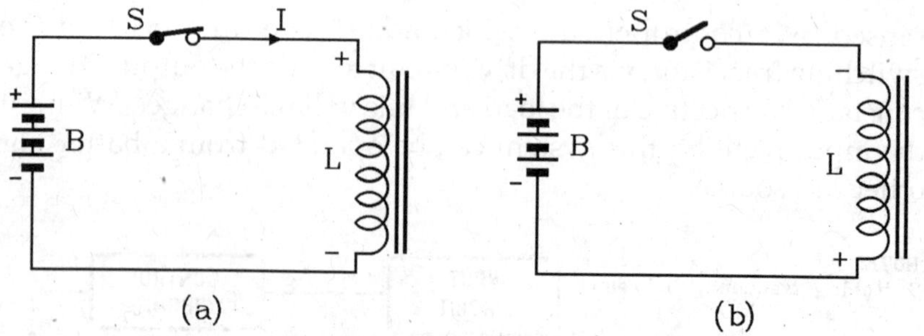

Figure 16.3: Effect of energy stored in an inductor.

Unlike a capacitor where the charge stored in it remains stored even after the removal of the source of voltage, the energy stored in an inductor remains stored only as long as the current is flowing. The moment switch S is opened, a very high voltage is induced in the coil as a back emf. The polarity of the back emf is shown in Figure 16.3b. This voltage adds up in series with the battery voltage and appears across the switch terminals. It can cause sparking or arcing across the contacts. In switching power supplies, as we shall see later, this energy is diverted to charge a capacitor which ultimately supplies d.c. to the load.

In Figure 16.4a, a battery is connected to a capacitor C and load resistor R in parallel. When the switch is closed, a current flows through L to supply power to the load as well as to charge the capacitor. The polarity of the voltage across L is shown in the Figure. When the switch is opened, a back emf is induced in the coil with the polarity shown in Figure 16.4b and the energy stored in the inductor charges the capacitor further through diode D. Diode is called a *free-wheeling* diode. This diode should be a fast one with very small turn-on and turn-off times. Schottky diodes have shorter switching on and off times.

At a current of 1 ampere, an inductance of 20 milli-henrys stores energy equal to $0.5\ LI^2 = 0.5 \times 20 \times 10^{-3} \times 1 \times 1 = 0.01$ joules. If this energy transfer takes 10,000 times in a second, the total energy transferred through the choke to the capacitor is $0.01 \times 10{,}000 = 100$ joules per second, or 100 watts.

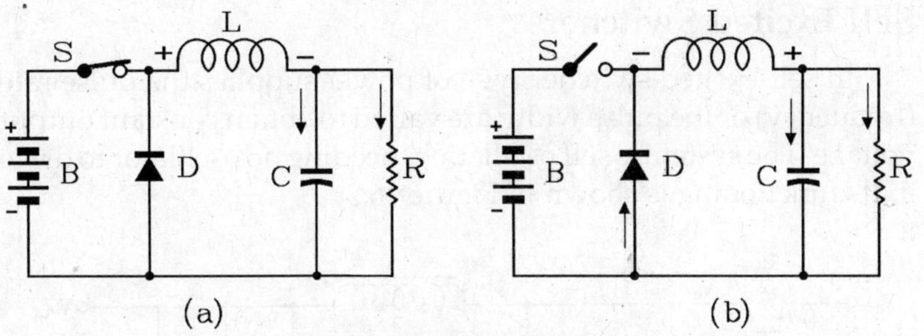

Figure 16.4: Effect of back emf and energy stored in an inductor.

This example illustrates that at higher frequencies a lot of power can be transferred even by a physically small coil. Switching type of power supplies, which use this method, are therefore compact and lightweight.

Since the inductances are small, they can be easily obtained by winding a small number of turns on a ferrite ring type core making it a toroidal inductor. Winding more independent coils on the same core makes it a toroidal transformer. For more informatioin refer to the book "Designing Coils and Transformers" by the Author and published by BPB Publications.

In switching type regulators, as we shall see later, transistors are used a switches. In such applications, the current gain of a transistor is not important. It is the switching speed that is important. In this kind of application, the transistor has only two states, fully conducting or fully non-conducting. In the fully conducting state, the base drive is hard enough to saturate the transistor. The voltage drop (V_{CE}) across the transistor in that condition is small and hence very little power gets dissipated in the transistor. When the transistor is cut off, the dissipation in the transistor depends on the leakage currents through the transistor. The working is similar to that of switch in Figures 15.2 and 15.3. This results in higher efficiency of regulated supplies because the power wasted in the switching transistors is a very small fraction of the total power handled.

Self Excited Switchers

In self-excited switcher type of power supplies the pulse rate (frequency) or the pulse width are varied to obtain constant output voltage. The system is self oscillating, needing no oscillator to drive it. Its functioning is shown in Figure 16.5

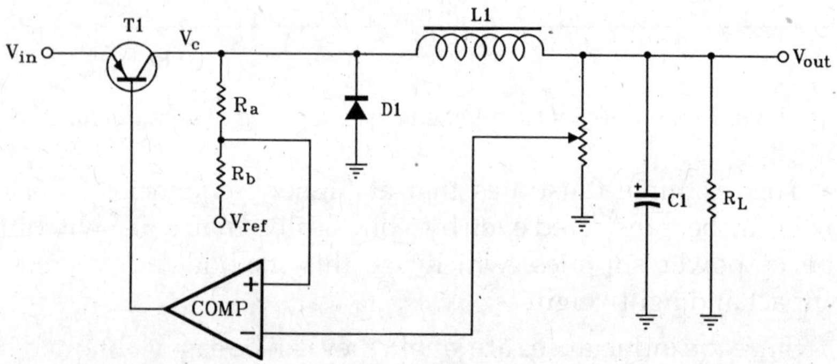

Figure 16.5: Basic self-excited switchers.

T1 is the switching transistor, D1 is the switching diode, while L1 and C1 form the filter. The comparator senses the voltage across L1, this being related to load curent by

$$V = L\frac{dI_L}{dI}$$

where L is the inductanceof L1.

Initially V_{out} will be low and T1 will be 'on'. This voltage at non-inverting input will approach $\beta_1 V_{in}$ when

$$\beta_1 V_{in} = V_{ref}\frac{R_a}{R_a + R_b} + V_c\frac{R_a}{R_a + R_b}$$

When the output voltage is reached, the comparator will turn T1 off. The diode D1 will become forward biased and supply a path for the inductive curent. This current and the sense voltage will start to decrease till V_{out} readches

$$\beta_2 V_{in} = V_{ref}\frac{R_a}{R_a + R_b}$$

where thecomparator will again turn T1 on and the cycle repeats. Thus, V = Vref + ripple. The frequency of oscillation f and capacitance C are

$$f = \frac{V_{out}\,(V_{in} - V_{out})}{L\,V_C\,[I_{max} - I_o]} \qquad C = \frac{(V_{in} - V_{out})}{8L\,f^2\,V_{in}\,(\Delta V)}$$

where ΔV is the ripple voltage, I_{max} and I_o are maximum and minimum inductor currents. Waveforms for V_C, I_L and V_{out} are shown in Figure 16.6.

The higher the value of f, the smaller will be value of L. The switching time of the comparator is quite fast, added by positive feedback. The diode and transistor should switch very fast or losses occur. Hence, Schottky diodes are used.

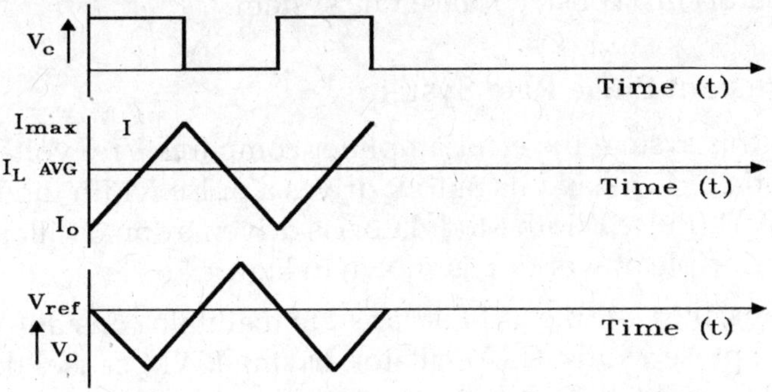

Figure 16.6: Waveforms in self excited switchers.

Also, inductor core saturation should not occur, otherwise peak transistor and diode currents will be very large, and power dissipation will increase. This can also lead to the destruction of the transistor and the diode. So, inductors with air gaps made of ferrite or permalloy should be chosen.

Low loss capacitors with adequate ripple current rating should be chosen since at 20-25 kHz even this can contribute to losses. Dry tantalum types are ideal.

As regards the selection of transistors, bipolar transistors were chosen in view of the frequency range of 20-30 kHz. Above this losses rise. Power MOSFETS can make things better and increase the frequency limit to 300-400 kHz.

Switching regulators can easily be implemented with linear regulators such as 723 or LM317 described earlier. Such switching regulators are described in chapters that follow.

Driven Switching Regulators

In driven switching regulators, the pulse width and rate are not varied randomly, but are controlled by an oscillator, keeping any one parameter constant and allowing the other to vary. The basic principle of operation is similar to self-excited switchers, except that the d.c. amplifier drives a VCO (Voltage Controlled Oscillator) in constant pulse width system, or a PWM (pulse width modulator) in a constant pulse rate system.

(a) Constant Pulse Rate System

In this system the error amplifier compares Vref voltage and a fraction of V_{out}, and its output drives a pulse width modulator. The PWM (Pulse Width Modulator) is driven by an oscillator. The basic principle of working is shown in Figure 16.7.

Thus, the ON time of T1 is varied to maintain constant V_{out} by varying pulse width. The oscillator driving PWM causes the transistor T1 to switch on at regular intervals.

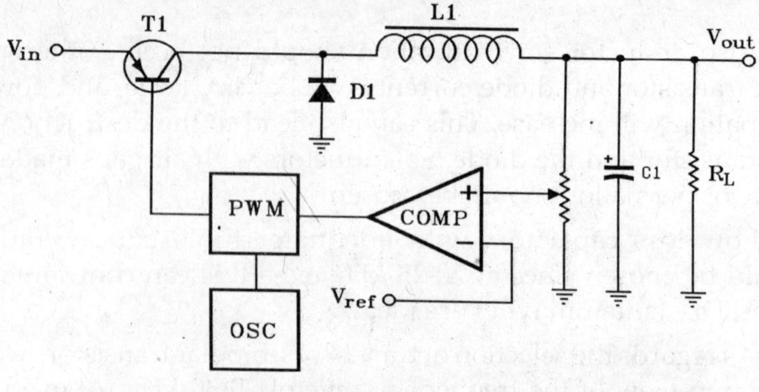

Figure 16.7: Basic pulse width modulator system of switching.

(b) Constant Pulse Width System

In this system, the error amplifier drives a VCO (Voltage Controlled Oscillator) which in turn drives a monostable multivibrator (MVB) of constant pulse width. Thus, the ON time of T1 is kept constant. This method of switching is shown in Figure 16.8.

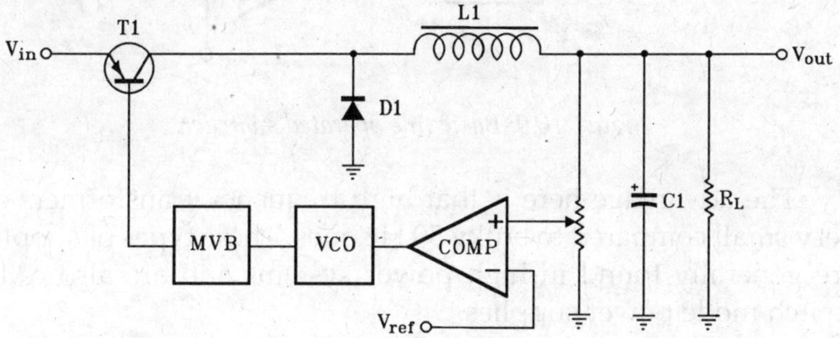

Figure 16.8: Basic pulse rate modulator system of switching.

Of the two systems described above, the constant pulse rate system is preferred since phenomenon like r.f.i. and line interference can be prevented by selecting the correct frequency. Driven switchers also offer more precise control of voltage.

Driven switchers can be implemented using separate PWM oscillators, comparators etc. but these often come packaged in a single IC—either with or without switching transistor.

(c) Line Operated Switchers

Line operated switchers (see Figure 16.9) operate on a slightly different principle. Here, the a.c. input is directly rectified and converted to a higher frequency (say, 20 kHz) by an inverter controlled by a PWM . The inverter's output is then stepped down by a ferrite transformer.

This output is then rectified, filtered and monitored by an error amplifier which drives the PWM. The inverter's output is thus controlled to get a constant V_{out}.

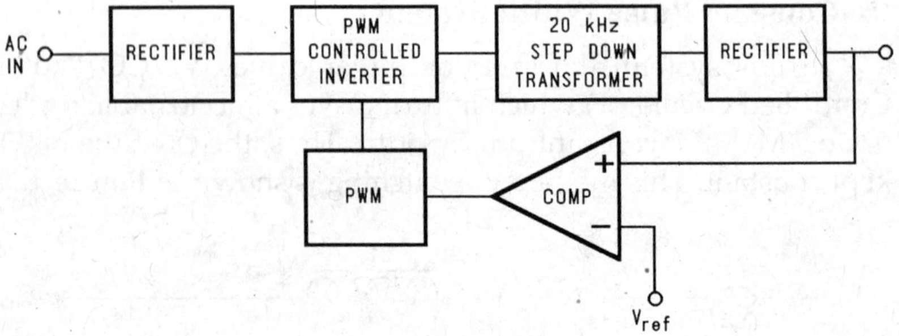

Figure 16.9: Basic line operated switcher.

The advantage here is that high frequency transformers are very small compared to bulky 50 Hz ones. These types of supplies are generally found in high power systems and are also called switch mode power supplies.

Thus, switching supplies have become viable alternatives to linear supplies. Recent developments in power semiconductors and those in magnetic materials and capacitors have brought switching supplies into the realms of the hobbyist from the design laboratories.

17

Switch Mode Power Supplies

Electronic equipment usually require power supplies that deliver large currents, say 5 A, at low voltages. Mains electric power is available at 230 V a.c. at 50 Hz. This has to be converted into low voltage d.c. using a step-down transformer and rectifiers, and ripples are filtered with capacitors and inductors as shown in Figure 18.1. Such supplies have been described in earlier chapters.

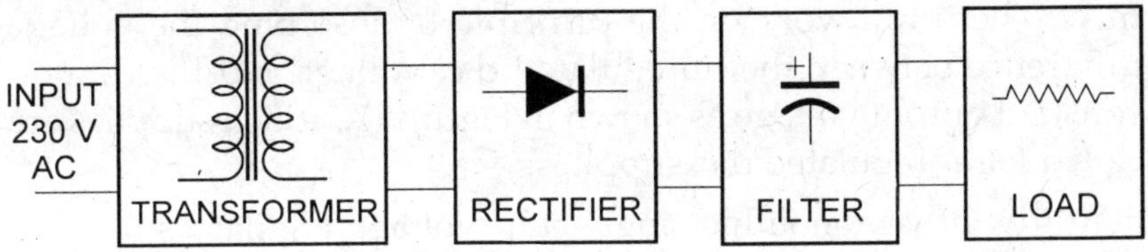

Figure 17.1: Block diagram of an unregulated d.c. supply.

Before 1972, this kind of supplies were quite suitable for most electronic equipment and instruments. But when integrated circuits came into existence, these d.c. supplies were required to be regulated against mains voltage variations and load current variations. For example, in digital circuits, TTL family of ICs needs a well regulated d.c. at +5 V level. In order to obtain stable voltage, a feedback control system was employed in the regulator ICs.

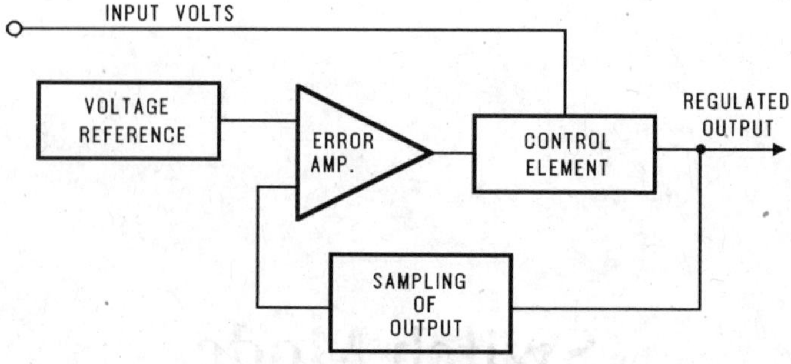

Figure 17.2: Principle of a normal linear mode regulated power supply.

In order to keep the output voltage constant, one must obtain a higher rectified voltage from the transformer and dissipate the excess power as heat. It means extra power loss and poor efficiency. A series-pass transistor is employed as a control element in such linear supplies to adjust the output voltage to the desired stable voltage. It drops the excess voltage across itself so as to maintain the output voltage equal to the desired value. If I_L is the current flowing, the power loss is I_L times ($V_{unregulated}$ - $V_{regulated}$)

Prior to 1975, the regulator ICs such as 723 and CA3085, were capable of voltage regulation by feedback at low currents only (100 mA). These ICs work on the principle of absorbing the voltage difference between the unregulated d.c. voltage and the desired constant output voltage as shown in Figure 17.2. Such supplies are called *linear* regulated d.c. supplies.

Presently monolithic single chip voltage regulators for currents up to 5 A are available. There are variable voltage regulators too. These are convenient, but have a poor efficiency, below 50 percent and dissipate heat at full load current.

A switching mode power supply (SMPS), on the other hand, is highly efficient and operates on entirely different principles. It appeared in 1980, when fast switching transistors became available at good current capacities. The output voltage of the switched mode power supply is controlled by varying either the duty cycle or frequency, or both, of the signal that turns the controlling transistor on and off.

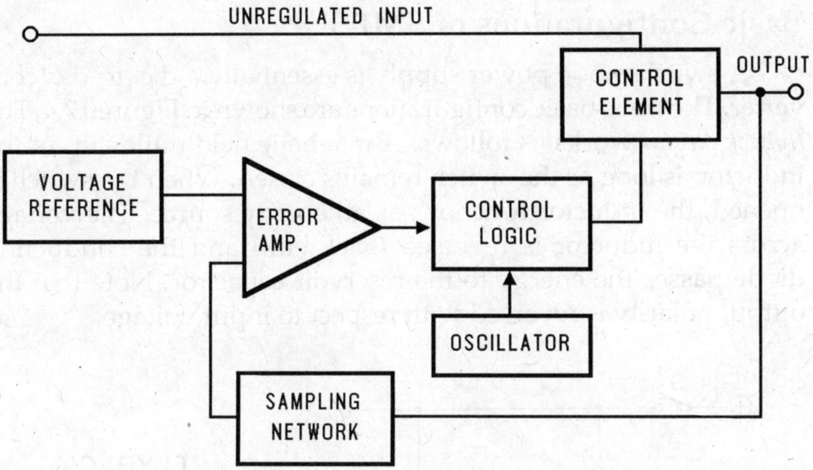

Figure 17.3: Block diagram of a switch mode regulator.

The block diagram in Figure 17.3 shows the concept of a switching regulator in simple form. The added elements are control logic and oscillator. The error amplifier controls the control logic and the control logic switches the control element on and off to maintain the output voltage. The output voltage is therefore kept regulated.

The control element usually consists of a transistor switch (a switching transistor or FET), an inductor and a diode. For each on, energy is pumped into the magnetic field associated with the inductor. This energy is then released to the load at the desired voltage. By varying the duty cycle or the frequency of the switching, or both, one can vary the stored energy in each cycle, and thus control the output voltage.

As a switch has only two positions, viz., on or off, it either allows energy to pass or stops it, but does not dissipate energy in itself as explained (see Figure 15.2). Since, the energy drawn is needed only to maintain the output voltage at a load current, there is no dissipation and hence a higher efficiency is maintained. Energy is pumped in discrete lumps, but the output voltage is kept steady by capacitor voltage.

Basic Configurations of SMPS

A switch mode power supply is essentially a d.c.-to-d.c. converter. The three basic configurations are shown in Figure 17.4. The *flyback circuit* works as follows. A magnetic field builds up in the inductor as long as the switch remains closed. When the switch is opened, the inductor functions as an energy source. The voltage across the inductor is reversed (back emf) and the conducting diode passes the energy to the reservoir capacitor. Note that the output polarity is reversed with respect to input voltage.

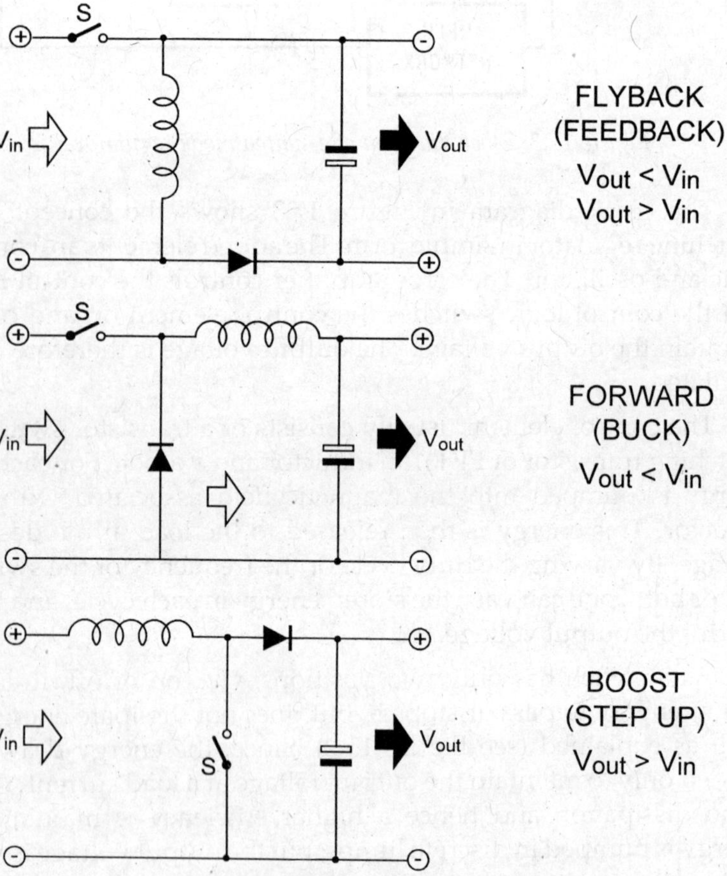

Figure 17.4 Basic configurations of a switch-mode power supply.

The *forward converter* does not reverse the polarity of the output voltage. The capacitor is charged via the inductor when the switch is closed. It also supplies power to the load. The difference between the input and the output voltage is available on the inductor. In contrast to that in the flyback converter, the switch is closed when the capacitor is being charged. When the switch is opened, the magnetic field of the inductor is weakened via the flyback or free wheeling diode. The switch is, of course, a power transistor, and the diode affords protection against induced voltage. In a forward converter, the output voltage is lower than the input voltage.

The third basic configuration is referred to as *boost* or *step up converter*. This circuit increases the output voltage, and is functionally similar to the flyback converter. Energy is stored in the inductor when the switch is closed. When the switch is opened, this energy is supplied to the load at the output via the diode.

The Continuous and Discontinuous Mode

Two modes of operation can be distinguished, depending on the current in the inductor—see Figure 17.5.

After closing the switch, the current in the inductor, increases linearly up to a specific maximum value (V_{in} = constant). After opening the switch, the current decreases linearly. The circuit operates in the *discontinuous current mode* if the current is zero in every period (Figure 17.5a). The capacitor supplies the load current during the remainder of the period. The discontinuous mode is characterized by the good response of the closed regulation circuit to fluctuations in the input voltage (line regulation), and the output load (load regulation). There is no energy in the inductor at the start of each period, and regulation can, therefore, take place on a period-to-period basis. It can, in fact, be argued that the inductor is not present in the regulator circuit. The maximum phase shift of 90° in the buffer capacitor ensures the stability of the closed regulation circuit. A disadvantage of the discontinuous mode is the relatively high peak current carried by the power switch. Flyback and step up converters usually operate in the discontinuous mode.

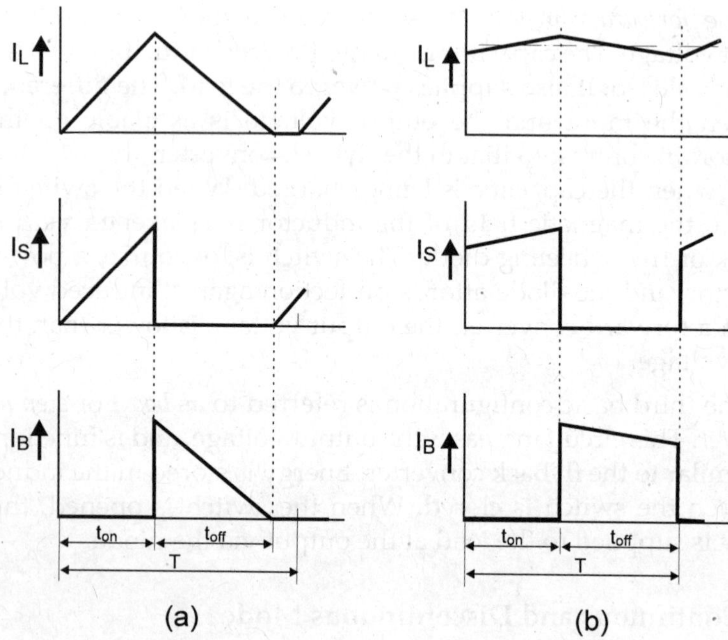

Figure 17.5: The discontinuous (a) and the continuous (b) modes differ in respect of the current carried by the inductor.

In the *continuous current mode*, the current through the inductor does not drop to zero at the end of every period (Figure 17.5b). The ripple current in the inductor is small relative to the load current, and this requires a fairly high self-inductance. The buffer capacitor, on the other hand, can be kept relatively small. The favourable shape factor of the current through the power transistor and the diode makes the continuous mode eminently suitable for high power applications. The response to load fluctuations is, however, worse than that of a circuit in the discontinuous mode. Each change in the output load current requires a corresponding change in the direct current through the (relatively large) self-inductance, and this process may take several periods to complete.

It is not possible for a system to automatically switch from continuous to discontinuous operation, or vice versa, because this would cause a considerable change in the open loop characteristics, giving rise to instability of the closed regulation system. This means that the load current of a system in the continuous mode should be higher than half the peak-to-peak value of the ripple current in the inductor. Forward converters usually operate in the continuous current mode.

Off-line Operation

In many cases, the input voltage is rectified and smoothed voltages obtained direct from the mains. The direct voltage so obtained (approx. 320 V for a 230 V mains supply) is rarely used for converting down to, say, 12 V, in view of the resultant low duty factor, and the need for large self inductances. Also, a direct connection to the mains is dangerous, and normally not permitted. This calls for a ferrite transformer, which, in an SMPS, has the advantage of being much smaller than a soft iron type used in the traditional 50 Hz mains supply. There are, however, a number of important considerations as to keeping the losses of the core material within acceptable limits. Ferrite is used instead of laminated iron because it offers a number of advantages. The construction of a ferrite is a light and insulating material.

The turns ratio of the ferrite transformer enables converting the high input voltage down to a value close to the desired output voltage. The conversion increases the duty factor, and hence it reduces peak currents in the power transistors.

Transformer Circuits

The simplest configuration of an SMPS is the single transistor flyback converter shown in Figure 17.6a. This circuits well known in low power supplies with an output rating up to about 250 W. In this application, the transformer is more properly referred to as a coupled self-inductance, because it assumes the function of the inductor shown in Figure 17.6a. This forward converter in continuous mode is more suitable for feeding relatively heavy loads. The most commonly found version is based on a single transistor and a demagnetization winding in the primary circuit as shown in Figure 17.6b.

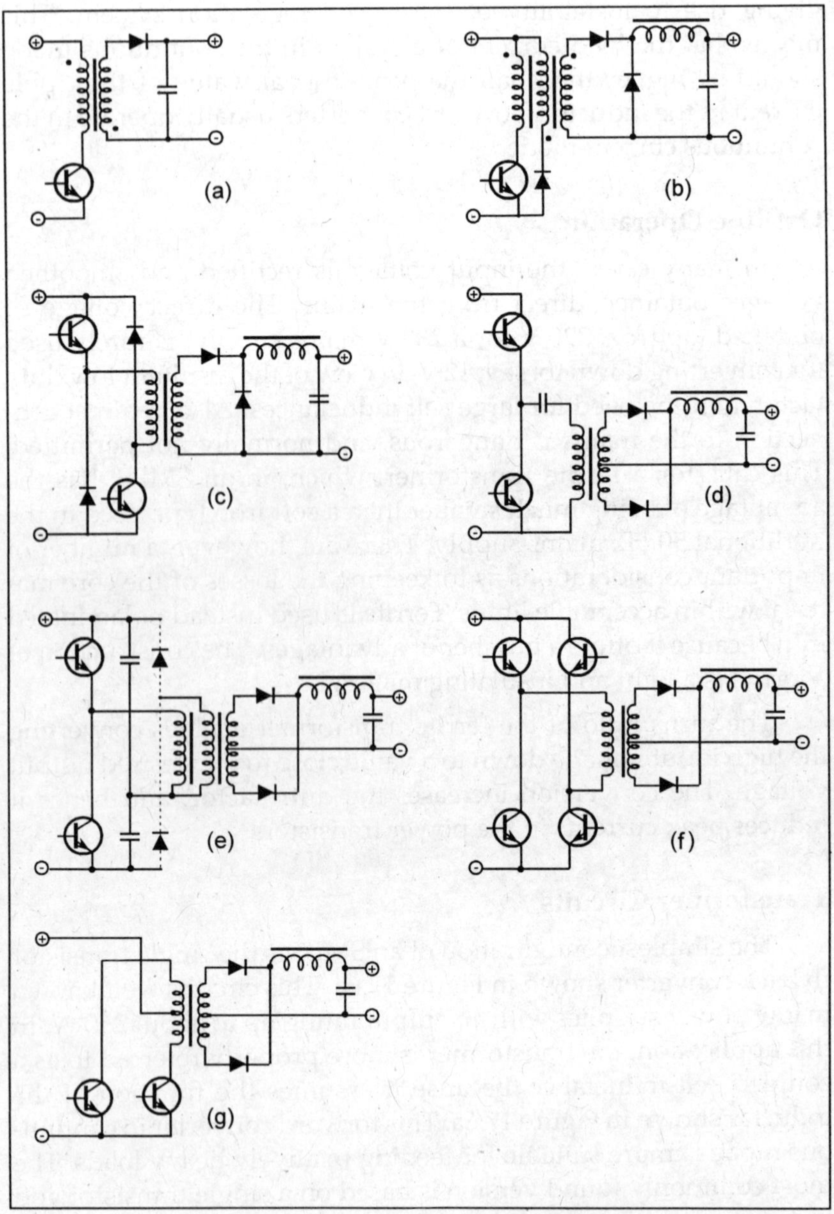

Figure 17.6: Various configurations of the switching power stages in SMPS.

The transistor must be able to handle at least twice the input voltage. The demagnetization winding can be omitted if the circuit is extended with a transistor and a flyback diode as shown in Figure 17.6c. In this circuit, the transistors need only withstand half the voltage. They are, however, driven with respect to different potentials, just as in the bridge circuits to be discussed.

Half or full bridge circuits are used mainly for high power applications. The full bridge variant is suitable for very heavy loads, thanks to the fact that the effective input voltage is doubled. The last variant, shown in Figure 17.6g, is also a bridge circuit, based on a center-tapped transformer that enables the transistors to be driven with respect to a common reference potential.

Bipolar transistors as well as power FETs can be used in the primary circuit. Bipolar technology is suitable for switching frequencies up to 50 or 100 kHz. Power FETs are faster, and can be used at higher frequencies without running into excessive switch losses. Consequently, the maximum usable frequency is about 1 MHz, and power FETs are more suitable in SMPSes in view of the ever increasing switching frequencies. Power FETs for relatively high voltages are quite expensive and more attractive for use in countries with 117 V mains supply, such as the U.S.A.

Core Saturation

Any transformer winding forms a self-inductance, and the average voltage across it should therefore be zero. When this is not so, the remaining direct voltage causes a linearly increasing direct current until the core is saturated. The H-field, and with it the current, then increases exponentially, in accordance with Faraday's law of constant increase of the magnetic flux per unit of time. This effect must be prevented because it can lead to destruction of the primary circuit.

In Figures 17.6b and 17.6c, the field in the transformer core is weakened with the aid of the flyback diode(s), but only as long as the duty factor remains below 50%. Problems owing to permanent magnetization are not expected to arise in the circuit of Figure

17.6d, where the coupling capacitor ensures the absence of direct current through the primary winding. The situation is more complex in Figure 17.6e. Although the primary winding is a.c. coupled, a direct voltage may still exist at the junction of the capacitors. This positive or negative potential may arise from less than perfect (i.e., unbalanced) driving of the power transistors which may have different recovery times also. Unbalancing of the primary circuit can be prevented with the aid of a compensation winding and 2 flyback diodes.

A coupling capacitor for blocking the magnetizing direct current is rarely used in the circuit of Figure 17.6f, because this is rated for very high power. Both the positive and the negative current in the primary winding are measured and any difference between them is compensated by controlling the duty factor. This safety measure can be applied to the circuit of Figure 17.6g also.

Voltage Control

In a switch mode power supply, the output voltage is measured, compared to a reference, and kept constant by controlling the duty factor of the drive signal applied to the power switches (i.e., transistors). The regulating effect of the control circuit depends on the open-loop characteristics of the system. The simplicity of the flyback converter makes it less suitable for many purposes, since the duty factor depends primarily on the load at a constant output voltage. Good voltage regulation requires a high amplification of the measuring and control circuit. In a forward converter, the voltage control circuit need not have a strongly regulating effect because in essence the output voltage depends only on the turns ratio of the transformer.

There are three basic types of voltage control system:

- Direct duty factor control.

The error signal is amplified and drives a pulse-width modulator, which in turn adjusts the duty factor as required. A high overall amplification is needed, at the cost of some stability—notably in the case of converters operating in the continuous mode.

- Voltage feedforward

This is the most commonly used system. Preregulation of the duty factor is implemented as a function of the input voltage, enabling the output voltage of the open loop system to be made independent of the input voltage. The control circuit is therefore only required for compensating load fluctuations. The preregulation improves the line regulation, and so ensures sufficient suppression of the hum.

- Current mode control

A second control circuit (inner loop) inside the voltage control circuit (outer loop) enables switching off the power transistor at a more or less fixed peak value of the current. The effect so obtained is the (quasi) disappearance of the inductor from the output filter. The whole system is then essentially a first order network with the capacitor in the output filter as the only phase changing element. The stability of the whole supply, as well as the response of the closed loop system to fluctuations in the input voltage and the load current, is excellent. High power forward converters of the continuous type are often equipped with a current mode control for obvious reasons.

Location of the Control Circuit

The voltage control circuit can be located at the primary or the secondary side of the supply. A circuit of the primary side (Figure 17.7b) makes it possible to drive the power stage direct from the IC, while it is a relatively simple matter to implement circuits for primary current monitoring and preregulation. A disadvantage of the primary location is the need for an insulating device in the control loop for transmitting an analog signal from the primary to the secondary side. This is usually done with the aid of an optocoupler. Only the error signal is transmitted to rule out instability of the output voltage owing to aging effects of the optocoupler.

A control circuit on the secondary side (Figure 17.7a) enables direct coupling of the voltage control circuit. Also, it allows the use of the reference circuit built into many of the available SMPS

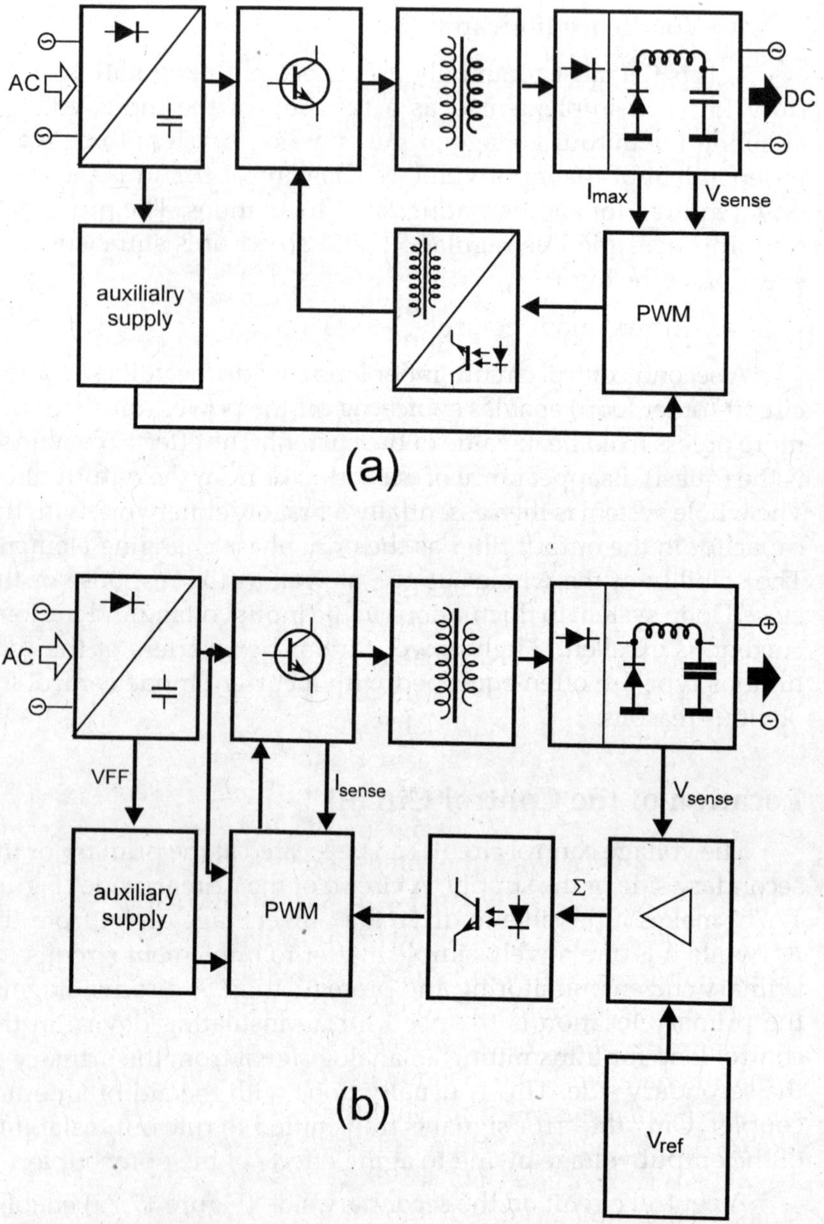

Figure 17.7: Functional diagram of a SMPS with the control electronics located at the secondary side (a) or the primary side (b).

controller integrated circuits. The PWM signal is fed to the power stage via a fast optocoupler, or a special pulse transformer. Differences in the drive applied to several power transistors are relatively simple to monitor and correct, but the primary location makes it difficult to keep tabs on the primary current.

Whatever the control circuit is used, it must have its own (start) supply. The self-generated voltage is, of course, only usable after the supply is fully operational.

SMPS Controllers

A wide variety of integrated circuits are available for controlling switch-mode power supplies. These ICs are essentially very similar, and a general block diagram is therefore given in Figure 17.8 to explain their operation.

The pulse width modulator is composed of a sawtooth generator, a voltage comparator and a set-reset bistable. A second input on the comparator is connected to the output of an op amp that amplifies the difference between the real and the required (set) output voltage. An accurate, and temperature compensated voltage reference is often included for adjusting a specific output voltage.

The remainder of the circuits in the chip shown have auxiliary functions, and serve for various types of protection. Voltage feed-forward or current mode control is possible by changing the slope of the sawtooth signal. Special control inputs make it possible to set a duty factor of zero. An analog input is activated above a predefined voltage level, and can be used for making a 2-level short-circuit protection. When the output current reaches the first level, the duty factor is held constant, and the current supplies a constant, maximum current. The duty factor is made zero above level 2. A digital input enables remote control of the supply (computer controlled test sites, etc.).

An input for setting the maximum duty factor is standard on most SMPS controllers. Properly driven, it prevents saturation of the transformer core, and hence an exponentially rising primary

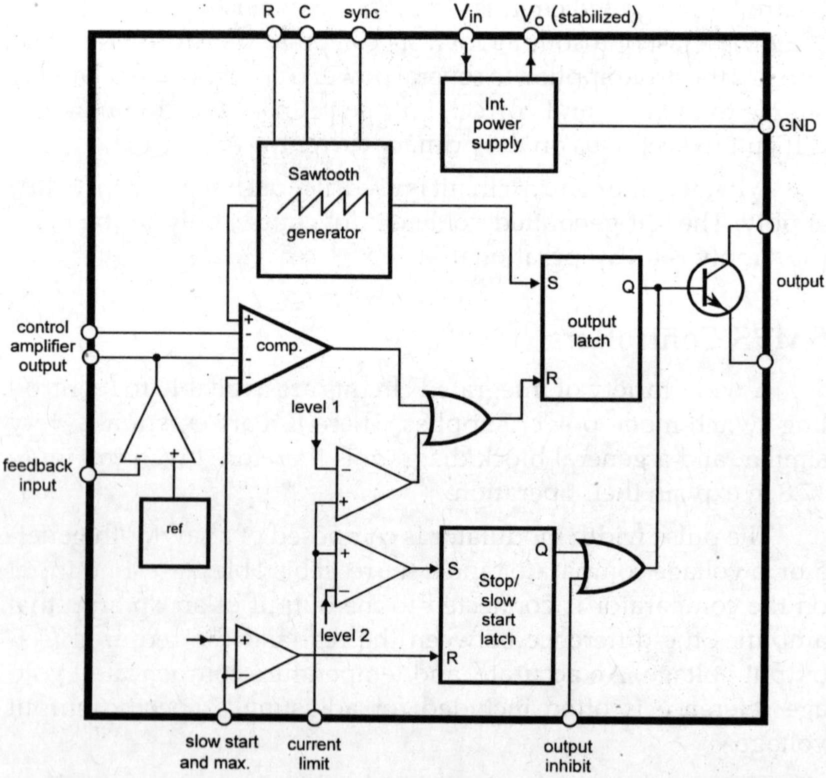

Figure 17.8: General block diagram of an integrated SMPS controller.

current. This safety measure is especially useful for supplies operating in the continuous current mode. In those, the duty factor has a tendency to rise to a maximum value at each change in the output current, in an effort to correct the direct current through the inductor in the output filter in accordance with the new load.

Components of Supplies

The bare-bone circuits of simple switched mode power supplies have been discussed so far. They merely describe the essential principles of energy conversion through storage of energy in the magnetic field of an inductor.

Energy is stored only during the conduction period of the switching transistor. The amount of energy decides the power capacity of an SMPS. In order to increase energy, large current must be pumped into high inductance. The inductance L is a measure of the flux per ampere of current in the coil. So, the energy stored in a coil is given by $0.5LF = 0.5 (\phi/I)F = 0.5 \phi I = 0.5 (LI) I = 0.5 LI^2$ where L is the inductance of the coil in henries, ϕ is the flux linking in the coil in webers, and I is the current in amperes.

Core

The flux ϕ is supported by the magnetic core of the coil. The core material chosen is generally ferrite as it can work at the frequencies which are employed in SMPS. Any frequency above 15 kHz is employed in these supplies, but as frequency increases the core material must have low energy loss due to hysteresis. At lower frequencies, i.e., below 10 kHz, it is not advantageous to make use of switched mode power supplies. The size of core reduces inversely with frequency. Some companies make SMPS even at 300 kHz using special ferrite material for the core.

Core size decides the value of flux which can be supported. Flux density is defined as flux per unit area. Ferrite cores can support up to 0.35 - 0.4 tesla (tesla = weber/m^2) of flux before saturation. The core should not be allowed to get saturated by passing too much direct current around its coil. It is advisable to work well below saturation level.

Ferrite cores of several types, the double-E, the E-I, pot core, cup core and the simple double limb type used in TVs. The saturation of a ferrite core is not a constant flux density as in iron cores. Usually, one has to operate ferrite at well below saturation level and provide thin air gaps in the core joints to avoid saturation. Too much air gap will require too many turns of wire for a given inductance and will increase copper losses.

For more information on ferrite and other core materials, see author's book "Designing Coils and Transformers" published by BPB Publications.

Transistor

Used as a switch, a transistor is inherently has a fast switch-on time, ample current rating to withstand surges and a high breakdown voltage rating to withstand the peak induced voltage in the coil at switch off. Such transistors are available. Switching time should be very small, below 1 μs, comprising turn on delay, rise time and fall times of current.

Diodes

Used at frequencies of 15 kHz and above, diodes are generally different from the common diffused junction types used at mains frequency. In SMPS, generally,one employs switching diodes with either gold dopes junctions or the Schottky barrier diodes which have metal semiconductor junctions. There is not much charge storage in the junction in such switching diodes. The difference between an ordinary diode and such a diode is apparent in the rectifier waveforms shown in Figure 17.9.

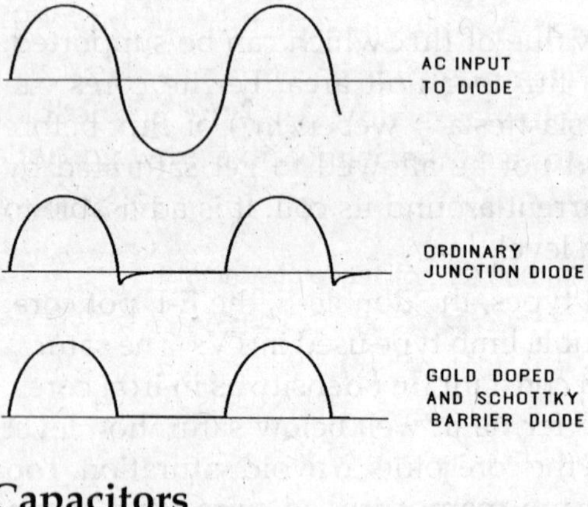

Figure 17.9: Behaviour of ordinary junction and gold doped diodes at high frequencies.

Capacitors

Electrolytic capacitors used for filtering ripples are necessary for any type of power supply. With power frequency rectifiers, the ripples are at 100 Hz but in switch mode power supplies, the frequency of switching is high, of the order of several kHz.

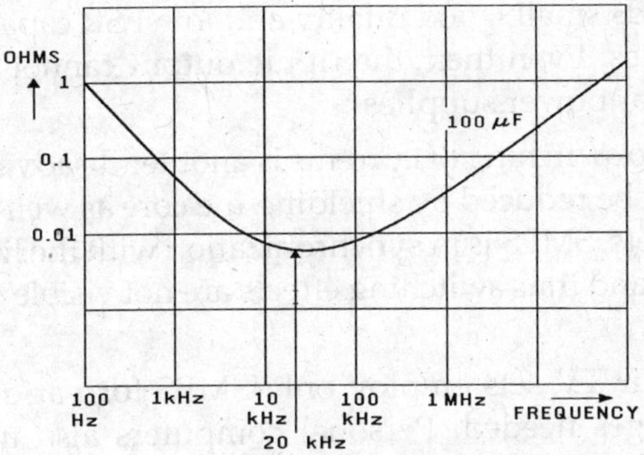

Figure 17.10: Variation of impedance of capacitor.

As such frequencies, electrolytic capacitors invariably have a series inductance and resistance effect. The equivalent series resistance (ESR) of a capacitor is about 0.05 to 0.1 ohms. It grows with frequencies above 10 kHz. The impedance of a capacitor becomes minimum at about 20 kHz, whereupon it increases further as shown in Figure 17.10.

In order to decrease the resistance, it is advisable to employ two or three capacitors of similar type in parallel rather than use a single one. This reduces the ESR.

Comparison with Linear Supply

Today, the switched mode power supply has become advantageous for most applications because of the availability of better material and components, as mentioned above. Primarily, the factors in favour of an SMPS are high efficiency as well as small size and weight.

High efficiency is obtained because of less wastage of power when supply is regulated. Smaller weight and size are due to smaller and light ferrite core transformers. operation at high frequencies (15-20 kHz) reduces the size of a transformer for the same power and iron core cannot be used above 400 Hz due to increased core heating.

For these advantages, a high price is paid for switching large currents at such high frequencies and for rectifying high frequency voltages. Filtering needs small, good quality and low ESR capacitors at these frequencies. Even then, the ripple output cannot be made as low as in linear power supplies.

RF radiation due to switching of currents is another disadvantage in SMPS. This can be reduced by shielding the core as well as the entire unit. In TV sets, SMPS is in synchronization with the line frequency (15,625 Hz) and thus switching effects are not visible on the screen.

Today, most colour TV sets employ only SMPS for various tube and circuit voltages needed. Personal computers also use SMPS for 5V, 12V. and 24 V supplies for large currents. Instruments employ only SMPS so as to make them lighter in weight. Pocket calculators with green fluorescent type display use SMPS for high tension display voltages from 3 Batteries. A comparison of the two types of power supplies is given in Table 18.1.

Table 18.1: Comparison between SMPS and Linear Supplies

Feature	Switch Types	Linear Types
Efficiency	65-75 percent	25-50 percent
Temperature rise	20 to 40°C	55 to 100°C
Ripple volts	Higher, 25 to 50 mV	Even 5 mV possible
Overall regulation (percentage drop in volts on load)	0.3 percent is common, tighter regulation is difficult to get	Even 0.1 percent is possible
Isolation from mains	Quite good	Not as good.
RF interference	Can be a problem unless properly shielded.	None
Magnetic material	Uses ferrite core	Uses stalloy or CRGO core
Weight	About 60 watts/ kg	20-30 watts / kg
Volume of unit	About 10-15 cc per watt	About two to three times higher
Cost	More parts, special ones increase the cost	Advantage for smaller units
Reliability	Depends upon availability of suitable transistors	More reliable

18

SMPS Circuits

We have already learnt how ready made commercially available integrated circuits can be used for designing of linear regulators. A number of such integrated circuits can also be used to make switch mode power supplies. Circuits using different types of such ICs for making switch mode regulators, as suggested by the manufacturers of those integrated circuits are described in this chapter. Practical circuits using integrated circuits specially designed for making switch mode regulators are also described in this chapter.

Switching Regulator Using CA3085

Figure 18.1 shows the circuit of a switch mode regulator using the CA3085 integrated circuit (see Figure 10.1). The unregulated d.c. input is applied to pin 3 (unregulated input) of the IC as well as to the emitter of a p-n-p power transistor. The collector current of the transistor supplies the load current via an inductor of 1.5 mH through a current limiting resistance R1.

A fraction of the regulated output, as decided by the voltage divider network comprising R2 and R3, is applied to pin 6 (inverting input of error amplifier in the IC) to sense the output. The positive input of the error amplifier within the IC is connected to the reference voltage.

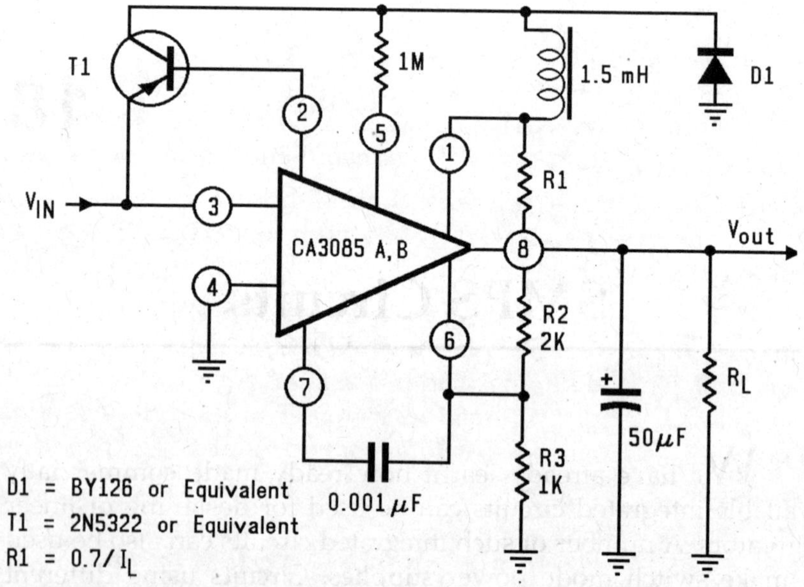

Figure 18.1: Switching regulator using CA3085.

Because of the presence of inductance, the current in the load grows slowly and charges the 50 μF output capacitor. Energy equal to $0.5LI^2$ is stored in the inductor as magnetic energy. As soon as the output voltage exceeds its correct voltage (determined by the voltage divider R2 and R3), the IC is shut off. This sudden stopping of current in the inductor induces a back emf in the coil. The polarity of this back emf is such that the current continues to flow in the same direction and charge the 50 μF capacitor through diode D1 to maintain the voltage on the capacitor which supplies power to the load R_L. When all the energy stored in the inductor is exhausted, the voltage across the 50 μF starts falling. When the voltage has fallen below a limit, the IC is turned on and current starts flowing through the inductor again. The process goes on repeating endlessly. A large value of the output capacitor ensures that the ripple content in the output voltage is within acceptable limits.

Positive Voltage Switching Regulator using 723 IC

The circuit shown in Figure 18.2 is a positive voltage switching regulator using the popular type 723 IC. The circuit resembles the positive voltage regulator using an external p-n-p transistor shown in Figure 10.17. The current drawn by the IC drops a voltage across R5 and this voltage applies a forward bias to transistor T1, which in turn, applies bias current to T2. The amplified collector current flows through a 1.2 mH inductor to the load, and charges the output filter capacitor C2. The output voltage is compared with a fraction of the reference voltage. When the output voltage exceeds the set output voltage, the voltage at the Current Sense (CS) input, the IC is switched off but the back emf in the inductor continues to charge the capacitor via diode D1. When the inductor's energy is exhausted, the voltage falls and the IC is turned on again.

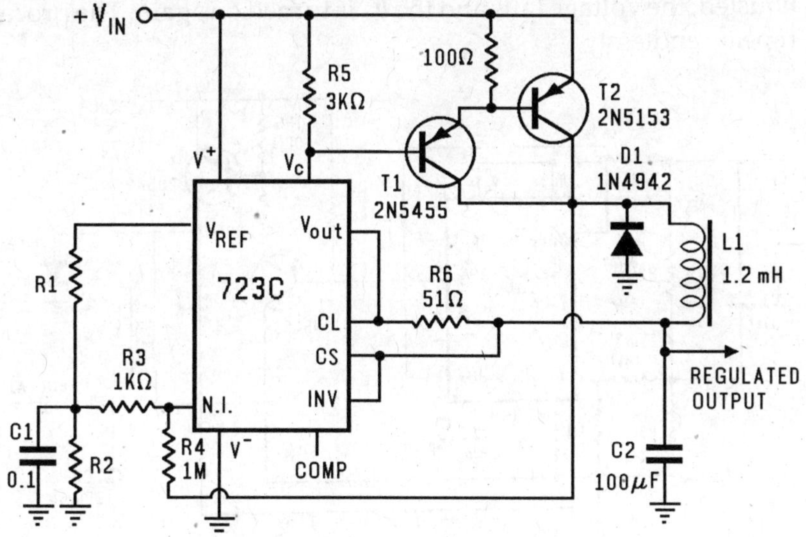

TYPICAL PERFORMANCE

Regulated Output Voltage	+5V
Line regulation (ΔV_{IN} = 30 V)	10 mV
Load Regulation (ΔI_L = 2A)	80 mV

Figure 18.2: Positive voltage switching regulator using 723.

Negative Voltage Switching Regulator using 723 IC

Figure 18.3 shows a negative voltage switching regulator using a 723 IC. The circuit is similar to the negative voltage regulator using a p-n-p transistor shown in Figure 10.20.

The working of this one is similar to the two circuits described earlier. As before, the output load current flows through an inductor L1 and charges the output filter capacitor C2. When the voltage across C2 exceeds the preset voltage (decided by the voltage dividing network comprising R1 and R2), the IC is turned off. This sudden shut down of current through the inductor induces a back emf in the inductor and this back emf continues to charge the capacitor C2 via diode D1. The path of current is from the load to ground and then through D1 back to the other end of the inductor.

When all the magnetic energy stored in the inductor is exhausted, the voltage falls and the IC is turned on again. The process repeats endlessly.

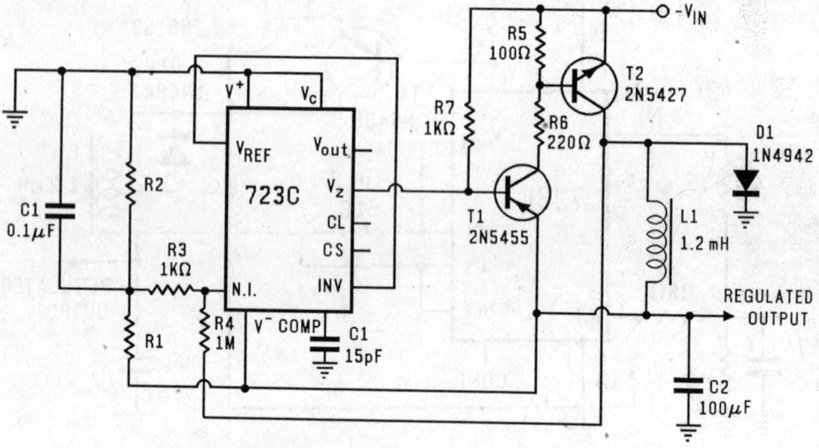

TYPICAL PERFORMANCE

Regulated Output Voltage	-15V
Line regulation (ΔV_{IN} = 20 V)	8 mV
Load Regulation (ΔI_L = 2A)	6 mV

Figure 18.3: Negative voltage switching regulator using 723.

Switch-Mode Voltage Regulators LM2575/LM2577 (National Semiconductor)

The devices in the LM2575 series are step-down regulators, and those in the LM2577 series are step-up regulators. Offering an efficiency of over 80%, these ICs are a good alternative to the 1-A regulators in the 78xx series. These step-up and step-down regulators are based on switch-mode voltage regulation, which gives them a higher efficiency (70-90%) than linear regulators (30-60%). By virtue of the on-chip power stages, the external component count remains low—basically, only a choke, a power diode and a few decoupling capacitors are required. The adjustable regulators in addition requires a voltage divider that determines the output voltage. All devices in the LM2575 and LM2577 series are available in a 5-pin TO-220 case, and a 4-pin TO-3 case. The case of these devices is always connected to ground. Fixed voltage as well as variable voltage devices are available—see Table 18.1.

The LM1575 and LM1577 should be used in heavy duty applications where an extended temperature range (T_j up to 150°C) is required.

The main technical data of the regulator family are summarized in Table 18.2.

Table 18.1: Overview of devices in LM2575/LM2577

TYPE	Output Voltage	TO 220 Series	TO3 Case
Step-down	5 V	LM2575T-5.0	LM2575K-5.0
	12 V	LM2575T-12	LM2575K-12
	15 V	LM2575T-15	LM2575K-15
	adjustable	LM2575T-ADJ	LM2575K-ADJ
Step-up	12 V	LM2577K-12	LM2577K-12
	15 V	LM2577K-15	LM2577K-15
	adjustable	LM2577K-ADJ	LM2577K-ADJ

Table 18.2: LM2575/LM2577 main technical data

Parameter		LM2575	LM2577
Input voltage	V_{IN}	3.5 – 35 V	3.5 – 40 V
Output voltage	V_{OUT}	1.23 – 30 V (always < V_{IN})	max. 60 V (always > V_{IN})
Output current	I_{OUT}	1 A	2 A (depends on VOUT)
Current limit (typ)	I_{CL}	2.2 A	4.3 A
Switch saturation voltage	V_{SAT}	0.9 V	0.5 V
Quiescent current	I_Q	10 mA	10 mA
Reference voltage	V_{FB}	1.23 V(± 2%)	1.23 V(± 2%)
Input current at V_{FB}	I_{FB}	50 nA	100 nA
Oscillator frequency	f_{OSC}	52 kHz (± 10%)	52 kHz (± 10%)
Efficiency		80%	80%
Operating temperature		-40°C < +125°C	-40°C < +125°C
Protectors		Internal current limit thermal shutdown	
Special features		ON/OFF input undervoltage detector	Soft start

Since the switch-mode regulators are based on a fixed oscillator frequency with a variable duty factor, they do not require the usual minimum load current to operate correctly. The ICs feature an internal thermal overload protection which is actuated at Tj = 125 °C, and an output current limiting circuit. A further advantage of these ICs is that they are inexpensive and uncritical of the choke type used.

Step-down Regulator LM2575

The operating principle of the so-called buck (step-down) regulator is illustrated in Figure 18.4. Switches S1 and S2 close in alternate fashion at a certain rate. When S1 is closed, the self inductance of the charge choke, L, causes a slowly rising current to flow. This current causes energy to be stored in the form of a magnetic field in the core of the choke. When the switches toggle (S1 is opened and S2 is closed) the self-inductance first causes a current I_L to be fed through L. As a result, the magnetic energy stored in the choke, and with it I_L, decreases gradually. This process is repeated as S1 and S2 close and open alternately. Capacitor C smoothes the output current, I_{OUT}, because it is charged as I_L rises, and discharged by the load as I_L drops again. By controlling the mark-space ratio of individual pulses, I_{OUT} can be set such that the desired output voltage V_{OUT} appears across the load.

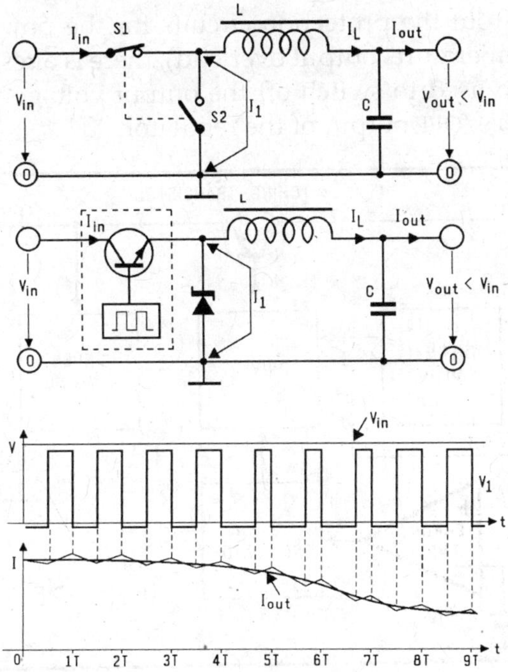

Figure 18.4: Basic operation of LM2575 step-down regulator.

The drawing in Figure 18.4 illustrates the process when the output current drops. Switch S1 is actually a power transistor controlled by a rectangular wave generator (Figure 18.4 middle). The other switch, S2, is formed by a diode that enables I_L to continue flowing when the power transistor is switched off. To prevent losses as a result of relatively high currents, the diode must have a fast switching characteristic. This requirement may be met by the use of a Schottky power diode.

The internal circuit of the step-down regulator Type LM2575 is shown in Figure 18.5. The power transistor is an n-p-n type that switches the input voltage, V_{IN}, to the output pin marked OUT. The transistor is driven by an oscillator with a fixed output frequency of 52 kHz. The pulse width of the oscillator output signals is controlled by an error amplifier, which compares the voltage supplied by a potential divider connected across the output with a fixed voltage supplied by an internal 1.23-V reference.

Apart from the protection circuits for the power transistor (junction temperature, output overload), there is a reset generator, which is also used to switch off the output voltage via the TTL-compatible ON/OFF output of the regulator.

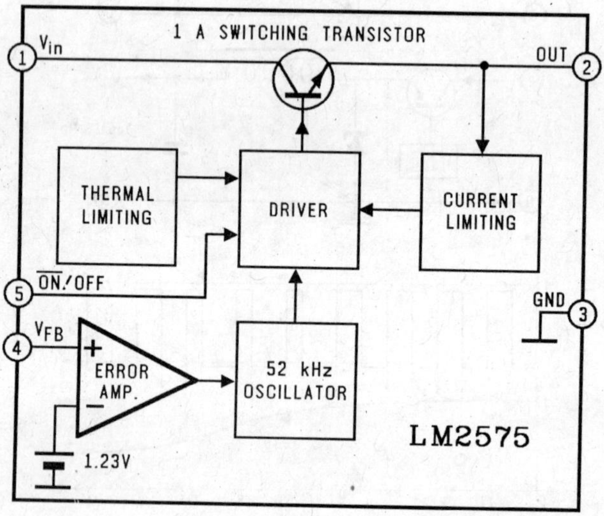

Figure 18.5: Internal schematic diagram of the LM2575.

A typical application circuit of the LM2575 is shown in Figure 18.6. Here, the adjustable device Type LM2575-ADJ is used. The output voltage is set with the aid of an external circuit. When a fixed-voltage device is used in this circuit, pin 4 must be connected to the output. A 3-A Schottky diode Type 1N5821 is used as the previously mentioned power diode, Components R2, D2 and C5 supply a voltage of +5 V for the ON/OFF control input, whose function may be tested with the aid of jumper J1. When the ON/OFF function is not required, the 5-V supply and the jumper may be omitted, and pin 5 connected permanently to ground.

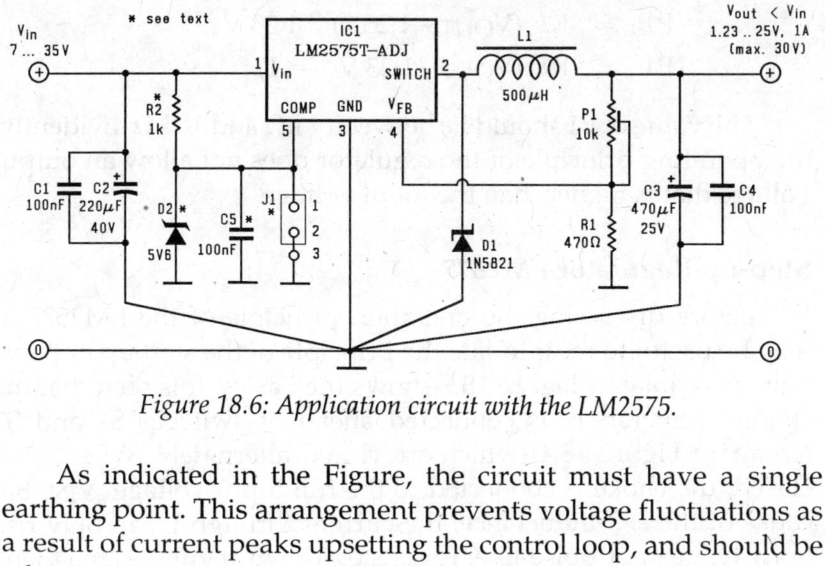

Figure 18.6: Application circuit with the LM2575.

As indicated in the Figure, the circuit must have a single earthing point. This arrangement prevents voltage fluctuations as a result of current peaks upsetting the control loop, and should be taken into account when a PCB is designed for the supply. Spurious pulses are suppressed by electrolytic capacitors C2 and C3 at the input and output.of the regulator. In some cases, the a.c. resistance of C3 may be too high, when the capacitor is best replaced by a number of capacitors in parallel with the same total capacitance (small electrolytic capacitors have a lower parasitic inductance than larger ones). The value of capacitor C3 must always be greater than 330 µF. The data sheet advises against the use of tantalum capacitors,w which cause instability of the control loop. When the output voltage is higher than 25 V, C3 must be replaced by at a type with a working voltage of 40 V or more. Capacitors C1 and C4 form a virtual short-circuit for the fast pulse edges, and so secure the necessary r.f. suppression.

The output voltage, V_{OUT}, of the regulator is simple to calculate. The equation for the current, I, allows us to determine the values of P1 and R1.

$$I = 1.23\ V / R1 = (V_{OUT} - 1.23V) / P$$

This allows the value of P1 to be determined when R1 is known.:

$$P1 = R1\ (V_{OUT} - 1.23V) / 1.23\ V$$
$$P1 = R1\ (V_{OUT} / 1.23\ V - 1)$$

The value of P1 should lie between 1 kΩ and 10 kΩ. Evidently, the operating principle of the regulator does not allow an output voltage that is higher than the input voltage.

Step-up Regulator LM2577

Before discussing the operation principle of the LM2577, it may be useful to recapitulate the principle of the step-up or boost voltage regulator. Figure 18.7 shows the basics. It is seen that the charge inductor, L, is connected ahead of switches S1 and S2 (compare Figure 18.4) which are closed alternately. When S2 is closed, the choke is connected to the full input voltage, V_{IN}. Because of the self-inductance, the current through L can only rise linearly, i.e., not pulse like. As in step-down regulator, magnetic energy is stored in the core of the choke. As soon as S2 is opened and S1 is closed, the self-inductance of the choke forces the current I_L to keep flowing in the same direction. The voltage developed by the self-inductance is polarized such that it is added to the input voltage. Hence, the output voltage is higher than the input voltage. Capacitor C smoothes the output current I_{OUT}. In the LM2577 (Figure 18.7), switch S2 is formed by a power transistor, and S1 by a fast switching diode. The diode is reverse biased when the transistor conducts. The voltage and current waveforms that occur when the output current is reduced are shown in Figure 18.7.

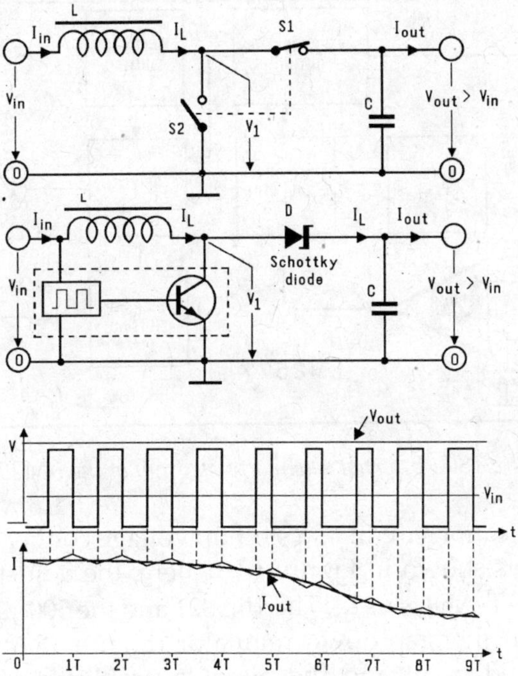

Figure 18.7: Basic operation of the LM2577 step-up regulator.

Figure 18.8 gives the block diagram of the LM2577. Like the LM2575, the LM2577 contains a 52-kHz pulse-width modulated oscillator driven by an error amplifier. The error amplifier obtains its control information from the voltage at the V_{FB} input.

In addition to the thermal and current overload protection circuits, the LM2577 features a soft-start generator that prevents a high rush-in current through choke L.

Another preventive function is that provided by the under-voltage limit circuit. When the input voltage drops below 2.9 V (typ.), the power stage is automatically switched off to prevent it from conducting almost continuously. When the under-voltage detector is actuated, the output voltage drops to virtually the input voltage.

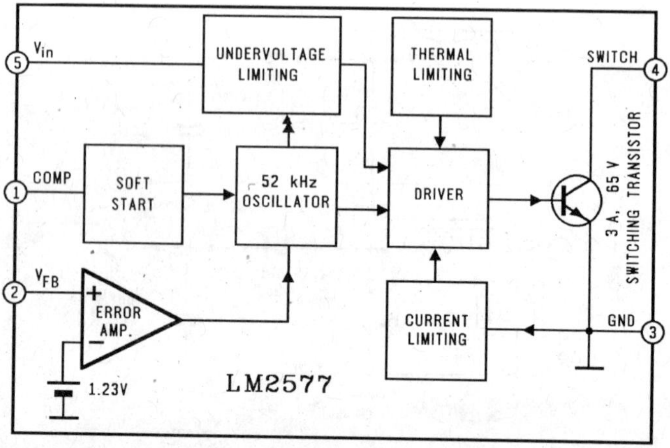

Figure 18.8: Internal schematic diagram of the LM2577.

An application circuit of a step-up voltage converter based on the LM2577 is shown in Figure 18.9. Here, the adjustable device Type LM2577T-ADJ is used. The 1N5821 and the 500-µH choke are familiar from the step-down regulator (Figure 18.6), as are the central ground point and the possible problems with a single electrolytic capacitor in position C3.

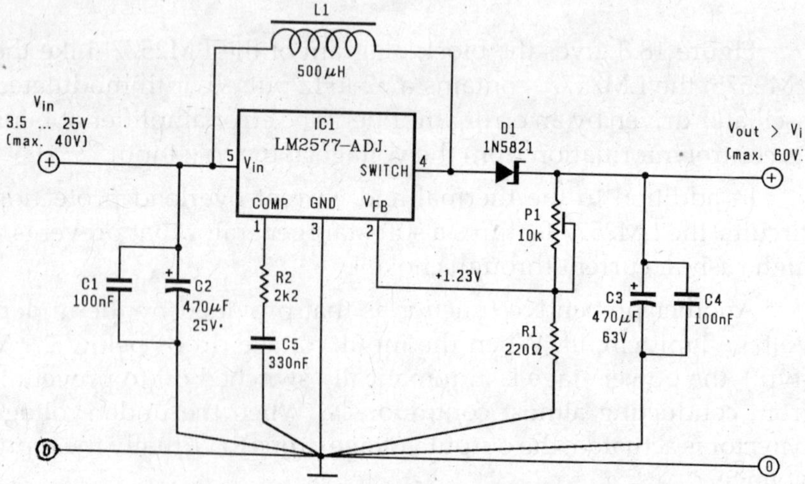

Figure 18.9: Application circuit with the LM2577.

Since the same reference voltage of 1.23 V is used, the calculation of the output voltage is identical to that shown for the LM2575. It should be noted, though, that the circuit in Figure 18.6 cannot supply output voltage lower than the input voltage. The maximum output current depends on the properties of choke L1, the ratio (V_{OUT}/V_{IN}), and the maximum permissible current through the power transistor in the LM2577 (4.3 A typ.). The data sheet provides the following information on I_{OUT} and V_{OUT}:

V_{OUT} < 60 V

V_{OUT} < 10 V_{IN}

I_{OUT} < 2.1 A V_{IN}/V_{OUT}

Also, when the output voltage is higher than 30 V, the 1N5821 should be replaced by a type with a higher rated voltage.

LM2574 Switch Mode Power Supply

When a small, reliable and inexpensive switch-mode power supply is needed, National Semiconductor's LM257x series of switch-mode power-supply controller ICs have several advantages over competitive products. The circuit shown in Figure 18.10 largely follows the NS recommended application configuration. The only time-consuming part is inductor L1. It is a triac suppressor coil whose self-inductance depends on the output voltage and maximum anticipated output current. The required inductance may be found on the graph in Figure 10.11. Most triac suppressor coils have an inductance of about 100 μH. Since the self-inductance of a coil is roughly proportional to the square of the number of turns, a 100 μH coil may be modified into a 470 μH version by multiplying its number of turns with $\sqrt{(470.100)} = 2.17$. The new winding may be of thinner enamelled wire to ensure that it fits in.

The second LC circuit behind the regulator L2-C5 provides additional ripple suppression, It a ripple voltage of a few millivolts is acceptable, L2 may be replaced by a wire link.

The output voltage is set with preset P1, where

$$P1 = R1[(V_O/1.23) - 1]$$

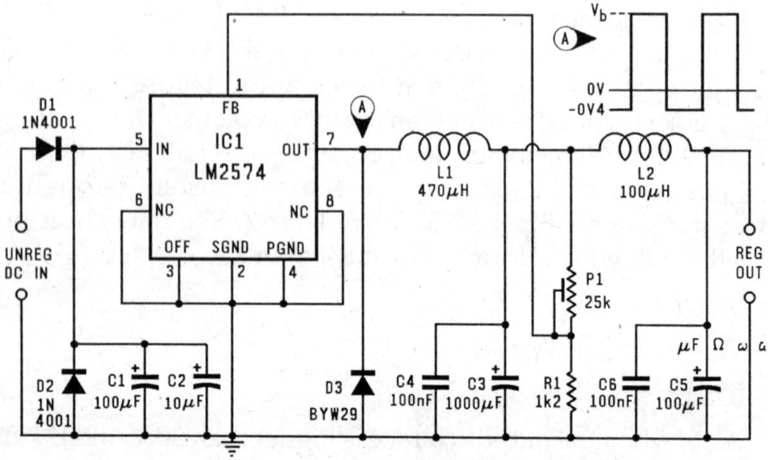

Figure 18.10: LM2574 switch-mode power supply.

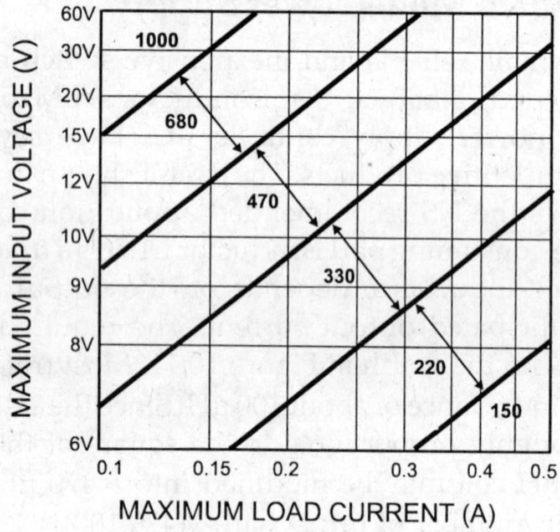

Figure 18.11: Value of inductance L1 for Figure 18.10.

Diode D1 at the input provides polarity reversal protection. If D1 is fitted, D2 is not required. However, if the voltage across D1 becomes unacceptably large, replace the diode by a wire link and fit D2 which will present a virtual short circuit to reverse input voltages.

Switched 5 V Power Supply

National Semiconductor's LM1575 and LM2575 are switched 5 V regulators. They are ideal for making +5V switched regulated supplies. They need only a choke, a freewheeling diode and a couple of electrolytic capacitors to meet your requirements. They provide power conversion with high efficiency. The only thing that is still not as good as in linear power supplies is the ripple.

The diode used in the circuit in Figure 18.12 is 1N5822 but a BYW29 or any fast diode will do, however, but do not use a common rectifier type 1N4001 etc.

The choke must be able to handle currents of at least 1 A.

The input voltage can lie between 8 V and 32 V. The maximum current the regulator can provide is about 1 A.

Pin 4 is the feedback input. Pin 5 may be used for switching the supply on/off. If that is wanted, connect this pin to a 2.4–15 V potential instead of to earth.

Build this supply with a central, star-shaped common earthing point as indicated in the diagram. Use good quality electrolytic capacitors (perhaps two in parallel). The more compact you keep the supply, the better it will work.

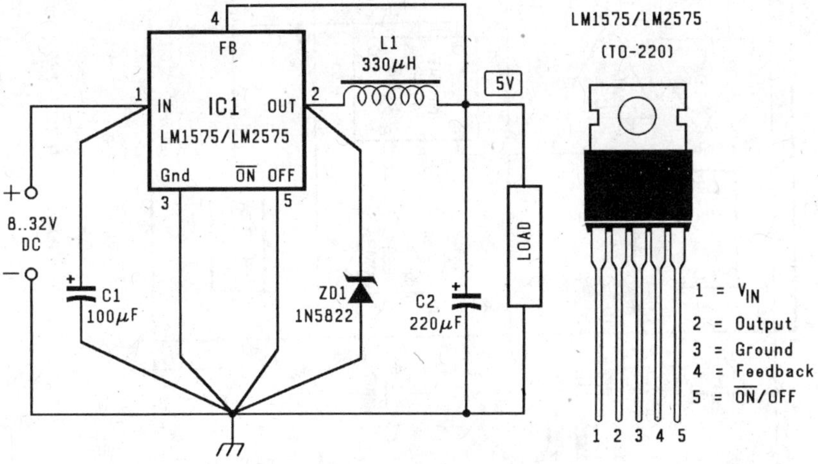

Figure 18.12: Switched 5 V supply.

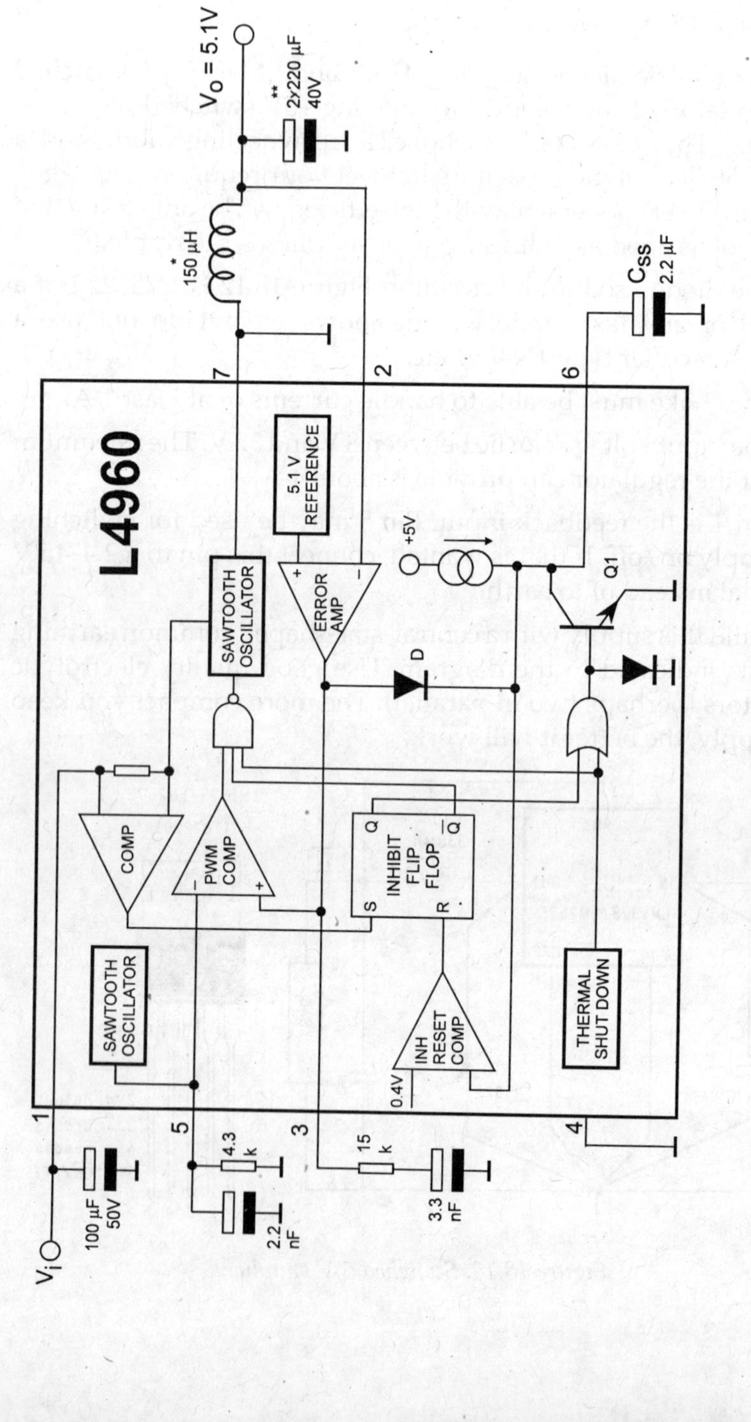

Figure 18.13: Switching regulator, 2.5 A, with current limiter and protection functions.

High Current Switching Regulator L4960

The SGS 4960 high current switching regulator IC offers all the advantages of a switching regulator yet is little more complex to use than a linear regulator. A 2.5A/5V regulator can be built with one L4960 and just eight components as shown in Figure 18.13. This diagram also shows a block diagram of the inside circuitry of the IC as well. Higher output voltages are obtained by adding two resistors. Moreover, the device includes current limiting and thermal protection circuits, eliminating the need to add extra circuitry.

A further advantage of L4960's source-sink output stage can switch in about 70 ns, allowing efficient operation (up to 90%) at switching frequencies of 100 kHz. The IC itself is assembled in Heptawatt 7-lead package—both horizontal and vertical mounting versions are available—and requires only a small heat sink. A versatile device, the L4890 may be used in a number of different ways.

The most obvious is a basic dc-to-dc converter configurations where a 50/60 Hz transformer, rectifier bridge and filter capacitor feed the input of the device with an unstabilized voltage.

How it Works?

The SGS L4960 is a monolithic step-down regulator providing output voltages from 5.1 V to 40 V, and delivering up to 2.5 A of output current.

At the heart of the device is a regulation loop consisting of a sawtooth oscillator, error amplifier, comparator and a source-sink output stage. An error signal is produced by comparing the output voltage with a precise 5.1 V on-chip reference which zener diode trimmed to ± 2%. This error signal is than compared with the sawtooth signal to generate the fixed frequency pulse-width modulated pulses which drive the output stage. Gain and frequency of the output loop are adjusted by an RC network connected to pin 3.

When the loop is closed directly by connecting the supply output to feedback input (pin 2) an output voltage of 5.1 V is produced. Higher output voltages are obtained by inserting a voltage divider in this feedback path.

Output overcurrents at switch-on are prevented by the soft-start function. The error amplifier output is initially clamped by the external capacitor C_{SS} and allowed to rise, linearly, as this capacitor is charged by a constant current source.

Output overload protection is provided in the form of a current limiter. The load current is sensed by an internal metal resistor connected to a comparator. When the load current exceeds a preset threshold, this comparator sets a flip flop which disables the output stage and discharges the soft start capacitor.

A second comparator resets the flip flop when the voltage across the soft-start capacitor has fallen to 0.4 V. The output stage is thus re-enabled and the output voltage rises under control of the soft-start network. If the overload conditions still present, the limiter will trigger again when the threshold current is reached.

The average short-circuit current is limited to a safe value by the dead time introduced in the soft-start network.

The thermal overload circuit disables circuit operation when the junction temperature reaches about 150°C and has hysteresis to prevent instability.

Multiple Outputs

A number of devices may be combined to produce a multiple output supply, permitting a building block approach in supply design. In multi-chip supplies, it is desirable to synchronize the switching frequencies and this can be done by connecting the oscillator pins to one RC network as shown in Figure 18.14.

Very often, low current auxiliary outputs such as ± 12 V are required. In this case it may be necessary to use several devices since an auxiliary output can be produced by adding an extra winding on the output inductor as shown in Figure 18.15. In this example the secondary winding is not isolated—the bottom end is at +5 V. This means that fewer turns are necessary than if it were isolated, and load regulation is improved.

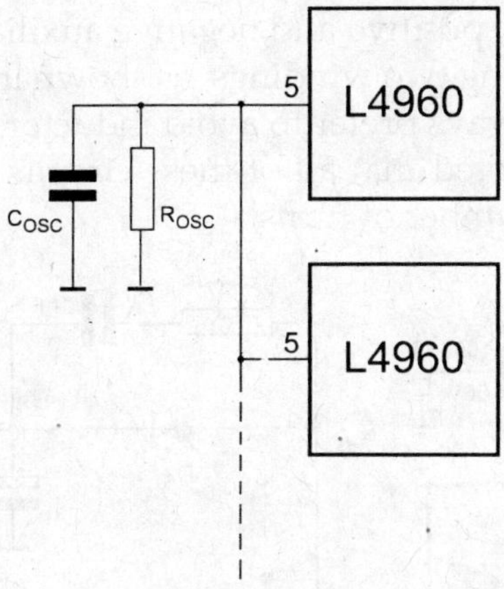

Figure 18.14: Synchronizing switching frequencies.

Circuits of this type have advantage of low cost and simplicity, and yield satisfactory performance as long as the power drain on the auxiliary output is no more than about 20% of the power delivered by the main output.

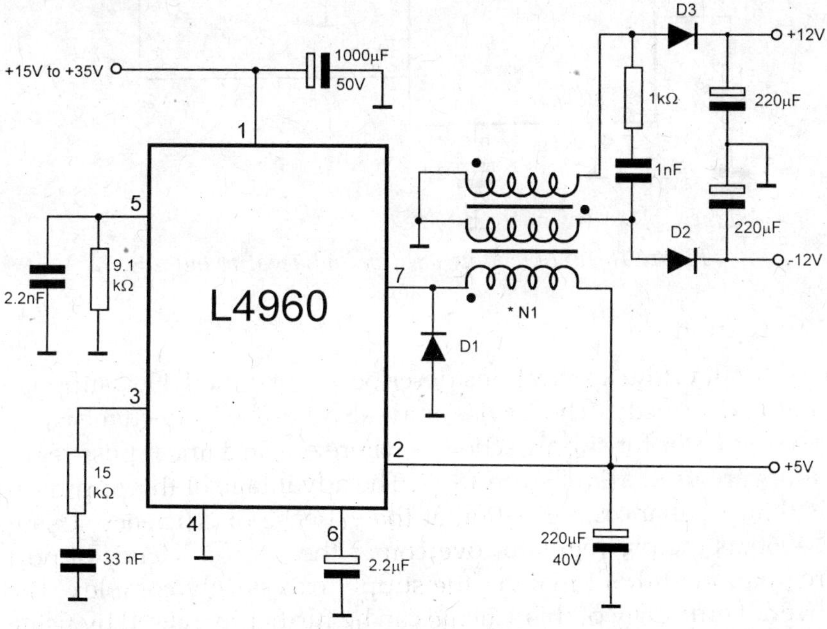

Figure 18.15: Getting auxiliary outputs from a switched regulator.

Extending this principle, positive and negative auxiliary outputs can be obtained by adding two windings, as shown in Figure 18.16. Since the designers always prefer to avoid inductors whenever possible it should be noted that all of these circuits require small toroids with a small number of turns.

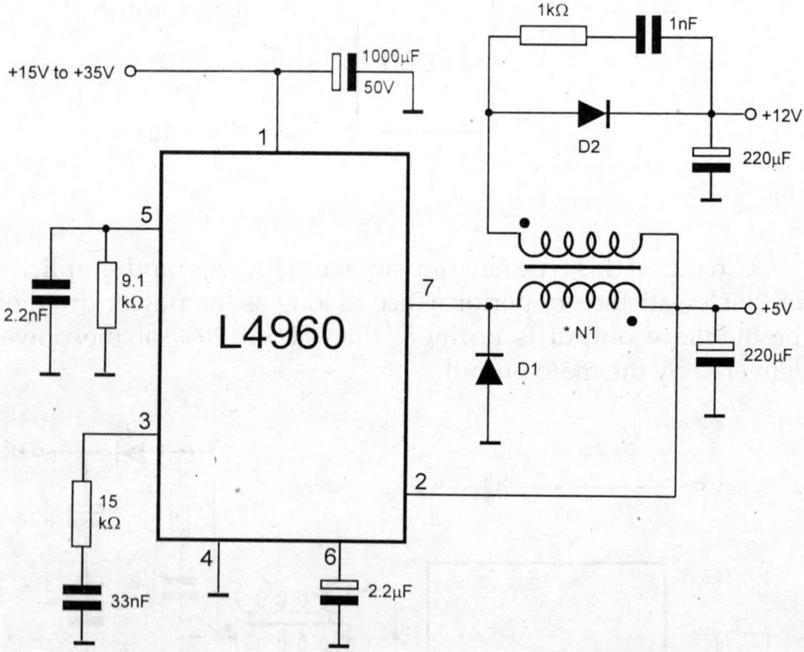

Figure 18.16: Extending positive and negative outputs.

Pre-regulation

In all of the applications described above the L4960 supplies the load directly. The device may also be used effectively as a pre-regulator for supply schemes where on-card linear post-regulators are used as in Figure 18.17. The advantage of this approach is that it enhances regulation at the expense of efficiency. Using L4960 as the pre-regulator overcomes the poor efficiency of post regulation without making the supply excessively complex. The overall efficiency of this scheme can be further increased by using new-generation low-drop linear regulators for final regulation.

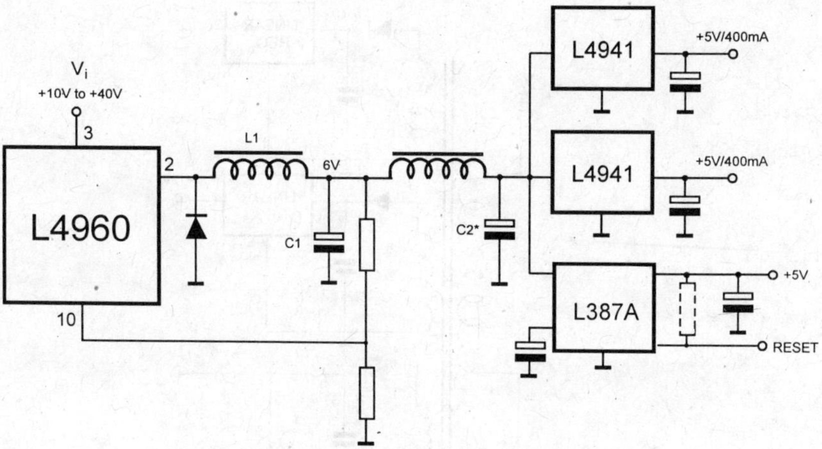

Figure 18.17: Using L4960 as a pre-regulator.

Off-line Switching Supplies

Another important application is where the L4950 is used to regulate auxiliary outputs in high power supplies where the mains transformer has been replaced by an off-line switching regulator to reduce size and weight.

To understand the advantage of using switching regulator chips in this application it is necessary to examine the drawbacks of the conventional circuit, illustrated in Figure 18.18. In this example, a feedback circuit guarantees the necessary precision for the main 5 V output while conventional linear regulators stabilize the other outputs. A major drawback of this supply design is the cost of the transformer. A separate secondary is needed for each output and each must be optimized to obtain a low drop-out voltage, otherwise power dissipation will be excessive.

The linear regulators used in these supplies are either ICs (for currents up to 2-3A) or discrete circuits. Either way, power dissipation becomes a problem in overload and short-circuits conditions since the current limiters are of the constant-current type. It is therefore necessary to complicate the circuit further by adding thermal protection circuits or to use a large, and costly heat sink.

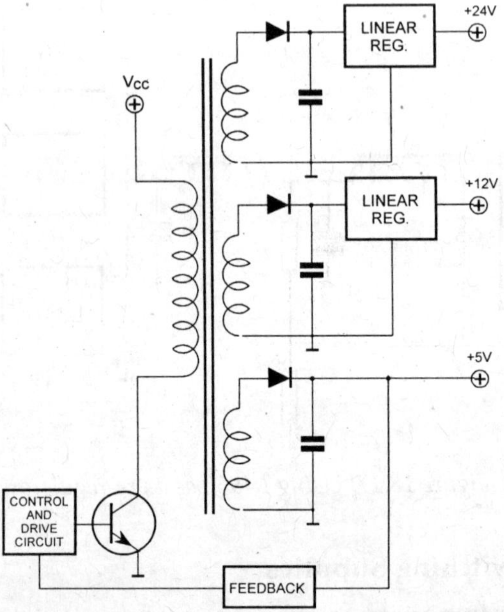

Figure 18.18: A conventional off-line switching supply.

A final consideration concerns the rectifier diodes between the secondary windings (one for each output voltage) and the filter capacitors. With a linear post regulators, the input current is always equal to the output current so the diodes must be dimensioned accordingly.

All of these problems are eliminated by using the L4960 as a post regulator as shown in Figure 18.19. Note that for all the auxiliary outputs only one secondary is needed, simplifying the transformer. Cross regulation is no longer a problem and the power dissipated in the stage depends mainly on load current and is almost unaffected by dropout. Moreover, the diode on the secondary can be smaller since the input current of a stepdown switching regulator is always less than the output current.

Finally, short-circuit protection is provided for all the auxiliary outputs by the chip's internal current limiter and internal protection circuit.

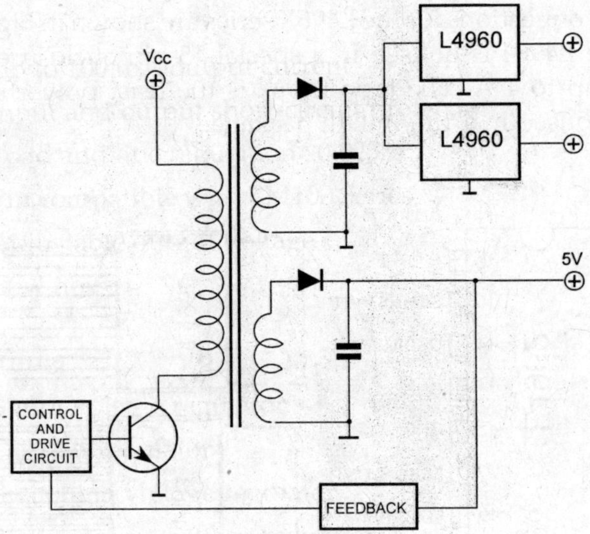

Figure 18.19: Using DC-DC converter circuits in place of linear regulators.

Switch Mode Power Supply

The power supply circuit described here is based around the integrated switch mode power supply controller type L4960 from SGS. This IC has the following features:

- ☐ Input voltage range : 9 — 50 V d.c.
- ☐ Output voltage adjustable between 5 and 40 volts
- ☐ Maximum output current : 2.5 A
- ☐ Maximum output power : 100 W
- ☐ Built-in soft-start circuitry
- ☐ Stability of internal reference : ± 4%
- ☐ Requires very few external components
- ☐ Duty factor 0—1
- ☐ High efficiency : up to 90%
- ☐ Built-in thermal overload protection
- ☐ Built-in current limiter for short-circuit protection.

Pin connections for the L496X series are shown in Figure 18.20. The type L4964 is housed in a special 15-pin enclosure, and can supply up to 4 A. The type L4962 is the least powerful version (1.5A), housed in a 16-pin DIL package.

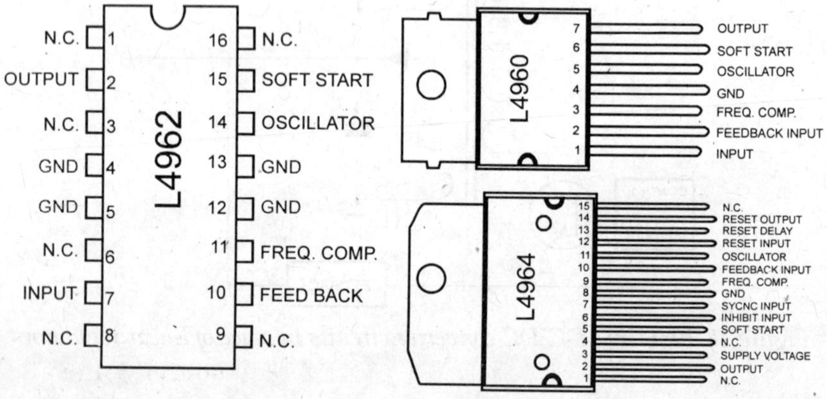

Figure 18.20: Pin connections of L496x series of integrated circuits.

The internal organization of the Type L4960 has been discussed with reference to Figure 18.13. The operation of the on-chip soft-start circuitry, and the current limiter, is illustrated by waveform diagrams in Figure 18.21a and 18.21b respectively. The thermal shut down circuit in the L4960 is activated when the chip junction temperature exceeds 150°C

For the sake of safety, the proposed SMPS unit is a transformer based design. The alternating input voltage to the board is obtained from the secondary of a mains transformer, which ensures that the d.c. input to the regulator is at least 3 V higher than the required output voltage at the maximum output current. It goes without saying that the transformer is preferably a toroidal type. For more information on toroidal transformers refer to Author's book "Designing Coils and Transformers" published by BPB Publications.

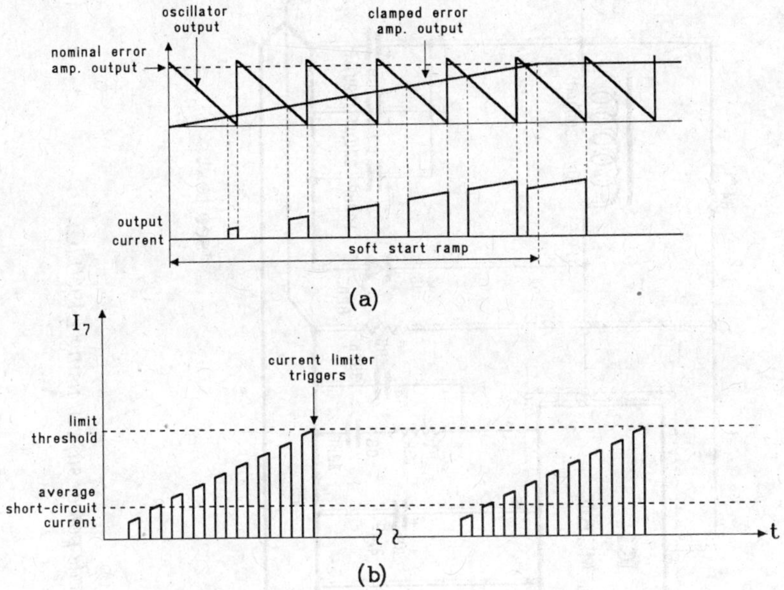

Figure 18.21: Waveforms for soft starting and current limiting operations.

Circuit Description

Figure 18.22 shows the circuit diagram of the mains section and the d.c. power supply. The alternating voltage from the secondary winding is applied to the respective inputs on the supply board, while the center tap is connected to ground. The unregulated input voltage V_{in} for the L4960 is supplied by a full-wave rectifier circuit composed of two 3 A diodes Type 1N5404, D1 and D2 and an electrolytic reservoir capacitor C1. Network R1-C3-C4 defines the gain of the closed regulation loop. A second network, C2-R2, is dimensioned for an oscillator frequency of about 100 kHz. The function of capacitor C5 is twofold: it defines the period of the soft start ramp (see Figure 18.21) as well as the average short circuit current. The feedback input of the regulator is connected to the junction of the output voltage divider R3-R4.

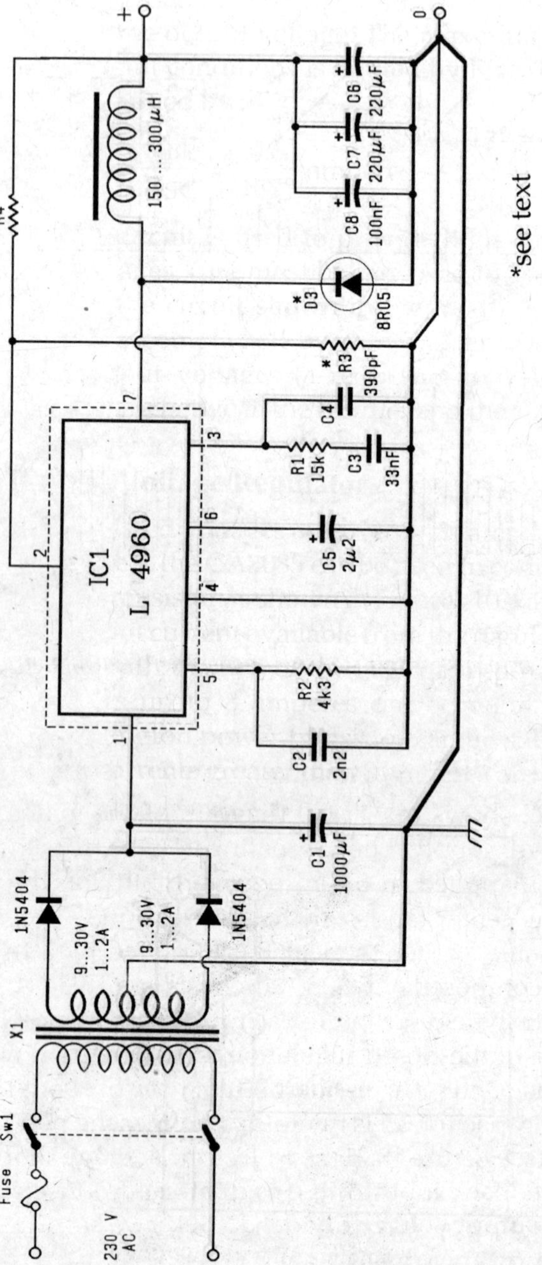

Figure 18.22: Mains section and switch mode power supply using L4960 IC.

The output voltage Vout of the L4960 is calculated from

$$V_{out} = 5.1\ [(R3 + R4)/R3]$$

assuming that $V_{in} > V_{out}$ by at least 3 V.

Remember that the minimum value of Vin is 9 V. A fixed output voltage of 5.1 V(± 4%) is obtained when R3 is omitted and R4 is replaced with a wire link. When R3 has the fixed value of 5k6, R4 alone determines the output voltage.

V_{out} = 9 V; R4 = 4k28

V_{out} = 12 V; R4 = 7k58

V_{out} = 15 V; R4 = 10k87

V_{out} = 18 V; R4 = 14k16

V_{out} = 24 V; R4 = 20k75

These resistor values are, of course, theoretical, and require ascertaining in practice. The output voltage is readily made variable by fitting R3 = 6k8 and replacing R4 with a 25k potentiometer. Power diode D3 is included as a safety measure. This fast rectifier limits the negative potential at the input of choke to a safe -0.6 V to -1 V when the output transistor in the regulator is switched off. In the absence of D3, the voltage at pin 7 would increase dangerously to several volts below the ground potential. Choke L1 is an essential part in the L-C network for noise and ripple suppression.

Testing

It should be noted that the supply requires a load at all times for proper operation. With 30 V a.c. input and loaded with 2 A at an output voltage of 5V, the temperature of the heat sink should not exceed about 60 °C at room temperature.

1.5 A Step-down Switching Regulator

Step-down regulators are frequently used to derive a low voltage from a high voltage without incurring losses. Owing to steadily increasing switching rates, the suppression of interference and noise that are unavoidable by-products of switching circuits has become much simpler.

The L4971 (from SGS Thompson) is a step-down monolithic power switching regulator delivering 1.5 A at a voltage between +3.3 V to +40 V (selected by a simple external divider). The regulator is housed in a DIL8 or SO16-SMD enclosure. A block diagram of the internal working of this integrated circuit is shown in Figure 18.23, and a practical circuit is shown in Figure 18.24.

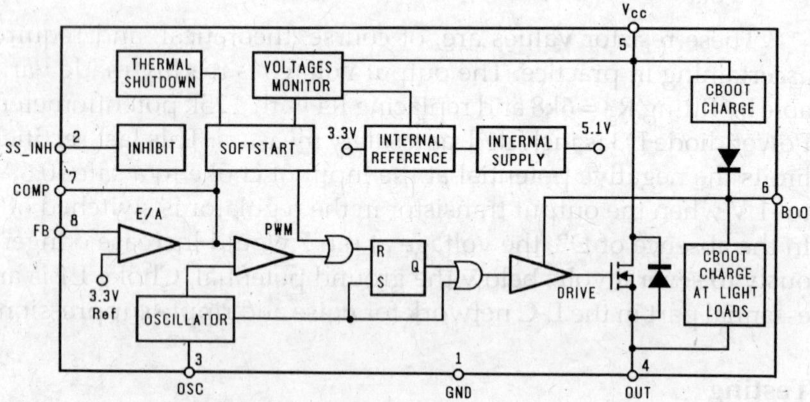

Figure 18.23: Block diagram of L4971 switching regulator.

The input voltage ranges from 8 V to 55 V, while the efficiency is 85%, rising to 95% when the input voltage is only a few volts higher than the output voltage.

The external divider to set the output voltage is formed by R3-R4. For an output of 3.3 V, V_{out} may be linked directly to FB (pin 8). Resistor values for some commonly needed output voltages are given in the Table given on the following page.

Values of resistors for different output voltages.		
Vout (V)	R3 (kΩ)	R4 (kΩ)
3.3	0	-
5.1	2.7	4.4
9.0	8.2	4.7
12	12	4.7
15	16	4.7
18	20	4.7
24	30	4.7

The normal switching frequency of the L4971 is 100 kHz, but this may be increased to 500 kHz by making R1 = 12 kΩ and C2 = 0.001 μF.

The soft start/inhibit input (SS_INH pin 2) may be used with an open collector output to inhibit the controller or capacitor C5 for the soft start function.

Other features include pulse-by-pulse current limit, hiccup mode for short-circuit protection, voltage feed forward regulation, protection against feedback loop disconnection, inhibit for zero current drain and thermal shut down.

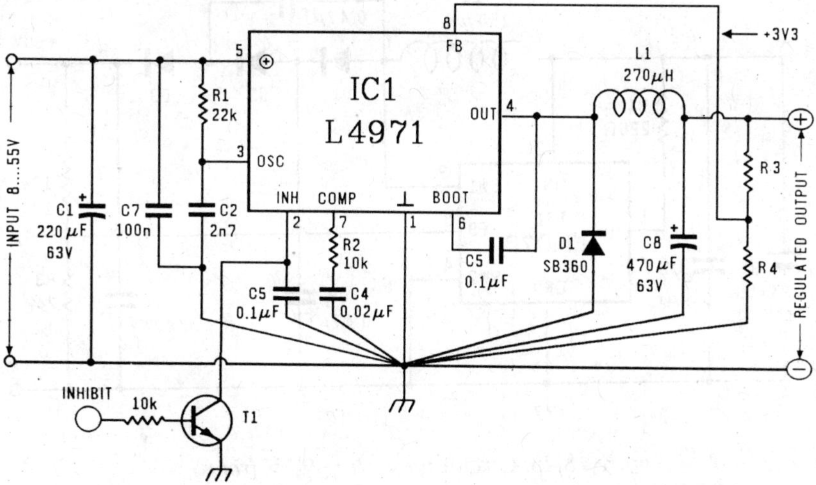

Figure 18.24: Step-down switching regulator using L4971.

High Voltage Converter: 90 V from 1.5 V

The circuit shown in Figure 18.25 shows one way of obtaining a voltage of 90 V from a 1.5 V battery supply.

The LT1073 switching regulator (Linear Technology) operates in boost mode and can work with an input voltage as low as 1.0 V. The switching transistor, which is hidden behind connections SW1 and SW2, briefly takes one end of choke to ground. A magnetic field builds up in the choke, which collapses when the transistor stops conducting. This produces a current in diode D1 which charges C3. The diode cascade comprising D1, D2, D3, C2, C3 and C4 multiplies the output voltage of the regulator (see Figure 7.13) by four, the pumping of C2 causing the voltage developed across C4 via C3, D2 and D3 to rise.

Finally, the regulator control loop is closed via the potential divider (10 MΩ and 24 kΩ). These resistors should be 1% tolerance metal film types. With the given component values, fast diodes with a reverse voltage of 200 V and a choke an output voltage of 90 V will be obtained. The output of the circuit can deliver a few milliamperes of current.

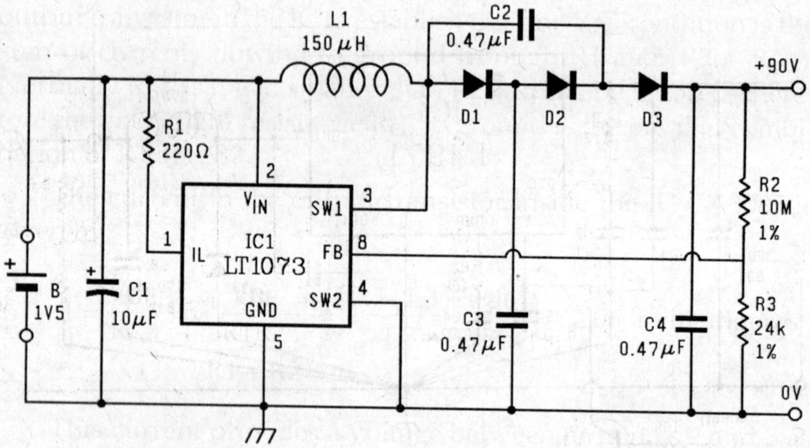

Figure 18.25: Circuit for getting 90 V from 1.5 V

12V to 3V Adapter

The 3 V car adapter shown in Figure 18.26 provides an output of 3V from a car battery of 12 V. This output is useful to operate devices such as games or music players that operate on 3V. Such gadgets are useful for kids or other passengers on the rear seat during a long drive.

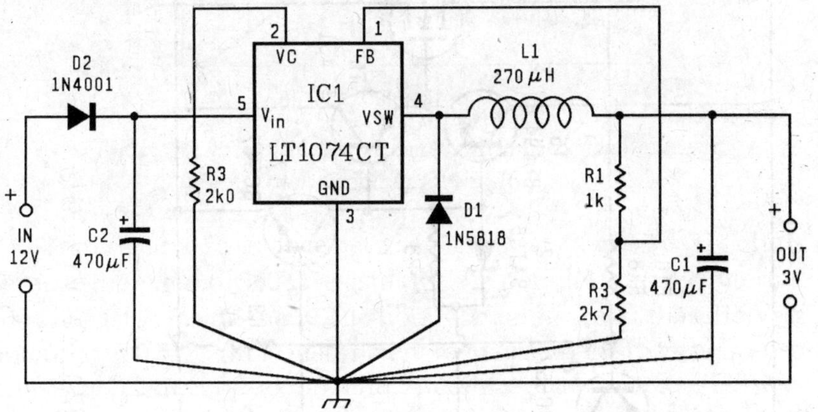

Figure 18.26: Circuit for getting 90 V from 1.5 V

The circuit is based on a standard LT1074CT switching regulator IC which is used here as a down converter. Diode D2 provides a reverse voltage protection while C2 decouples the input to the switcher IC. The IC briskly switches the supply voltage on and off in response to the input signal applied to its FB input, to the extent that the average output voltage is at the required level. The values of potential divider R1 - R3 have been chosen to attenuate the output voltage to that there is 2.5 V at the FB pin. The difference between the attnuated output and the internal 2.5-V reference is used to control the modulation effect of the switcher. Inductor L1 along with the IC form the main switching components while C1 provides decoupling for the output load. Resistor R2 provides frequency stabilization.

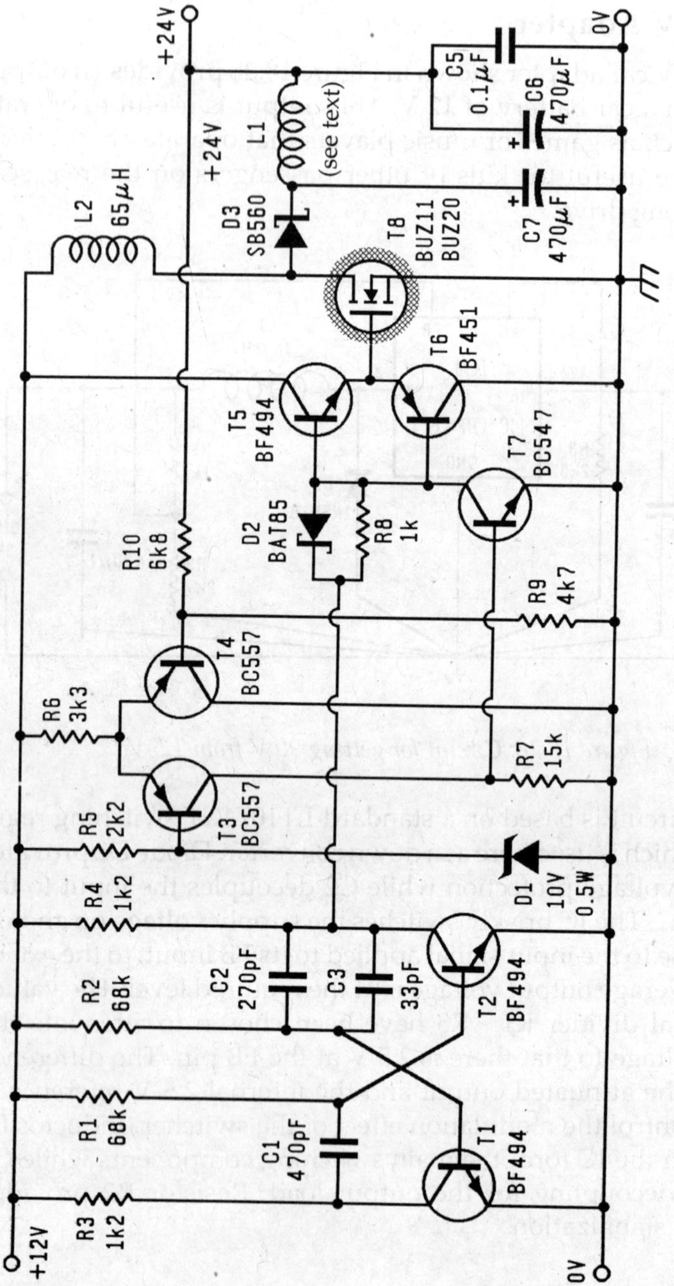

Figure 18.27: A 12-V to 24-V Converter.

12 V to 24 V Converter

The dc-to-dc converter shown in Figure 18.27 delivers a maximum power of about 36 watts at an efficiency of about 90%. Apart from a power FET and a Schottky diode, this circuit is comprised entirely of familiar and inexpensive parts. In spite of this, the specifications are excellent:

Efficiency	approx. 90 %
Ripple voltage	max. 10 mV
Output current	max. 1.5 A
Switching frequency	40 kHz
Input voltage	12 V
Output voltage	24 V regulated

The switching element is a fast power FET (T8). This FET has a relatively high input capacitance and is switched on and off by a push-pull stage consisting of two r.f. transistors (T5 and T6). Schottky diode D2 increases turn-off speed even further, which is crucial here because we are aiming to get the highest possible efficiency.

The switching signal is provided by a simple multivibrator, which is also made from two r.f. transistors (T1 and T2). Differential amplifier (T3/T4) has been added to get a regulated output voltage of 24 V.

L2 is a 5 A suppressor choke with a self inductance of 65 μH. L1 is a part of the output filter, the purpose of which is is to eliminate r.f. noise. This is an air core coil which you can make by winding 25 turns of 0.5 mm dia. enamelled copper wire around a 10 mm diameter drill. Because of the high efficiency, the dissipation of T8 remains smaller than about 3.6 W, so a modest heat sink of 10K/W will suffice. It is advisable that the 12 V input supply includes a fast fuse, rated about 3.5 A.

Considering that the duty cycle has a substantial effect on the efficiency, a second capacitor (C3) has been added in parallel with C2. The optimum setting can be determined by varying this additional capacitor.

The remaining components are not at all critical. Any 5 A suppresser choke will work for L2, any Schottky diode for D3 and just about any power MOSFET for T8 (BUZ10, BUZ20, BUZ100).

Parts List

Capacitors

C1	470 pF
C2	270 pF
C3	33 pF
C4	100 nF
C5	1000 μF 16 V
C6, C7	470 μF 35 V

Resistors

R1, R2	68 kΩ
R3, R4	1kΩ2
R5	2kΩ2
R6	3kΩ3
R7	15 kΩ
R8	1 kΩ
R10	6 kΩ8

Inductors

L1	25 turns 0.5 ma dia enamelled copper wire, 10 mm dia. no core
L2	65 μH/5A suppressor coil (ring core)

Semiconductors

D1	zener diode 10 V, 500 mW
D2	BAT85
D3	SB650
T1, T2, T5	BF494
T3, T4	BC557
T6	BF450 (BF451)
T7	BC547
T8	BUZ11 (BUZ20)

19

SCR Controlled Power Supplies

The use of SCRs (Silicon Controlled Rectifiers) for controlling a.c power was explained in Chapter 4. Although it is posible to design circuits to delay the phase angle for firing an SCR by using discrete components only, it is better to use ready-made integrated circuits designed for this purpose. That makes the design simple, easy and reliable.

L-120 AB Phase Controller

Electrical Characteristics

Power supply	12 V
Power dissipation P_D	800 mW
Temperature range	20°C — 100°C
Function	Phase control zero-crossing trigger zero voltage switch/ ramp generator
Outline	DIP 16 pins (see Figure 19.1)

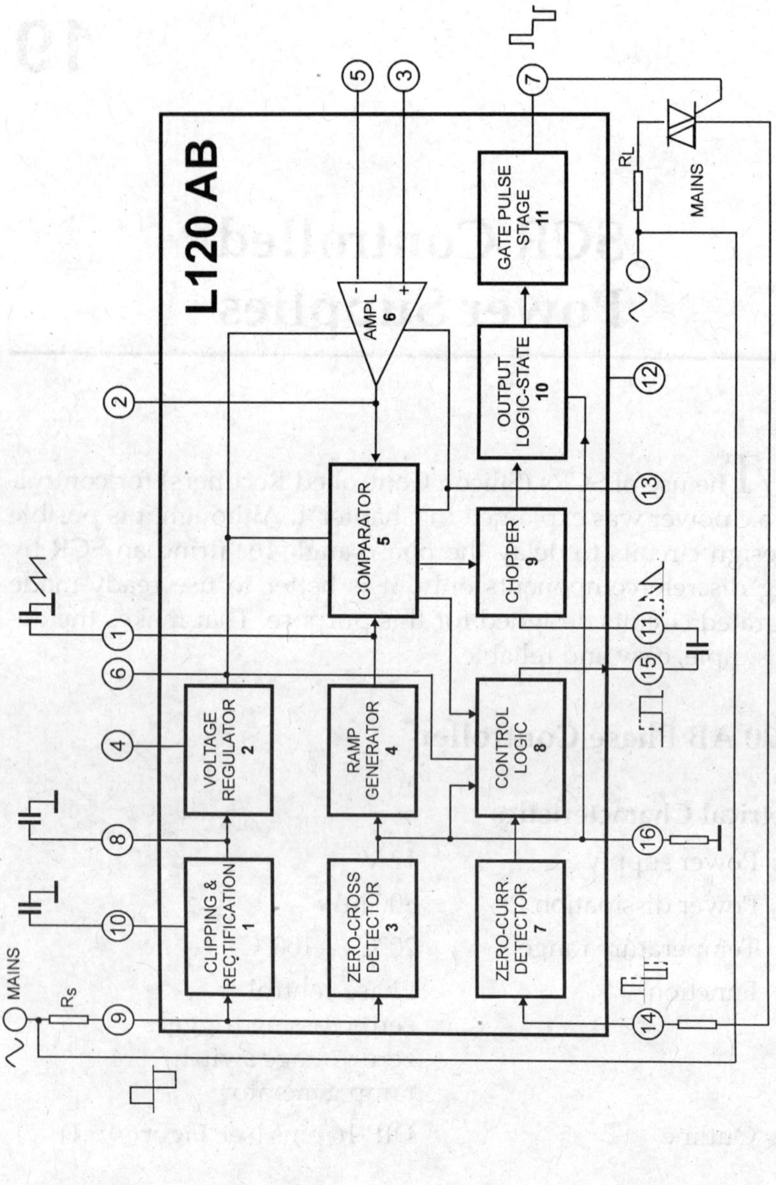

Figure 19.1: Block diagram and pin connections for L-120 phase controller.

Mains Power Supply With Primary Regulation

Figure 19.2 shows an example of a practical circuit using control of power to the primary of a transformer to maintain a regulated d.c. output. Since this reduces the heat lost in the regulating element IC2, the overall efficiency is increased.

In this regulated power supply, the d.c. output is regulated by IC2 (LM317). The output voltage can be varied from 1.2 volts to 25 volts, and the current can be 1.5 A at any of these voltages, provided IC2 is mounted on a suitable heat sink. The heat dissipated in the IC is minimized by primary regulation.

Another advantage of primary regulation is that the power supply is protected against variations in the mains supply. This aspect is normally ignored with secondary regulation, so it is assumed that the primary fluctuations have no effect on the secondary regulation. The present circuit is therefore of particular importance for use where the mains supply is subject to large variations.

The regulation functions so that the voltage drop across the regulator IC2 is held constant. This voltage is transferred by the current source T1 into a current through the LED in the opto-coupler. When the voltage drop diminishes, the current through the LED is smaller. The transistor in the opto-coupler gets less drive, and the voltage at pin 3 of IC1 drops. Voltage regulator IC1 contains a complete circuit for phase gating control with Triac Tri1. The gating angle of this triac depends on the comparison between the direct voltage at pin 3 and an on-chip generated sawtooth signal, the frequency of which is determined by capacitor C1, (= 100 nF). In this example, the triac switches the mains voltage earlier so that buffer capacitor C2 receives more energy.

Noise caused by the phase gating circuit must be prevented of entering the mains supply by a mains noise filter. This is necessary because it can cause false triggering. In Figure 19.2, inductors L1, L2, L3, L4 and capacitor C9 form the noise filter for this purpose.

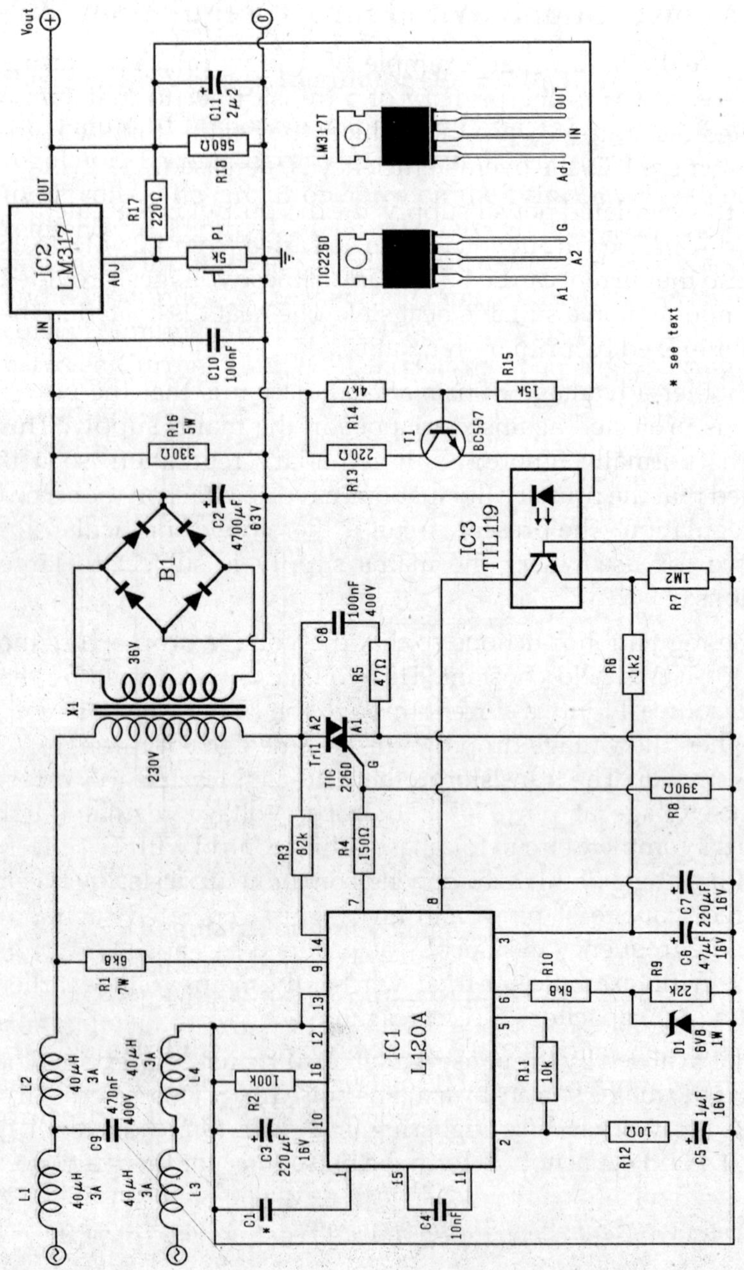

Figure 19.2: Power supply with primary regulation.

Parts List

B1	Bridge rectifier, 50 V3, 5A	P1	5 k
		R1	6k8, 7W
C1	see text	R2	100 k
C2	4700 μF, 63V	R3	82 k, 1W
C3	220 μF, 16V	R4	150 Ω
C4	10 nF	R5	47 Ω
C5	1 μF, 16V	R6	1k2
C6	47 μF, 16V	R7	1M2
C7	220 μF, 16V	R8	390 Ω
C8	100 nF, 400V	R9	22 k
C9	470 nF, 400V	R10	6k8 k
C10	100 nF	R11	10k
C11	2μ2, 40V	R12	10 Ω
IC1	L120A	R13	220 Ω
IC2	LM317	R14	4k7
IC3	Optocoupler TIL119	R15	15 k
		R16	330 Ω, 5W
		R17	220 Ω
L1	40 μH, 3A	R18	560 Ω
L2	40 μH, 3A	T1	BC557
L3	40 μH, 3A	Triac	TIC226D
L4	40 μH, 3A	X1	Transformer 230V/36V

20

Television Power Supplies

Television receiver power supplies provide low voltage (LV), high voltage (HV) and E.H.T (Extra High Tension) d.c. sources. In addition, filament power source for vacuum tubes (if used) and for cathode ray tube also forms part of the power supply.

Low Voltage Supply

The low voltage d.c. that is needed for feeding various sections of the receiver circuit is obtained by rectifying and filtering the mains supply. It may be classified as unregulated and regulated d.c. supply. Most tube type receiver power supplies are unregulated. Small changes in d.c. output voltage due to normal fluctuations in the mains supply do not affect the performance of the receiver drastically because the circuits are designed to operate satisfactorily for reasonable variations in the B+ voltage. Solid state receivers employ a regulated low voltage d.c. source because transistors and ICs are very sensitive to over voltage.

In receivers that employ only tubes, B+ voltage is generally between 100 and 350 volts with moderate current demands. For transistor receivers, a V_{CC} source between 15 and 100 volts is necessary, but the current drain is quite high. Receivers that use hybrid (tubes as well as transistors) circuitry need a d.c. supply from 15 to 250 volts.

For the horizontal output stage and other auxiliary circuits a d.c. supply over 600 V is required for vacuum tubes. This is met from the 'boosted' supply that is economically generated in the horizontal output circuit by augmenting the existing B+ supply.

E.H.T. Supply

In monochrome receivers and E.H.T. supply of 10 to 16 kV is required at the final anode of the picture tube. The colour picture tube needs a still higher d.c. voltage, that ranges from 15 kilovolts to 25 kilovolts. The current drain is, however, very small and seldom exceeds one milliampere. The E.H.T. voltage is generated by stepping up and rectifying high voltage retrace pulses produced in the horizontal output transformer of the deflection circuits. The design of the horizontal output stage determines regulation of the high voltage supply. In colour receivers, besides providing a regulated low (B+) supply, a regulator is also used for high voltage supply to prevent changes in focus and convergence in the colour picture tube. Note that if only one power supply of the usual type is decided upon for both low and high voltage requirements, it would turn out to be bulky, very expensive and quite out of proportion with other sections of the receiver. Hence, the low voltage and E.H.T. supplies are designed separately.

The power drawn from the a.c. source is about 100 watts in hybrid monochrome receivers but rises to nearly 200 watts in colour receivers of that type. Present day monochrome receivers which employ only transistors and ICs consume as low as 60 watts of mains power.

Power Supply Input Circuits

A typical circuit arrangement of the input section of a power supply employing a transformer is shown in Figure 20.1. Besides the features already discussed, a hash filter us used at the mains supply input points.

The power line acts as an antenna and may pick up unwanted signals from motors, neon signs or any other sparking contacts and feed them into the receiver. Such a disturbance is called *hash* because it does not have a specific frequency. The hash may be seen or heard as unwanted noise. The inductors in the hash filter block any line transients and other disturbing frequencies, while the capacitors act as a short to such line disturbances. In addition, the hash filter also helps in confining to itself the signals generated within the receiver. These signals would otherwise interfere with other sets through mains.

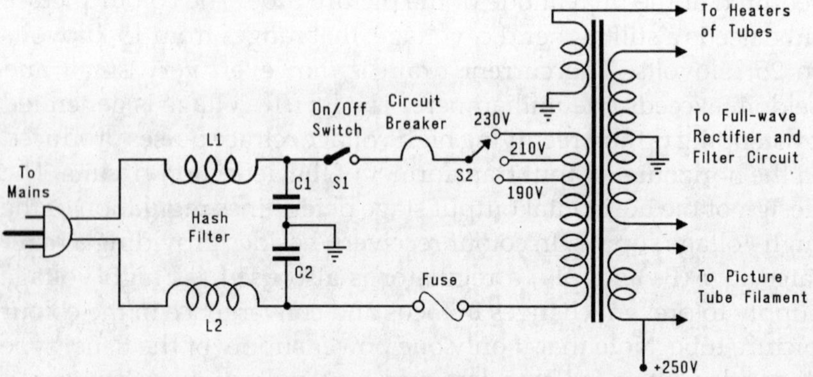

Figure 20.1: Input circuit of a TV power supply.

Figure 20.1 also shows switches, fuses and a circuit breaker. The switch S1 is the receiver on/off switch and S2 is a rear switch which selects a tapping on the primary of the transformer. It is set in accordance with the available line voltage in a locality.

Fuses used in TV receivers are of two types—fast acting and time delay. A fast acting fuse is used in circuits which do not have surge and transient currents. It is constructed of a single strand of wire that has no heat sink to dissipate momentary overloads. Such fuses are normally rated at 200 to 300% of the full load current.

Time delay fuses are designed to pass transients and surges without opening. Such fuses will, however, blow with sustained overloads or short-circuits. These are constructed out of a fuse wire that is under spring tension. The spring is embedded in a solder alloy which acts like a heat sink for momentary current surges. With sustained overload, the solder melts and the fuse opens.

The circuit breaker is for overload protection. The contact in the circuit breaker is through a spring loaded bimetallic strip which bends when heated. During normal operation the strip stays in the closed position and current flowing through the switch will not warm it sufficiently to cause it to open. This level of current is called the 'hold current'. If there is an overload, say a partial short in the transformer, the current through the bimetallic strip will be sufficient to bend it by heating and snap open the switch contact. Thus any excessive power drain will trip it, preventing further damage to the receiver and possible to the house wiring.

Once the circuit breaker opens, it will remain so, until it is closed by pushing a button protruding from the circuit breaker. Occasionally the circuit breaker will trip due to temporary surge voltages. It is a good practice to reset the breaker and observe further trippings, if any. If the breaker trips again as soon as the receiver is turned on, it is an indication that the power supply section is defective.

Low Voltage Power Supplies

Low voltage power supplies either operate from a transformer or are supplied directly from the line. The latter type of receivers are sometimes referred to as 'line connected' while those using a transformer are said to be 'line isolated'.

The line connected receiver is so termed because one side of the a.c. supply line is connected to the receiver chassis. This could be dangerous if the user came in contact simultaneously with the receiver chassis and some external ground, unless the mains cord is plugged to connect neutral or earth wire to the chassis. Voltage doublers are used in such receivers to obtain higher d.c. voltages.

Most of the high quality receivers employ a mains transformer because, though bulky and expensive, it gives many advantages. The main merits of using a transformer are:

(a) Since the a.c. source is completely isolated by separate primary and secondary windings, there is no shock hazard in touching the chassis or other exposed metal parts.

(b) The B+ voltage generation is efficient because full-wave rectification can be employed by using a center tapped high voltage secondary.

(c) A separate filament winding can be provided for parallel connections of heater circuits of vacuum tubes in a vacuum tube type of receiver, and so tubes having different filament current ratings can be used.

(d) The d.c. voltage bias that is often needed on the heaters of picture and damper tubes can be easily fed to the isolated filament winding.

(e) With a mains input transformer, if tappings are provided on the primary, the a.c. voltage to the receiver can be changed to counteract persistent low or high a.c. mains voltage changes. Such a provision, if made, is referred to as an in-built voltage regulator.

Figure 20.2 shows the circuit of the low voltage side of a transformer type of power supply.

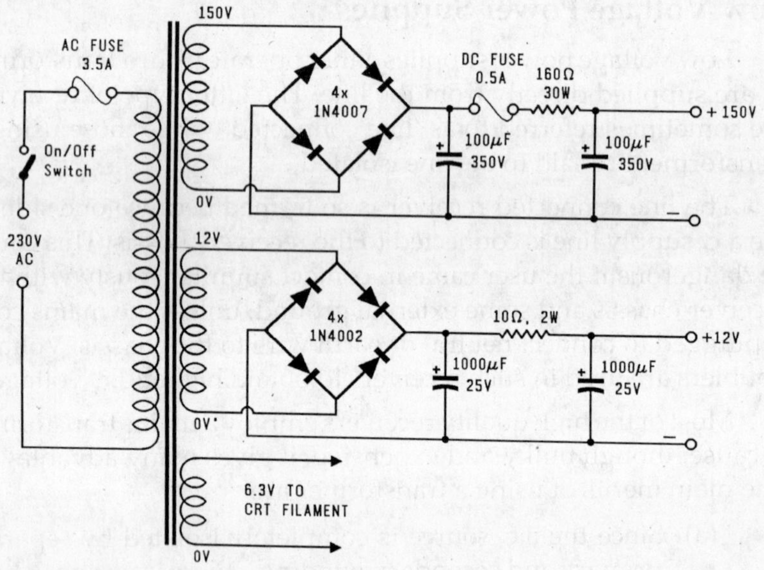

Figure 20.2: Low voltage side of a transformer type of TV power supply.

In some receiver designs an autotransformer is employed for stepping up the voltage. Though, this is economical in transformer cost, it does not provide isolation from the a.c. mains, unless it is made sure that the 'hot' side of the line is not connected to the chassis.

As stated earlier, a TV receiver needs two levels of low voltage supply. In solid state receivers, one supply is at about 20 volts for supplying power to the integrated circuits, and the other is at about 110 volts to feed horizontal and vertical amplifiers for providing beam deflection.

In the design shown in Figures 20.3 and 20.4, the power drawing stages (e.g., horizontal amplifier) are designed to operate at 110 volts. This leaves plenty of margin for the regulator to maintain 110V in spite of wide variations in the mains supply voltage.

In general, the negative side line in a regulated power supply is earthed to the chassis ground and the positive side is given to the regulator circuits. But it is equally possible to insert the regulator on the negative side (see Figure 5.15). In Figure 20.3 also, the regulator circuit is inserted on the negative line. The positive line is connected to the output terminal.

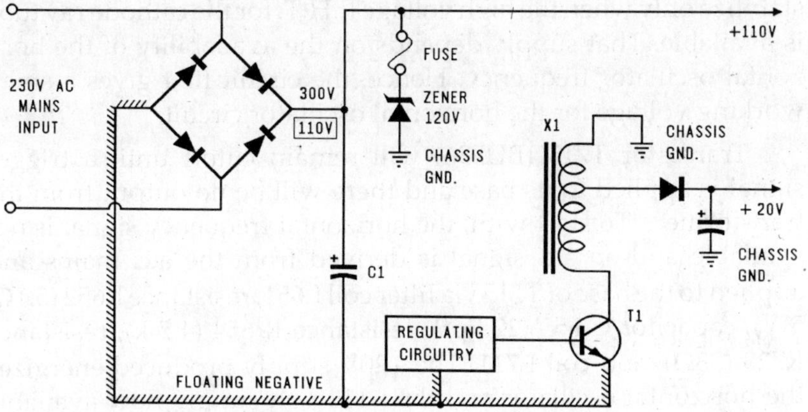

Figure 20.3: One method of obtaining low voltage supply in a TV receiver.

The d.c. supply obtained by rectifying the 230 volts a.c. is about 300 volts. The regulator works in such a way the the positive side is maintained at 110 volts with respect to the chassis ground. The difference between the negative side and chassis ground is lost in the regulator circuit. The negative side of the rectified d.c. is *floating* and this fact is represented in circuit diagrams by a hatched line as shown in Figure 20.3 which also explains working of this type of circuit arrangement.

The a.c. voltage available at the mains is rectified by a bridge rectifier and that delivers about 300V d.c. The emitter of the transistor is connected to the floating negative and its collector is connected to the chassis ground through a winding of the transformer as shown in Figure 20.3. Transistor T1 and transformer X1 form a regulator which drops the difference between the desired regulated output (110V d.c.) and the full rectified output which can vary along with variations in the line voltage. As a safety measure, a 120V zener diode is also connected through a fuse so that if the regulator circuitry fails, the voltage of the 110 V line does not exceed 120V. A part of the a.c. voltage available across X1 is rectified and filtered to provide +20V regulated. Detailed circuit is shown in Figure 20.4.

The design of this power supply is such that the circuit will stabilize only when the high voltage E.H.T. for the cathode ray tube is available. That supply depends on the availability of the horizontal oscillator frequency. Hence, the circuit first gives a small working voltage for the horizontal oscillator circuit.

Transistor T715 (BU536) will remain cutoff until a trigger signal is applied to its base and there will be no output from the transformer. To start with, the horizontal frequency signal is not available and an a.c. signal is derived from the a.c. mains and applied to the base of T715 via filter coil L651, resistance R652 (5.1Ω, 5W), capacitor C 653 (22 kpf), resistance R 654 (4.7 k), resistance R723 (33Ω), and coil L711. The +20V supply produced energizes the horizontal oscillator and the +110 volt supply now available supplies energy to the horizontal amplifier to start generating E.H.T.

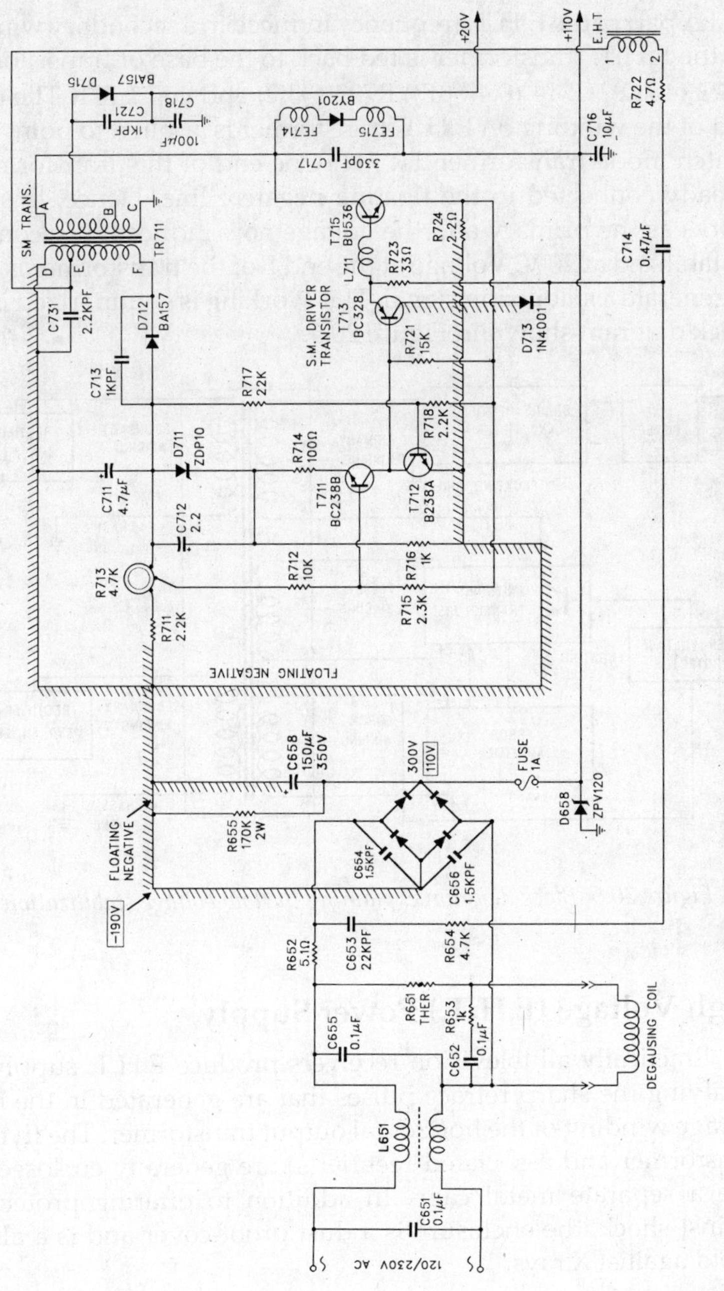

Figure 20.4: Power supply section of a colour television receiver.

A part of the E.H.T. frequency induced in a secondary winding on the E.H.T. transformer is fed back to the base of transistor via R722 (4.7Ω), C714 (0.47 μF), R723 (33Ω) and coil L711. The other end of the winding on E.H.T. transformer is applied to point D on switch mode transformer TR 711. One end of this transformer is already connected to the floating negative line. Hence, this part works as the primary and the voltage now induced in secondary is stabilized at 20 V. Voltage at the end F of the transformer is used to generate a stabilizing signal. This working is summarized in the block diagram shown in Figure 20.5.

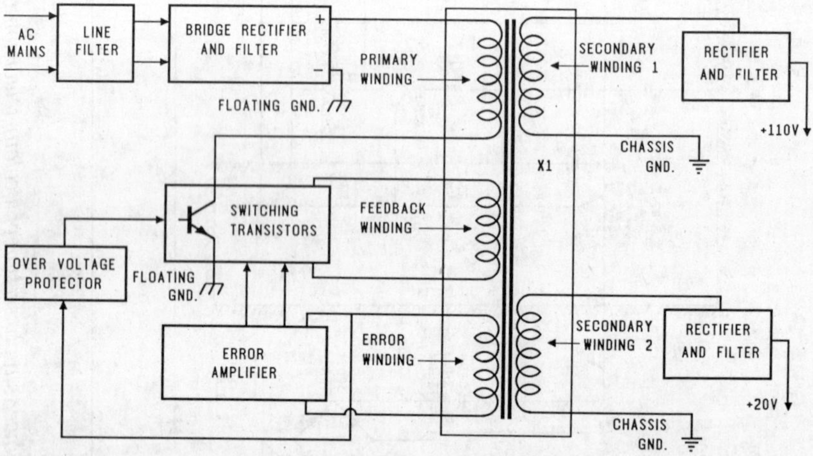

Figure 20.5: Block diagram explaining the low voltage stabilization.

High Voltage (E.H.T.) Power Supply

Practically all television receivers produce E.H.T. supply by rectifying the sharp retrace pulses that are generated in the high voltage winding of the horizontal output transformer. The flyback transformer and associated rectifier(s) are generally enclosed inside a separate metal cage. In addition to offering protection against shock, the enclosure is a dust proof cover and is a also a shield against X rays.

We know that to be efficient, switch mode supplies work at about 20 kHz or more. However, the the case of TV receivers, the SMPS operates at 15,625 Hz, which is the frequency generated by the horizontal deflection oscillator. In any other frequency is used, it is likely to generate some kind of interference in the picture display.

In a television receiver, the cathode ray beam is deflected horizontally at a linear rate. When the beam reaches the right edge of the picture display, it is quickly deflected to the left side. This type of deflection is obtained by feeding the horizontal deflection coils on the yoke with a sawtooth type of current that rises linearly and then falls abruptly. The sharp falling of this current induces a high voltage in the transformer winding, which is stepped up and rectified to generate E.H.T. supply for the anode of the CRT. The current drain is small and the capacitance formed by the coating on the inside and outside of the CRT forms enough capacitance to work as a filter capacitor. This method of generation of E.H.T is shown by a block diagram in Figure 20.6.

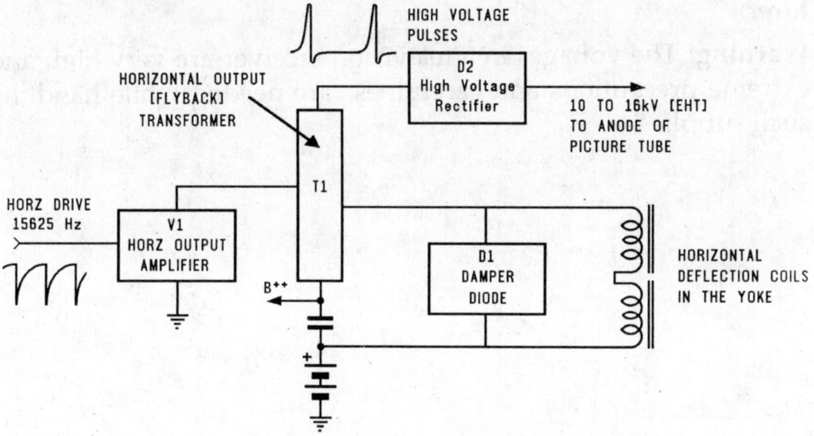

Figure 20.6: Block diagram of horizontal output stage.

The main sections of the horizontal output stage are (a) output amplifier V1, (b) output transformer X1 (commonly known as flyback transformer), (c) damper diode D1, (d) high voltage rectifier D2, and (e) deflection windings in the yoke.

The amplifier operates with self-bias and is designed to perform as a switch which closes for a fixed period during each horizontal deflection cycle to supply power to the output transformer. The transformer's main functions are to provide impedance match between the output stage and deflection windings and to act as a step-up high voltage transformer for D2, the E.H.T. rectifier. The damper diode conducts soon after the flyback stroke to suppress shock excited oscillations. While doing so the damper also causes horizontal scan of the beam on the left-half of the raster.

In many designs, the rectifiers are embedded in the winding itself as shown in Figure 7.17.

Power supply designs outlined in Figures 203 and 20.4 do not use a power transformer, and that makes the design light-weight and economical. We have described the basic principles of such designs. Actual circuitry varies from manufacturer to manufacturer.

Warning: The voltages in a television receiver are very high and extreme precautions and carefulness are needed while handling such supplies.

21

Uninterruptible Power Supplies

Power protection is imperative for all sensitive electronic equipment because power supply in India is erratic and poor in quality. this happens because the demand is more than the supply of electric power, and overloading of mains supply lines results in power failure, power cuts, voltage fluctuations, power brown outs and a host of other problems.

The intermittent power plays havoc with sensitive electronic equipment by interrupting a smooth flow of electric current and bringing all processes to a sudden and uncontrolled stop. The damage is more noticeable with computers where sudden power failure can result in loss of valuable data and corruption of many important files.

Voltage regulating devices like Constant Voltage Transformers (CVT) and Servo stabilizers offer some protection while the power is there, but when the power fails, such stabilizers are rendered useless.

Only an Uninterruptible Power Supply (UPS) can provide complete power protection. A UPS provides clean regulated power regardless of whether the mains power is present, is fluctuating, or has failed completely.

The UPS

As stated above, the word **UPS** stands for **Uninterruptible Power Supply**. This kind of supply is designed to provide a continuous source of electrical power to a computer in the event of an ac-power-line failure because sudden interruption of power to a computer can cause heavy data loss. The UPS plugs into the house a.c. current socket and the computer plugs into the UPS. Surge protection and a.c. power conditioning circuitry is included in the UPS to protect the computer from voltage spikes and other electrical conditions that could cause damage the computer.

Earlier, all electronic equipment used to use linear supplies to convert the a.c. to the required d.c. voltages. In order to reduce the size of electronic equipment manufacturers developed a new kind of supply—SMPS. The SMPS eliminated the bulky power transformer thus saving considerable space and cost. Today, all good electronic equipment manufacturers use SMPS because they are cheaper, compact, highly efficient, reliable and work with a wide input voltage range. For example, most colour TV's today use SMPS and work uncomplainingly as long as the voltage is between 90 and 300 volts.

Well designed SMPSes work with a wide input frequency of 40 to 60 Hz and wide input voltage of 180 to 250 volts. SMPSes also work with square and quasi-square wave forms and are fairly tolerant of harmonic distortion.

What is a UPS?

In general, a UPS (Uninterruptible Power Supply) contains a rechargeable battery and dc-to-ac inverter circuitry. The battery charging circuitry in the UPS will maintain the battery at a full charge during normal operation. If the a.c. power falls low or goes off for any reason, the inverter automatically turns on to maintain the 230 V a.c. supply. You can continue to work on the computer, either until the UPS battery discharges, or until the a.c. power comes back. At the very least, this will give you enough time to save files and shut down your computer system normally.

Following is the description of some of the methods used for generating the output a.c. in a UPS system. This part of the UPS called the inverter is the most critical and should be analyzed carefully before selecting a UPS.

(a) Sine Wave Amplification

The type of UPS is often called the Classic UPS as sits method of inversion is considered obsolete for UPS use and is not employed anywhere else. This type of UPS uses a low level sine wave generator and amplifies that signal by using a large number of power transistors in parallel. This functioning is similar to that of an audio amplifier being run at full volume.

This type of UPS is simple and inexpensive to manufacture. However, the average efficiency is below 45%, which means that larger and more expensive batteries are required to run the UPS for a short period of time. The cost of running the UPS is very high as nearly 55% extra energy is wasted. A large amount of the energy is dissipated as heat and will overload any type of cooling or air conditioning system. Since the transistors are used in their active region (as amplifiers), they are more prone to failure. The extra cost of running this type of UPS is many times more than the amount saved in the initial purchase price.

(b) Thyristor/Transistor Square Wave Inverters

This is one of the crudest form of generating a.c. and is suitable for general purpose use like powering lights and fans. The output waverform is so full of noise, it is almost certain to harm sensitive equipment if connected. There is no voltage regulation either and the output voltage rises and falls with the battery voltage.

(c) Step Wave Inverter

This type of UPS uses a number of inverters to produce a series of square waves. These waves are phase shifted and then added to get a step-wave that approximately resembles a sine wave. The larger the number of steps, the smaller the output filter.

The driving circuitry is fairly complex and is generally used for UPS systems above 10 KVA. This type of UPS, though bulky and expensive, is fairly reliable and is popular with some UPS manufacturers.

(d) Ferroresonant Inverters

This type of UPS unit is common in use. They have been manufactured over the last 30 years and control circuit design is simple and thus lends to the reliability of the UPS. However, recent studies have shown that the ferroresonant transformer responds poorly to step changes in the load and the output voltage oscillates before settling down. These UPS units tend to consume considerable power even when not fully loaded. They are noisy and bulky, but are also very reliable due to the simplicity of the circuit design.

(e) PWM Inverters

Pulse Width Modulation (PWM) is probably the most commonly used terminology by UPS manufacturers. True PWM UPS systems are of two types—the state-of-the-art method employs high frequency switching above 20 kHz and this makes the UPS virtually silent since the frequency is beyond the hearing range of human beings. Also, the higher the frequency, the better is the response to change in load or battery voltage. The size of the filter becomes proportionately smaller as the frequency increases. Many manufacturers have dispensed with the bulky iron core transformers thus making the UPS systems unbelievably compact, silent and cost effective.

Another version of the high frequency UPS system uses a reversible bridge which doubles as a rectifier, charger and inverter. This results in a substantial saving in the cost of the UPS. The reversible bridge however, has one serious limitation: the input frequency range is only ±5% as comparable to ±10% available from other UPS manufacturers. This means that the UPS will keep switching to battery whenever the frequency drifts by more than 2 Hz. This type of UPS cannot be used with a generator unless the generator frequency is closely regulated.

The more commonly available UPS systems employ a switching frequency between 2 to 10 kHz. Since the frequency is well within the audio range, the UPS systems emit an irritating high pitch squeal.

There are many variations to the low frequency PWM UPS systems. Some use a simpler single pulse width modulation while others use specially selected frequency to cancel out harmonics.

On-line UPS

On-line UPS systems are the ultimate form of power conditioning and power protection. A true On-line UPS completely isolates the attached equipment form the power line. The UPS protects against spikes, surges, brownouts, blackouts, line noise, change in frequency and even allows the electronic equipment to be used in remote locations where mains power is not available.

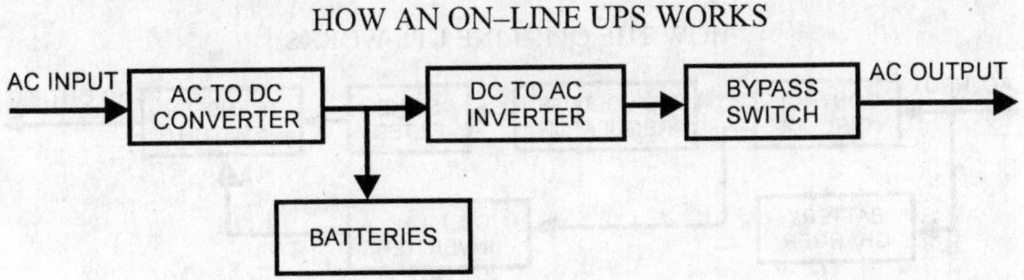

Figure 16.1: An on-line uninterruptible power supply.

The On-line system consists of a rectifier-battery charger-inverter and control circuitry as shown in Figure 16.1. The rectifier unit converts a.c. power to a d.c. of the same voltage as the attached batteries. While a.c. is present, the battery charger charges the batteries and the rectifier supplies regulated d.c. power to the inverter. The inverter with the help of control circuitry, inverts the d.c. back to a.c. By converting the main line power to d.c. and then back to a.c., the UPS effectively isolates the attached equipment from all kinds of power pollution.

In the event of power failure, even for a few picoseconds, the inverter in the UPS simply draws its power from the batteries instead of from the rectifier. There is no break, switching or interruption in the power supply. And when the power returns, the batteries are automatically recharged.

Even though the role of the UPS is clearly defined, the methods used by various UPS manufacturers differ. The reliability and the efficiency of the different types of UPS need to be examined closely.

Off-line UPS Systems

Another development is the off-line for standby UPS, more often used by micro and mini computers. The off-line UPS was developed with a view to provide protection at a reasonable cost. The continuing drop in computer prices forced the on-line UPS manufacturers to develop economically priced power back-up system. Figure 16.2 shows the block diagram of an off-line UPS.

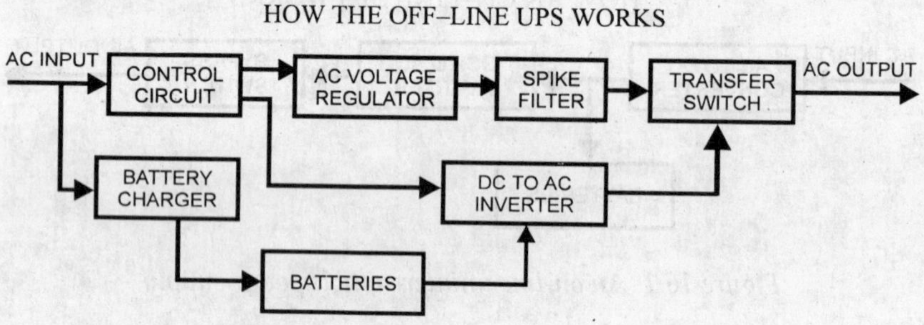

Figure 16.2: An off-line uninterruptible power supply.

There are many types of off-line UPS systems, some offer protection only against power failure, while others regulate and filter the voltage and provide comprehensive protection against brownout surges and power failure.

A good off-line UPS monitors and regulates the a.c. power to the attached equipment. The moment the power goes below or above the set limits, it transfers the load to the inverter which draws its power from the attached batteries. There is a time lag when the off-line UPS switches from mains power to the inverter. The time taken to switch is normally between 1 to 15 milliseconds, but differs drastically from manufacturer to manufacturer. The SMPS or power supply in most electronic equipment uses capacitors that store electrical charge and the equipment continues to work without interruption while the off-line UPS transfers to inverter or back to a.c. when the power returns.

However, due to the large variety of electronic equipment and off-line UPS systems it is recommenced to test the off-line UPS with the electronic equipment before purchase.

What to Look For in a UPS

Look for a UPS that can be forced into the inverter mode without the need to disconnect it from a.c. active a.c. line. Most medium sized units have an on/off switch and a test/alarm disable switch.

Higher grade UPSes have a single on/off power push-button These models can usually be forced into the inverter mode by pressing and holding the push button for a few seconds.

Most medium or small UPSes use a single 12-V gel cell. Larger supplies may use two 6-V batteries in series. Some use two 12-V batteries in series to form a 24-V system for maximum utility. A 12-V system can easily be configured to work with an external deep-discharge battery.

Technical Considerations

1. Input voltage range (± 25% recommended)
2. Input frequency range—ability to run on generator
3. Efficiency of the UPS system—operational cost
4. Mean time between failure and mean time to repair.
5. Size of unit—(space today is expensive)
6. Number of hours UPS can run on batteries without overheating and shutting down.

7. Noise—whether the level is acceptable for the proposed location of the UPS.
8. Power rating—verify the power rating of the UPS by attaching pure resistive load like bulbs of the specified VA rating to output of the UPS. Many unscrupulous manufacturers sell UPS systems of less than half their stated rating to unsuspecting customers.
9. If possible, ask for a demo machine for testing before finalizing purchase.

Using Your UPS

The UPS is intended for medium power output for short term use. Don't expect an internal battery driven inverter to supply heavy loads for a long time.

How Long Will it Last?

A test may help you estimate how long you can expect to use the UPS's battery-driven inverter with a given load. Connect a lamp with a 60-W light bulb. Use a voltmeter to record the battery voltage at the beginning of the test and every 5 to 15 minutes. Plotting the data on a graph will show the results.

Appendix A

Colour Codes

All information about a component cannot be printed in text on the component. For this reason, colour codes are used. For example, the different leads of a transformer are identified by their colours, but this varies from manufacturer to manufacturer. However, in the case of resistors and capacitors, the codes have been accepted universally. These codes are described here.

Resistor Colour Codes

Carbon and metal oxide resistors are usually marked with colour codes which indicate their ohmic value and tolerance. Two methods of colour coding are in common use; one that involves four coloured bands whilst the other that uses five colour bands. These methods are described on the following pages.

Standard Resistance Values

The most commonly used resistors are available in tolerences of 20%, 10%, and 5%. Standard values have been chosen to cover all values within tolerence. These are given below: Resistors are available in multiples or submultiples of these values like 1.2M, 120k, 12k, 1.2k, 120, and 1.2 ohms.

20% Tol.	10				15				22				33				47				68			
10% Tol.	10		12		15		18		22		27		33		39		47		56		68		82	
5% Tol	10	11	12	13	15	16	18	20	22	24	27	30	33	36	39	43	47	51	56	63	68	75	82	91

Resistor Colour Codes

Four Colour Bands System

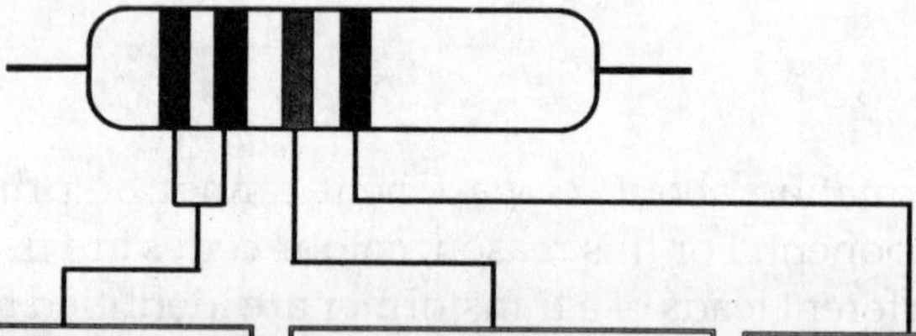

1st and 2nd COLOUR BANDS (VALUE)	
BLACK	0
BROWN	1
RED	2
ORANGE	3
YELLOW	4
GREEN	5
BLUE	6
VIOLET	7
GREY	8
WHITE	9

3rd COLOUR BAND (MULTIPLIER)	
SILVER	0.01
GOLD	0.1
BLACK	1
BROWN	10
RED	100
ORANGE	1000
YELLOW	10,000
GREEN	100,000
BLUE	1000,000

4th COLOUR BAND (TOLERANCE)	
RED	± 2%
GOLD	± 5%
SILVER	± 10%
NO BAND	± 20%

Examples:

1st band	blue = 6
2nd band	gray = 8
3rd band	red = 2 × 100
4th band	gold = ± 5%
Value	68 × 100 = 6.8k
Tolerance	± 5%

1st band	orange = 3
2nd band	orange = 3
3rd band	silver = × 0.01
4th band	silver = ±10%
Value	33 × 0.01 = 0.33Ω
Tolerance	± 10%

Five Colour Bands System

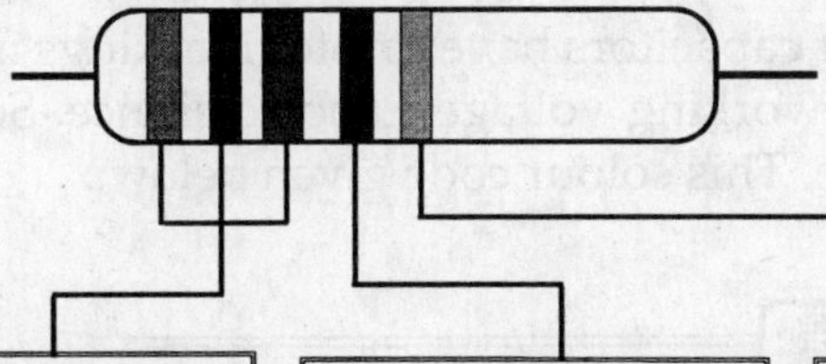

1st, 2nd, 3rd COLOUR BANDS (VALUE)	
BLACK	0
BROWN	1
RED	2
ORANGE	3
YELLOW	4
GREEN	5
BLUE	6
VIOLET	7
GRAY	8
WHITE	9

4th COLOUR BAND (MULTIPLIER)	
SILVER	001
GOLD	0.1
BLACK	1
BROWN	10
RED	100
ORANGE	1000
YELLOW	10,000
GREEN	100,000
BLUE	1000,000

5th COLOUR BAND (TOLERANCE)	
BROWN	± 1%
RED	± 2%
GOLD	± 5%
SILVER	± 10%
No BAND	± 20%

BS Coding

Some types of resistors have values marked using the system of coding defined in BS 1852. This system involves marking the position of the decimal point with a letter to indicate the multiplier concerned. A further letter is then appended to indicate the tolerance as shown in following tables:

Letter	Multiplier
R	1
K	1,000
M	1,000,000

Letter	Tolerance
F	±1%
G	±2%
J	±5%
K	±10%
M	±20%

For example: 1M2 = 1.2 megohm
33k = 33kΩ; 4R7 = 4.7Ω , and
R22 = 0.22 Ω ; ± 20%

Capacitor Colour Codes

A vast majority of capacitors have printer markings in text to indicate their values, working voltages, and tolerance. Some are marked in colour code. This colour code given below:

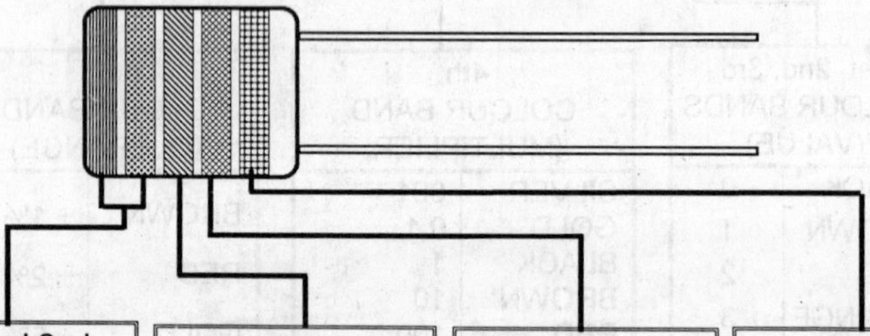

1st and 2nd COLOUR BANDS (VALUE)	
BLACK	0
BROWN	1
RED	2
ORANGE	3
YELLOW	4
GREEN	5
BLUE	6
VIOLET	7
GRAY	8
WHITE	9

3rd COLOUR BAND (MULTIPLIER)	
ORANGE	1000
YELLOW	10,000
GREEN	100,000

4th COLOUR BAND (VALUE)	
WHITE	± 10%
BLACK	± 20%

5th COLOUR BAND (VALUE)	
RED	250V
YELLOW	400V

The more modern method of marking resin doped polyester (and other) types of capacitors involves quoting the value (in μF, nF or pF), the working tolerance (often either ± 10% or ± 20%), and the working voltage using _ and ~ symbols to indicate d.c. and a.c. respectively. Several manufacturers use two separate lines for their markings and these have the following meanings:

First line: capacitance (in pF or μF) and tolerance (K = ± 10%, and M = ± 20%)

Second line: rated d.c voltage and code for dielectric material.

Appendix B

Battery Characteristics

Traditional batteries are now being replaced by ones with better outputs and functions. There has been an increase in the variety, size, and applications of primary and secondary batteries.

Parmeters of Primary Battery Types

Battery Type	Anode	Cathode	Nominal voltage (V)	Energy density (Wh/kg)
Leclanche	Zn	MnO_2	1.5	85
Alkaline-MNO_2	Zn	MnO_2	1.5	125
Zinc-Air	Zn	O_2 (Air)	1.5	340
Li/SO_2	Li	SO_2	3.0	260
Li/$SOCl_2$	Li	$SOCl_2$	3.6	320
Li/MNO_2	Li	MnO_2	3.0	230

Parmeters of Secondary Battery Types

Battery Type	Anode	Cathode	Nominal voltage (V)	Energy density (Wh/kg)
Lead acid	Pb	PbO_2	2.0	35
Nickel-cadmium	Cd	Ni Oxide	1.2	35
Nicker-metal hydride	(MH)	Ni Oxide	1.2	50
Zinc-Air	Zn	O_2 (Air)	1.5	150
Lithium-ion	C	Li_2CoO_2	4.0	90

Primary Portable Batteries

Brand	Composition	Type	Voltage (V)	Capacity (mAH)	Dimension (mm)	Weight (g)	Applications
Varta	Li-MnO_2 Flat cell	LFP 7	3.0	7	0.4 × 9 × 29	0.3	Smart cards
		LFP 25	3.0	25	0.4 × 22 × 29	0.5	
					(H × W × L)		
Varta	Lithium Cylindrical Cells	CR 1/2 AA	3.0	950	14.75, 25.1	11.5	Defence
		CR 2/3 AA	3.0	1350	14.75, 33.5	15,0	
		CR 2/3 A	3.0	1350	17.0 , 33.5	17.0	
		CR AA	3.0	2000	14.75, 50.5	21.5	
		CR 2 NP	3.0	1400	11.6 , 60	13.0	
		CR 2/3 AH	3.0	1350	17.0 , 34.5	16.0	
					(D, H)		
Varta	Zinc-air cells Zn-O_2	V5	1.4	33	5.75, 2.15	0.18	Hearing aids
		V10	1.4	70	5.8 , 3.6	0.3	
		V13	1.4	240	7.9 , 5.4	0.83	
		V312	1.4	140	7.9 , 3.6	0.58	
		V675	1.4	600	11.6, 5.4	1.85	

Brand	Composition	Type	Voltage (V)	Capacity (mAH)	Dimension (mm)	Weight (g)	Applications
Varta	Lithium Button Cells	CR 1216	3.0	25	12.5, 1.6	0.7	Memory protection, digital diaries, power source
		CR 1220	3.0	35	12.5, 2.0	0.8	
		CR 1616	3.0	55	16.0, 1.6	1.2	
		CR 1620	3.0	60	16.0, 2.0	1.2	
		CR 2016	3.0	90	20.0, 1.8	1.8	
		CR 2025	3.0	170	20.0, 2.5	2.5	
		CR 2032	3.0	230	20.0, 3.2	3.0	
		CR 2320	3.0	135	23.0, 2.0	2.9	
		CR 2430	3.0	280	24.5, 3.0	4.0	
		CR 2450	3.0	560	24.5, 5.0	6.2	
		CR 1/3 N	3.0	170	11.0, 10.8	3.0	
		V 28 PXL	6.0	170	13.0, 25.1	8.8	
Varta	Oxygen Sensor Cells	V 675 O	1.4	400	11.6, 5.4	1.85	
Varta	Hydrogen Generating Cells	V 675 H	0.4	130	11.6, 5.4	1.85	

Brand	Composition	Type	Voltage (V)	Capacity (mAH)	Dimension (mm)	Weight (g)	Applications
Varta	Alkaline	V3 GA	1.5	30	7.9, 3.6	0.6	Digital
	Manganese	V8 GA	1.5	25	11.6, 2.1	0.8	cameras,
	Cells	V10 GA	1.5	50	11.6, 3.05	1.1	flash lights,
	(Zn-MNO_2)	V12 GA	1.5	80	11.6; 4.2	1.6	mobile
		V13 GA	1.5	125	11.6, 5.4	1.8	
		V74 PX	15.0	45	16.0, 35.0	14.0	
		V4034 PX	6.0	100	13.0, 25.2	10.4	
		V625 U	1.5	200	15.5, 6.1	3.3	
		V23 GA	12.0	38	10.3, 28.5	7.5	
Varta	Silver Cells	V 301	1.55	115	11.6, 4.6	1.78	Watches
	(AgO_2)	V 303	1.55	170	11.6, 5.4	2.33	
		V 309	1.55	70	7.9, 5.4	1.08	
		V 315	1.55	20	7.9, 1.65	0.4	
		V 317	1.55	8	5.8, 1.6	0.18	
		V 341	1.55	11	7.9, 1.4	0.27	
		V 344	1.55	100	11.6, 3.6	1.49	
		V 357	1.55	180	11.6, 5.4	2.33	High
		V 364	1.55	20	6.8. 2.15	0.33	Capacity
		V 371	1.55	44	9.5, 2.1	0.61	Version
		V 377	1.55	27	6.8, 2.6	0.39	(HCV)
		V 28 PX	6.2	145	13.0, 25.2	11.1	HCV
		V 76 PX	1.55	145	11.6, 5.4	2.4	HCV

Characteristics of Commonly Used Rechargeable Batteries

	Ni-Cd	Ni-MH	Lead-Acid	Li-ion	Li-ion polymer
Gravemetric energy density (Wh/kg)	45-80	60-120	30-50	110-160	100-130
Internal resistance (in mΩ) (includes peripheral circuits)	100 to 200 (6V pack)	200 to 300 (6V pack)	<100 (12V pack)	150 to 250 (7.2V pack)	200 to 300 (7.2V pack)
Cycle life (to 80 percent of initial capacity)	1500	300 to 500	200 to 300	500 to 1000	300 to 500
Fast charge time	1h typical	2 to 4h	8 to 16h	2 to 4h	2 to 4h
Overcharge tolerence	Moderate	Low	High	Very low	Low
Self-discharge per month (room temperature)	20%	30%	5%	10%	-10%
Cell voltage (nominal)	1.25V	1.25V	2V	3.6V	3.6V
Load current — Peak	20C	5C	5C	> 2C	> 2C
— Best result	1C	0.5C or lower	0.2C	1C or lower	1C or lower
Operating Temperature (discharge only)	-40°to 60°C	-20° to 60°C	-20° to 60°C	-20° to 60°C	0° to 60°C
Maintenance requirement	30 to 60 days	60 to 90 days	3 to 6 month	Not required	Not requird
Commercial use since	1950	1990	1970	1991	1999

Secondary Portables

Brand	Composition	Type	Voltage (V)	Capacity (mAH)	Dimensions (mm)	Weight (g)	Applications
Varta	Li Button	MC 614	3.0	0.7	6.8, 1.3	0.18	Memory
		MC 621	3.0	1.1	6.8, 2.15	0.24	
		ML 1216	3.0	7.0	12.5, 1.6	0.7	
		ML 2025	3.0	30.0	20.0, 2.5	2.5	
		ML 2032	3.0	40.0	20.0, 3.2	3.0	
					(D, H)		
Varta	Standard NiMH	V 7/8 H	8.4	150	15.7×48.5× 26.6	47.0	Memory
		V 6/8 H	7.2	150	48.5, 26.6	41.0	
		CP 2010 H	3.6	220	15.0×43.3×30.0	28.0	
Varta	NiMH button cells	V 15 H	1.2	16,	11.5, 3.1	1.3	Cordless phones, torch light, memories
		V 150 H	1.2	150	5.9×14.1×25.6	6.0	
		V 250 H	1.2	250	25.1, 6.7	10.0	
		(all high capacity series)					
Varta	NiMH button cells	V 65 HT	1.2	70	15.5, 6.0	4.0	Cordless phones, torch light, memories
		V 110 HT	1.2	120	5.9, 14.1, 25.6	6.0	
		(high temp series)					

Brand	Composition	Type	Voltage (V)	Capacity (mAH)	Dimensions (mm)	Weight (g)	Applications
Varta	NiMH button cells	V 20 HR	1.2	22	11.5, 2.2	0.9	Cordless phones, torch light, memories
		V 400 HR	1.2	400	4.68×24.1×34.1	11.0	
		V 600 HR (high rate series)	1.2	600	6.4×24.1×34.1	15.0	
Varta	NiMH cylindrical	VH 700 AAA	1.2	700	10.5, 43.6	13.0	Digital cameras, mobiles, flash lights
		VH 1200 AA	1.2	1280	14.5, 49.1	26.0	
		VH 1800 4/5A	1.2	1880	17.0, 43.0	34.0	
		VH 4000 4/3A	1.2	3800	17.0, 67.0	53.0	
Varta	NiMH prismatic	VH 600 FGM	1.2	600	34×6.6×16	12.5	Mobile phones, battery packs
		VH 860 F6	1.2	560	48×6.1×7	18.0	
Varta	NiCd	5001	1.2	150	12, 29	—	Cameras, toys, torch
		5003	1.2	300	10.5, 44	—	
		5014	1.2	2200	26, 50	—	
		5214	1.2	1400	25, 50	—	

Appendix C

Electrical Characteristics

At times, designers of electronic power supplies may need more details about the electrical characteristics of the devices to be used by them than the package outlines, pin connections and performance parameters etc. described along with the device in different chapters in this book. In this appendix, we give detailed electrical characteristics of some of the more popular types of integrated circuits.

CA3085, CA3085A, CA3085B types

Characteristics	Symbol	Test Conditions $T_A = 25°C$ (Unless indicated otherwise)	CA3085 Min	CA3085 Typ	CA3085 Max	CA3085A Min	CA3085A Typ	CA3085A Max	CA3085B Min	CA3085B Typ	CA3085B Max	Units
Reference Voltage	V_{REF}	$V^+_{in} = 15\ V$	1.4	1.6	1.8	1.5	1.6	1.7	1.5	1.6	1.7	V
Quiescent Regulator Current	I_{QUIESC}	$V^+_{in} = 30\ V$	-	3.3	4.5	-	-	-	-	-	-	mA
		$V^+_{in} = 40\ V$	-	-	-	-	4.65	5	-	-	-	
		$V^+_{in} = 50\ V$	-	-	-	-	-	-	-	4.05	7	
Input Voltage Range	$V_{in(range)}$	-	7.5	-	30	7.5	-	40	7.5	-	50	V
Maximum Output Voltage	$V_{o\ (max)}$	$V^+_{in} = 30, 40, 50\ V$, $R_L = 365\Omega$ Term No. 6 to Gnd.	26	27	-	36	37	-	46	47	-	V
Minimum Output Voltage	$V_{o\ (min)}$	$V^+_{in} = 30V$	-	1.6	1.8	-	1.6	1.7	-	1.6	1.7	V
Input Output Voltage Differential	$V_{in}-V_{out}$	-	4	-	28	4	-	38	3.5	-	48	V
Limiting Current	I_{LIMIT}	$V^+_{in} = 16V$; $V^+_{out} = 10V$ $R_{SCP} = 6\Omega$	-	96	120	-	96	120	-	96	120	mA
Load Regulation	-	$I_L = 1$ to 100 mA; $R_{SCP} = 0$	-	-	-	-	0.025	0.15	-	0.025	0.15	$\%V_{out}$
	-	$I_L = 1$ to 12 mA; $R_{SCP} = 0$	-	-	0.003	0.1	-	--	-	-	-	
Line Regulation	-	$I_L = 1$ mA; $R_{SCP} = 0$ $T_A = 0°C$ to $+70°C$	-	0.04	0.15	-	0.04	0.1	-	0.04	0.08	%/V
Output Resistance	r_o	$V^+_{IN} = 25$ V; f = 1 kHz	-	0.075	1.1	-	0.075	0.3	-	0.075	0.3	Ω

Characteristics	Symbol	Test Conditions: $T_A = 25°C$ (Unless indicated otherwise)		Limits: CA3085 Min	CA3085 Typ	CA3085 Max	CA3085A Min	CA3085A Typ	CA3085A Max	CA3085B Min	CA3085B Typ	CA3085B Max	Units
Equivalent Noise Output voltage	V_{NOISE}	$V^+_{IN} = 25$ V	$C_{REF} = 0$	-	0.5	-	-	0.5	-	-	0.5	-	mV_{p-p}
			$C_{REF} = 0.22\mu F$	-	0.3	-	-	0.3	-	-	0.3	-	
Ripple Rejection	-	$V^+_{IN} = 25$ V, f = 1 kHz	$C_{REF} = 0F$	-	50	-	-	50	-	45	5-	-	dB
			$C_{REF} = 2\mu F$	-	56	-	-	56	-	50	56	-	
Load Transient Recovery Time	t_{ON}	$V^+_{IN} = 25V$; +50 mA Step		-	-	1	-	1	-	-	1	-	µS
	t_{OFF}	$V^+_{IN} = 25V$; -50 mA Step		-	-	3	--	3	--	--	3	-	µS
Line Transient Recovery Time	t_{ON}	$V^+_{IN} = 25V$; f = 1 kHz; 2V Step		-	0.8	-	-	0.8	-	-	0.8	-	µS
	t_{OFF}			-	0.4	-	-	0.4	-	-	0.4	-	µS
Temperature Coefficient of Reference and Output Voltages	ΔV_{REF} ΔV_o	$I_L = 0$; $V_{REF} = 1.6$ V		-	0.0035	-	-	0.0035	-	-	0.0035	-	%/°C

Electrical Characteristics CA723

Parameter	Conditions	Min	Typ	Max	Units
Line Regulation	V_{IN} = 12V to V_{IN} = 15V		0.01	0.1	$\%V_{out}$
	V_{IN} = 12V to V_{IN} = 40V		0.1	0.5	$\%V_{out}$
	0°C ≤ T_A ≤ 70°C, V_{IN} = 12V to 15V			0.3	$\%V_{out}$
Load Regulation	I_L = 1 mA to I_L = 50 mA		0.3	0.2	$\%V_{out}$
	0°C ≤ T_A ≤ 70°C, I_L = 1 mA to I_L = 50mA			0.5	$\%V_{out}$
Ripple Rejection	f = 50 Hz to 10 kHz, C_{REF} = 0		74		dB
Average Temp. Coefficient of Output Voltage	0°C ≤ T_A ≤ 70°C		0.003	0.015	%/°C
Short Circuit Current Limit	R_{SC} = 10 Ω, V_{OUT} = 0		65		mA
Reference Voltage		6.80	7.15	7.50	V
Output Noise Voltage	BW = 100 Hz to 10 kHz, C_{REF} = 0		20		μV_{RMS}
Standby Current Drain	I_L = 0, V_{IN} = 30 V		2.3	4.0	mA
Input Voltage Range		9.5		40	V
Output Voltage Range		2.0		37	V
Input-Output Differential		3.0		38	V

Note 1: Ratings apply to ammbient temperature up to 25°C. Above that, derate linearly at 0.3 mW/°C

Note 2: V+ must be connected to a +3V or greater supply.

Electrical Characteristics - LM109, LM309

Parameter		Conditions	LM109			LM309			Units
			Min	Typ	Max	Min	Typ	Max	
Output Voltage		$T_j = 25°C$	4.7	5.05	5.3	4.8	5.05	5.2	V
Line Regulation		$T_j = 25°C$ $7.0V \le V_{IN} \le 25V$		4.0	50		4.0	50	mV
Load Regulation	LM 109H	$T_j = 25°C$ $5mA \le I_{OUT} \le 0.5A$		20	50		20	50	mV
	LM 109K	$5mA \le I_{OUT} \le 1.5A$		50	100		50	100	mV
Output Voltage		$7.0V \le V_{IN} \le 25V$ $5mA \le I_{OUT} \le I_{max}$ $P < P_{max}$	4.6		5.4	4.75		5.25	V
Quiescent Current		$7.0V \le V_{IN} \le 25V$		5.2	10		5.2	10	mA
Quiescent Current Change		$7.0V \le V_{IN} \le 25V$			0.5			0.5	mA
		$5mA \le I_{OUT} \le I_{max}$			0.8			0.8	
Output Noise Voltage		$T_A = 25°C$ $10Hz \le f \le 100kHz$		40			40		μV
Long Term Stability					10			20	mV

NOTE: Without a heat sink the thermal resistance of the TO-5 package is about 150°C/W, while that of TO-3 package is approximately 35°C/W.

Electrical Characteristics - LM123, LM223, LM323

Parameter	Conditions	LM 123/LM 223			LM 323			Units
		Min	Typ	Max	Min	Typ	Max	
Output Voltage	$T_j = 25°C$ $V_{in} = 7.5V$, $I_{out} = 0$	4.7	5	5.3	4.8	5	5.2	V
Output Voltage	$7.5V \le V_{IN} \le 15V$ $0 \le I_{OUT} \le 3A$, $P \le 30W$	4.6	-	5.4	4.75		5.25	V
Line Regulation	$T_j = 25°C$ $7.5V \le V_{IN} \le 15V$		5	25		5	25	mV
Load Regulation	$T_j = 25°C$, $V_{IN} = 7.5V$ $0 \le I_{OUT} \le 3A$		25	100		25	100	mV
Quiescent Current	$7.5V \le V_{IN} \le 15V$ $0 \le I_{OUT} \le 3A$		12	20		12	20	mA
Output Noise Voltage	$T_j = 25°C$ $10 Hz \le f \le 100 kHz$		40			40		μV_{rms}
Short Circuit Current Limit	$T_j = 25°C$ $V_{IN} = 15V$ $V_{IN} = 7.5V$		3 4	4.5 5		3 4	4.5 5	A A
Thermal Resistance Junction to Case			2			2		°C/W

NOTE: Without a heat sink the thermal resistance of the TO-3 package is about 35°C/W. With a heat sink, the effective thermal resistance can only approach the specified values of 2°C/W, depending on the efficiency of the heat sink.

Electrical Characteristics - LM125, LM126, LM325, LM326

Parameter	Conditions	Min	Typ	Max	Units
Output Voltage	$T_j = 25°C$				
LM125/LM325A		14.8	15	15.2	V
LM323		14.5	15	15.5	V
Output Voltage	$T_j = 25°C$				
LM126		11.8	12	12.2	V
LM326		11.5		12.5	V
Input-Output Differential		2.0			V
Line Regulation	V_{IN} 18V to 30V, $I_L = 20$ mA $T_j = 25°C$		2.0	10	mV
Line Regulation over Temp. Range	$V_{IN} = 18V$ to 30V, $I_L = 20$ mA		2.0	20	mV
Load Regulation	$I_L = 0$ to 50 mA, $V_{IN} = \pm 30V$ $T_j = 25°C$				
V_O^+			3.0	10	mV
V_O^-			5.0	10	mV
Output Voltage Balance	$T_j = 25°C$				
LM125, LM325A				± 150	mV
LM325				± 300	mV
Output Voltage Balance	$T_j = 25°C$				
LM126,				± 125	mV
LM326				± 250	mV
Temperature Stability of V_O			±0.3		%

Parameter	Conditions	Min	Typ	Max	Units
Short Circuit Current Limit	$T_j = 25°C$		260		mA
Output Noise Voltage	$T_j = 25°C$, BW = 100 - 10 kHz		100		μV_{rms}
Positive Standby Current	$T_j = 25°C$, $I_L = 0$		1.75	3.0	mA
Negative Standby Current	$T_j = 25°C$		3.1	5.0	mA
Long Term Stability			0.2		%/kHr
Thermal Resistance Junction to Case (Note 4) LM125H,LM325H LM126H, LM326H Junction to Ambient Junction to Ambient	 (Still Air) (400 Lt/min Air Flow)		 20 20 215 82		 °C/W °C/W °C/W °C/W
Junction to Ambient LM325AN, LM325N	(Still Air)		90		°C/W
Junction to Ambient LM326N			150		°C/W

NOTE 1: That voltage to which the output may be forced without damage to the device

NOTE 2: Unless otherwise specified, these specifications apply for $T_j = 55°C$ to +150°C on LM126. $T_j = 0°C$ to +125°C on LM326. $V_{IN} = \pm 20V$, $I_L = 0$ mA, $I_{MAX} = 100$ mA, $P_{MAX} = 2.0$ W for the TO-5 H package. $I_{MAX} = 100$ mA, $P_{MAX} = 1.0$ W for the DIP N package.

NOTE 3: If the junction temperature exceeds 150°C, the output short circuit duration is 30 seconds

NOTE 4: Without a heat sink, the thermal resistance junction to ambient of the TO-5 Package is about 215°C/W. With a heat sink, the effective resistance can only approach the junction to case values specified, depending on the efficiency of the sink.

Electrical Characteristics - LM150, LM350 (Note 1)

Parameter	Conditions	LM150			LM350			Units
		Min	Typ	Max	Min	Typ	Max	
Line Regulation	$T_A = 25°C$, $3V \leq (V_{IN} - V_{OUT}) \leq 35V$, $I_L = 0.010A$ (Note 2)		0.005	0.01		0.005	0.03	%/V
Load Regulation	$T_A = 25°C$, $10\ mA \leq I_{OUT} \leq 3A$		0.1	0.3		0.1	0.5	%
Adjustment Pin Current			50	100		50	100	μA
Reference Voltage	$3V \leq (V_{IN} - V_{OUT}) \leq 35V$ $10\ mA \leq I_{OUT} \leq 3A$, $P \leq 30W$	1.20	1.25	1.30	1.20	1.25	1.30	V
Line Regulation	$3V \leq (V_{IN} - V_{OUT}) \leq 35V$ (Note 2) $I_L = 0.010A$		0.02	0.05		0.02	0.07	%/V
Load Regulation	$10\ mA \leq IOUT \leq 3A$ (Note 2)		0.3	1		0.3	1.5	%
Minimum Load Current	$(V_{IN} - V_{OUT}) = 35V$		3.5	5		3.5	10	mA
Current Limit	$(V_{IN} - V_{OUT}) \leq 10V$ $(V_{IN} - V_{OUT}) = 30V$, $T_j = +25°C$	3.0 0.3	4.5 1		3.0 0.25	4.5 1		A A
RMS Output Noise. % of V_{OUT}	$Tj = 25°C$, $10\ Hz \leq f \leq 10\ kHz$		0.001			0.001		%

NOTE 1: Unless otherwise specified, these specifications apply $-55°C \leq Tj \leq +150°C$ for the LM150. $0°C \leq Tj \leq +125°C$ for the LM 350. V_{IN}-V_{OUT} = 50V, and I_{OUT} = 10 mA. These specifications are applicable for power dissipation up to 30W for the K package and 25W for the T package. Power dissipation is guaranteed to these values up to 15 volts input-output differential. Above 15V differential, power dissipation will be limited by internal protection circuitry.

Note 2: Regulation is measured at constant junction temperature. Changes in output voltage due to heating effects must be taken into account separately.

Electrical Characteristics - LM317 (Note 1)

Parameter	Conditions	LM317 Min	LM317 Typ	LM317 Max	Units
Line Regulation	$T_A = 25°C$, $3V \le V_{IN} - V_{OUT} \le 40V$ (Note 2)		0.01	0.04	%/V
Load Regulation	$T_A = 25°C$, $10\ mA \le I_{OUT} \le I_{MAX}$				
	$V_{OUT} \le 5V$ (Note 2)		5	25	mV
	$V_{OUT} \ge 5V$ (Note 2)		0.1	0.5	%
Adjustment Pin Current			50	100	μA
Reference Voltage	$3V \le (V_{IN} - V_{OUT}) \le 40V$ $10\ mA \le I_{OUT} \le I_{MAX}$, $P \le P_{MAX}$	1.20	1.25	1.30	V
Minimum Load Current	$V_{IN} - V_{OUT} = 40V$		3.5	10	mA
Current Limit	$V_{IN} - V_{OUT} \le 15V$				
	K and T Package	1.5	2.2		A
	H and P Package	0.5	0.8		A
	$V_{IN} - V_{OUT} = 40V$				
	K and T Package		0.4		A
	H and P Package		0.07		A
RMS Output Noise, of V_{OUT}	$T_A = 25°C$, $10H \le f \le 10\ kHz$		0.003		%

NOTE 1: Unless otherwise specified, these specifications apply: $0°C \le T_j \le +125°C$ for the LM317. V_{IN}-V_{OUT} =5V and I_{OUT} = 0.1A for the TO-5 and TO-202 packages and I_{OUT} = 0.5A for the TO-3 package and TO-220 package. Although power dissipation is internally limited, these specifications are applicable to power dissipation of 2W for TO-5 and TO-202 and 20W for the TO-3 and TO-220. I_{MAX} = 1.5A for TO-3 and TO-220 package, and 0.5A for TO-5 and TO-202 package.

NOTE 2: Dissipation is measured at constant junction temperature, using pulse testing with a low duty cycle.

Electrical Characteristics - LM337 (Note 1)

Parameter	Conditions	LM317			Units
		Min.	Typ.	Max.	
Line Regulation	$T_A = 25°C$, $3V \le V_{IN} - V_{OUT} \le 40V$ (Note 2)		0.01	0.04	%/V
Load Regulation	$T_A = 25°C$, $10\ mA \le I_{OUT} \le I_{MAX}$ $V_{OUT} \le 5V$ (Note 2) $V_{OUT} \ge 5V$ (Note 2)		 15 0.3	 50 1.0	 mV %
Adjustment Pin Current			65	100	μA
Reference Voltage	$3V \le (V_{IN} - V_{OUT}) \le 40V$ $10\ mA \le I_{OUT} \le I_{MAX}$, $P \le P_{MAX}$	-1.200	-1.250	-1.300	V
Minimum Load Current	$V_{IN} - V_{OUT} \le 40V$		2.5	10	mA
Current Limit	$V_{IN} - V_{OUT} \le 15V$ K and T Package H and P Package $V_{IN} - V_{OUT} = 40V$ K and T Package H and P Package	 1.5 0.5 	 2.2 0.8 0.4 0.17		 A A A A
RMS Output Noise, of V_{OUT}	$T_A = 25°C$, $10H \le f \le 10\ kHz$		0.003		%

NOTE 1: Unless otherwise specified, these specifications apply: $0°C \le T_j \le +125°C$ for the LM337. $V_{IN}-V_{OUT} = 5V$ and $I_{OUT} = 0.1A$ for the TO-5 package and TO-TO-202 package, and IOUT = 0.5A for TO-3 and TO-220 package. Although power dissipation is internally limited, these specifications are applicable to power dissipation of 2W for TO-5 and TO-202 and 20W for the TO-3 and TO-220. $I_{MAX} = 1.5A$ for TO-3 and TO-220 package, and 0.5 for TO-5 and TO-202 package.

NOTE 2: Dissipation is measured at constant junction temperature, using pulse testing with a low duty cycle.

Electrical Characteristics - L200 (Tj = -25°C to 150°C for L200C and Tj = -55°C to 150°C for L200 unless otherwise specified)

Parameter		Test Conditions		Min.	Typ.	Max.	Unit
GENERAL							
I_d	Quiescent Drain Current (Pin 3)	$V_i = 20V$	$T_j = 25°C$		4.2	8.2	mA
e_n	Output Noise Voltage	$V_o = V_{ref}$ $B = 1MHz$	$I_o = 10$ mA		80		μV
CURRENT REGULATION LOOP							
V_{sc}	Current Limit Sense Voltage between pins 5 and 2	$V_i = 10V$ $T_j = 25°C$	$V_o = V_{ref}$		0.45		V
	Average Temperature Coefficient of V_{sc}				0.03		%/°C
$\frac{\Delta I_o}{I_o}$	Current Load Regulation	$V_i = 10V$ $I_o = 0.5A$ $I_o = 1A$ $I_o = 1.5A$	$\Delta V_o = 3V$		 1.4 1 0.9		 % % %
I_{sc}	Short Circuit Current	$V_i - V_o = 14V$ (pins 2 and 5 short circuited)		2	2.5	3.6	A

NOTE 1: A load step of 2A can be applied provided that input-output differential voltage is lower than 20 V.

NOTE 2: The same performance can be maintained at higher output levels if a bypassing capacitor is privided between pins 2 and 4.

Parameter		Test Conditions	Min.	Typ.	Max.	Unit
VOLTAGE REGULATION LOOP						
V_{ref}	Reference Voltage (Pin 4)	$V_i = 20V$ $I_O = 10$ mA $T_j = 25°C$	2.65	2.77	2.85	V
$\frac{\Delta V_{ref}}{\Delta T}$	Average Temperature Coefficient of Reference Voltage	$V_i = 20V$ $I_O = 10$ mA for $T_j = -25°C$ to $125°C$ for $T_j = 125°C$ to $150°C$		-0.25 -1.5		mV/°C mV/°C
V_o	Output Voltage Range	$I_o = 10$ mA $T_j = 25°C$	2.85		36	V
$\frac{\Delta V_o}{V_o}$	Voltage Load Regulation (Note 1)	$\Delta I_o = 2A$ $\Delta I_o = 1.5A$		0.15 0.1	1 0.5	% %
SVR	Supply Voltage Rejection Ratio	$V_o = 5V$ $I_o = 500$ mA $\Delta V = 10V_{p\text{-}p}$ f = 100 Hz (note 2)	60	70		dB
$\frac{\Delta V_i}{\Delta V_o}$	Line RegulationCurrent Load Regulation	$V_o = 5V$ $I_o = 500$ mA $V_i = 8$ to 18V Tj = 25°C	54	70		dB
Z_o	Output Impedance	$V_i = 10V$ $V_o = V_{ref}$ $I_o = 0.5A$ f = 100 Hz		1.5×10^{-3}		Ω
	Drop-out Voltage between Pins 1 and 5	$I_o = 1.5A$ $\frac{\Delta V_o}{V_o} \leq 2\%$		2	2.5	V
I_4	Bias Current at Pin 4	$T_j = 25°C$		3	10	µA
$\frac{\Delta I_4}{\Delta T \bullet I_4}$	Average Temperature Coefficient (Pin 4)			-0.5		%/°C

Electrical Characteristics μA78G/78GC

Parameter	Condition	Min.	Typ.	Max.	Units
Input Voltage Range	$T_j = 25°C$	7.5		40	V
Output Voltage Range	$V_{IN} = V_{OUT} + 5V$ $V_{OUT} + 3V \le V_{IN} \le V_{OUT} + 15V$; $T_j = 25°C$	5.0		30	V
Output Voltage Tolerence	$5\text{ mA} \le I_{OUT} \le 1.0\text{ A}$ $P_D \le 15W$, $V_{IN(MAX)} = 38V$			5.0	$\%V_{OUT}$
Line Regulation	$V_{OUT} + 3V \le V_{IN} \le V_{OUT} + 15V$;			0.75	$\%V_{OUT}$
Load Regulation	$V_{IN} = V_{OUT} + 5V$; $5\text{ mA} \le I_{OUT} \le 1.5\text{ A}$			2.0	$\%V_{OUT}$
Control Pin Current	Tj = 25°C		1.0	5.0	μA
Quiescent Current	Tj = 25°C		3.2	5.0	mA
Ripple Rejection	$8V \le Vin \le 18V$; f = 120 Hz μA78G $V_{OUT} = 5V$, μA78GC	68 62	78 78		dB dB
Output Noise Voltage	Tj = 25° $10Hz \le f \le 100\text{ kHz}$, $V_{OUT} = 5V$		40		μV
Dropout voltage	μA78G μA78GC			3.0 2.5	V V
Short Circuit Current	Tj = 25°C $V_{IN} = 30V$		750		mA
Peak Output Current	Tj = 25°C		2.2		A
Control Pin Voltage (Reference)	Tj = 25°C	4.8	5.0	5.2	V

Electrical Characteristics μA78MG/78MGC

Parameter	Condition	Min.	Typ.	Max.	Units
Input Voltage Range	$T_j = 25°C$	7.5		40	V
Output Voltage Range	$V_{IN} = V_{OUT} + 5V$	5.0		30	V
Output Voltage Tolerence	$V_{OUT} + 3V \leq V_{IN} \leq V_{OUT} + 15V$; $T_j = 25°C$ $5\ mA \leq I_{OUT} \leq 350\ mA$ $P_D \leq 15W$, $V_{IN(MAX)} = 38V$			4.0 5.0	$\%V_{OUT}$ $\%V_{OUT}$
Line Regulation	$V_{OUT} + 3V \leq V_{IN} \leq V_{OUT} + 15V$;			0.75	$\%V_{OUT}$
Load Regulation	$5\ mA \leq I_{OUT} \leq 500\ mA$; $V_{IN} = V_{OUT} + 7V$			1.0	$\%V_{OUT}$
Control Pin Current	$T_j = 25°C$		1.0	5.0	μA
Quiescent Current	$T_j = 25°C$		2.8	4.0	mA
Ripple Rejection	$8V \leq V_{IN} \leq 18V$; $T_j = 25°C$, $I_{OUT} = 300\ mA$, $V_{OUT} = 5V$, $f = 120\ Hz$, $I_{OUT} = 100\ mA$	62 62	80		dB dB
Output Noise Voltage	$10Hz \leq f \leq 100\ kHz$, $V_{OUT} = 5V$		25		μV
Dropout Voltage	μA78MGHM μA78MG (H and C)			3.0 2.5	V V
Short Circuit Current	$V_{IN} = 35\ V$, $T_j = 25°C$		300		mA
Peak Output Current	$T_j = 25°C$		800		mA
Control Pin Voltage (Reference)	$T_j = 25°C$	4.8	5.0	5.2	V

Electrical Characteristics μA79G/79GC

Parameter	Condition	Min.	Typ.	Max.	Units
Input Voltage Range	$T_j = 25°C$	-40		-7.0	V
Output Voltage Range	$V_{IN} = V_{OUT} - 5V$	-30		-2.23	V
Output Voltage Tolerence	$V_{OUT} -15V \le V_{IN} \le V_{OUT} - 3V$; $T_j = 25°C$ $5\ mA \le I_{OUT} \le 1.0\ A$ $P_D \le 15W$, $V_{IN(MAX)} = -38V$			4.0	$\%V_{OUT}$
Line Regulation	$(V_{OUT} - 15V) \le V_{IN} \le (V_{OUT} - 3V)$			0.75	$\%V_{OUT}$
Load Regulation	$V_{IN} = V_{OUT} - 5V$; $5\ mA \le I_{OUT} \le 1.5\ A$			2.0	$\%V_{OUT}$
Control Pin Current	$T_j = 25°C$		0.4	2.0	μA
Quiescent Current	$T_j = 25°C$		0.5	1.5	mA
Ripple Rejection	$-18V \le V_{IN} \le -8V$; $f = 120\ Hz$ μA79G $V_{OUT} = -5V$, μA79GC	50 50	60 60		dB dB
Output Noise Voltage	$T_j = 25°$ $10Hz \le f \le 100\ kHz$, $V_{OUT} = -5V$		125		μV
Dropout Voltage	μA79G μA79GC			2.8 2.3	V V
Short Circuit Current	$T_j = 25°C$ $V_{IN} = -30V$		250		mA
Peak Output Current	$T_j = 25°C$		2.2		A
Control Pin Voltage (Reference)	$T_j = 25°C$	-2.32	-2.23	-2.14	V

Electrical Characteristics μA79MG/79MGC

Parameter	Condition	Min.	Typ.	Max.	Units
Input Voltage Range	$T_j = 25°C$	-40		-7.0	V
Output Voltage Range	$V_{IN} = V_{OUT} - 5V$	-30		-2.23	V
Output Voltage Tolerence	$V_{OUT} -15V \le V_{IN} \le V_{OUT} - 3V$; $T_j = 25°C$ $5\ mA \le I_{OUT} \le 150\ mA$ $P_D \le 15W$, $V_{IN(MAX)} = -38V$			4.0	$\%V_{OUT}$
Line Regulation	$V_{OUT} -20V \le V_{IN} \le V_{OUT} - 2.5V$;			1.0	$\%V_{OUT}$
Load Regulation	$V_{IN} = V_{OUT} - 7V$; $5\ mA \le I_{OUT} \le 500\ m\ A$			1.0	$\%V_{OUT}$
Control Pin Current	$T_j = 25°C$			3.0	μA
Quiescent Current	$T_j = 25°C$		0.5	1.5	mA
Ripple Rejection	$-18V \le V_{IN} \le -8V$; $T_j = 25°C$, $I_{OUT} = 300\ mA$ $V_{OUT} = -5V$, $f = 120\ Hz$, $I_{OUT} = 100\ mA$	54 50	65		dB dB
Output Noise Voltage	$10Hz \le f \le 100\ kHz$, $V_{OUT} = -5V$		125		μV
Dropout Voltage	μA79G μA79GC			2.5 2.3	V V
Short Circuit Current	$T_j = 25°C$ $V_{IN} = -30V$		100		mA
Peak Output Current	$T_j = 25°C$		650		mA
Control Pin Voltage (Reference)	$T_j = 25°C$	-2.32	-2.23	-2.14	V

Appendix D

Surge Suppressors

When sophisticated electronic equipments break down, whether at home or at plant, the natural reaction is to blame the dealers and manufacturers for supplying defective quality equipments. But in reality it may not be always true. Most equipment failures are caused by electrical interferences and disturbances arising out of surges/spikes ((sudden shootup of voltage for less than a microsecond) and voltage fluctuation. Therefore what is needed is to control/suppress/clean such conditioned power.

What are Surges?

Surges are sudden increases in current and voltage of the order of 1000 V to 6000 V and above for less than a few millionths of a second, resulting in very high electrical energies. Such electrical jolts can bombard the circuitry and components of sophisticated electrical and electronic equipment and reduce their life and reliability considerably.

Surges/spikes are mainly caused due to lightning, electrical substations switching on/off, nearby machinery such as electric motors, air conditioners, refrigerators, spot welding equipment and even tube light switching.

Surges generated at one place can travel quite a distance on the mains line, enter the equipment and cause damage. Repairs/replacements are costly and it is wise to take preventive actions in power supply designs. Figure D1 shows surges on mains supply voltages lines.

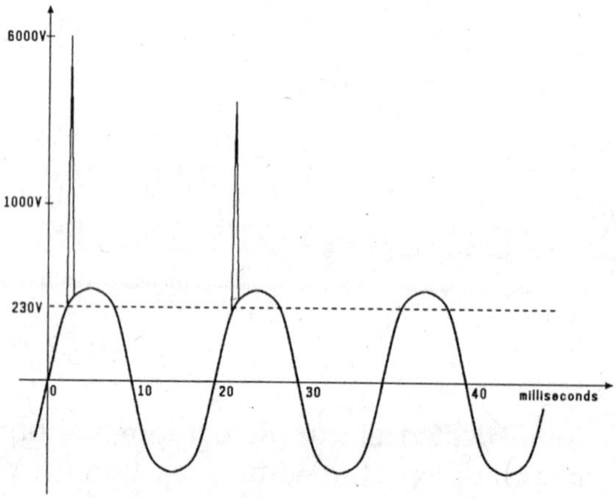

Figure D1: Spikes riding on mains voltage.

Effect of Surges

Computer's memory gets lost, and erroneous data are produced. Electronic cash registers can malfunction. The TV picture may jump. The photo-copier can stop in-between. Electro-medical equipment/EAPBX/FAX/TELEX/etc. may stop working. The VCR and Hi-Fi Systems can get damaged. In general, down time of the equipments increase and they need frequent maintenance and replacements of components.

The solution to the surge problem is to use protectors. A surge protector is a solid state device which works instantaneously.

Selenium Surge Suppressors

Selenium rectifier, a semiconductor device, has been modified to provide excellent protection for other semiconductor devices. One such device is KlipSel by Usha Rectifier Corporation. Either a non-polarized Klip-Sel or a polarized Klip-Sel connected as shown in Figure D2 is effective in clamping transients to safe values. The bipolar one breaks down in either direction and can be used directly in a.c. circuits. The polarized ones are more suited for use on rectified a.c. pulses.

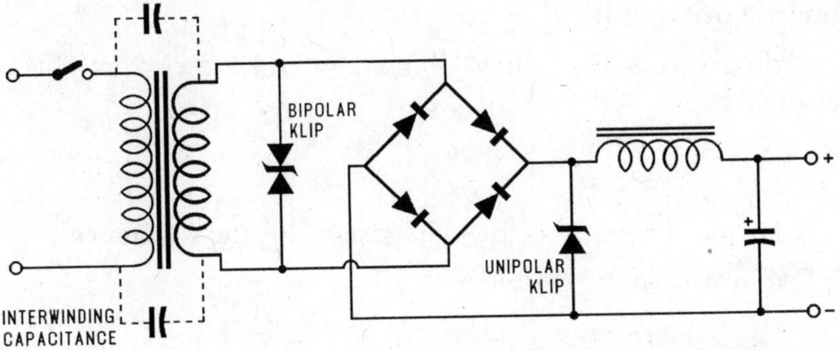

Figure D2: Protection from damage by surges using Klip-Sel.

Varistors

Varistors are devices that exhibit non-ohmic conduction. Ohm's law states that the ratio of voltage applied across a resistance to the current flowing through it is a constant at a given temperature. When the above ratio varies as a function of the applied voltage, then the material is called a variable resistor, or in short, a *varistor*.

The voltage-current characteristics of a varistor, therefore, is different from the well known straight line characteristic of a resistor. This deviant behaviour has advantages and makes varistors capable of protecting sensitive equipment from voltage transients. Present day varistors offer quick suppression of voltage spikes from the few volts level required for semiconductor devices to hundreds of kilovolts for power distribution systems.

Metal Oxide Varistors

Ceramic materials having deviant ohmic properties are now widely used for protection and are equally efficient in electronic equipment as well as high voltage transmission lines. These are called *varistors*. Silicon carbide (SiC) and titanium dioxide (TiO_2) were the early materials used in the manufacture of varistors, but nowadays, metal oxides like zinc oxide (ZnO) are more popular due to their superior characteristics.

Principle of Operation

Ordinary resistors follow Ohm's law and the current has a relation like

$$I = \frac{V}{R}$$

where I is the current, V is the voltage and R is the resistance.

For varistors the expression is

$$I = V^{\alpha}$$

where α is a positive number greater than unity.

From the above relation, the conductivity varies in proportion with the power of the applied voltage. For example, for $\alpha = 10$, the conductivity increases by more than 1000 times if the applied voltage is doubled. Since the response time of the devices is kept very small, even large voltage spikes are effectively bypassed when a varistor is connected at the input end of a circuit.

But how is it that the varistor has such a deviant characteristic? It is due to the material properties of metal oxides like ZnO used in the manufacture of varistors. The varistor action in a ZnO device is located around the boundary of ZnO grains.

When a varistor is connected across a voltage load, it offers a large impedance and draws negligible current from the supply lines in the order of microamperes (less than 100 microamperes) under normal supply conditions. Occurrence of a surge reduces the varistor impedance to a very low conducting value because of the presence of a very high voltage, and the surge current is diverted away from the load. As a result, the transient high voltage is suppressed and kept below a safe level and the load is protected from surge damage. The response time of varistors is about 100 nanoseconds. This is a very small time period, and it is this fast-acting characteristic which renders varistors suitable for protecting electronics circuitry.

Varistors are small in size and it is easy to install and use them in practical applications.

Application Areas

Metal oxide varistors (MOVs) find the commonest application in protection of power supplies of electronic equipments and control systems. Extended life of diodes used in rectifier circuitry and higher system reliability can be expected out of their use. Where the load being driven is inductive, it is better to provide a MOV across the load so that surges generated by the load as switching off are prevented from burning the driver transistors an/or associated devices. Various other applications are possible in control and power switching systems.

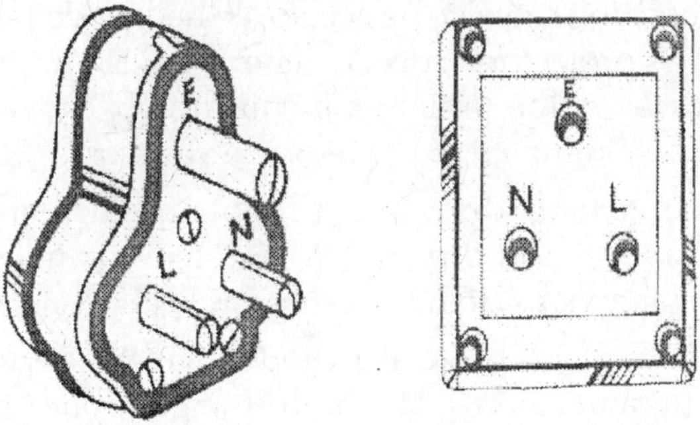

Figure D3: A three-pin plug with built-in surge suppressor.

The plug shown in Figure D3 has such protectors built inside. They protect on all three modes:-line, neutral and earth. Typical specifications for two types are given below:

Description	**5 Amps.**	**15 Amps**
Max. Voltage Rating	250 V RMS	250 V RMS
Max Load Current	6 amps	16 amps
Max Surge Current	1000 amps	2000 amps
Response Time	< 10 nano-seconds	< 10 nanoseconds
Total Energy Rating	30 Joules	75 Joules
Resetting	Automatic	Automatic

Contact Protection

If a circuit has an inductor (e.g., primary of a transformer), then a high voltage is induced when the switch is turned off. This can cause sparking or arcing at the switch contacts and reduce its life. How the contact erosion can be reduced by using contact protectors was explained with reference to Figure 6.2. Selenium contact protectors, when connected across the contacts of a switch absorb the high voltage transient and increase the life of a switch.

Performance Comparison

Surge protection can also be accomplished by devices like gas discharge surge protectors, special zener diodes and RC or LRC networks. However, the limitations and drawbacks of these devices have been eliminated in metal oxide varistors.

Gas discharge tubes draw heavy follow-on currents and have a switching time of the order of a few tens of microseconds. Moreover, they are not suitable for d.c. applications.

Solid-state avalanche (zener) diodes have a very small response time but cannot be used for high voltage applications. Zener breakdown occurs only in the reverse biased condition and hence two devices connected back to back are required for a.c. application. Another disadvantage of avalanche diodes lies in their small energy absorption capability.

RC or LRC networks do not offer uniform response to all types of surges. They are basically damping circuits and turn out to be sources of surges due to resonance oscillations. They are often bulky and hence occupy too much space.

Varistors, in general, have quick response time and are small in size. They have large surge absorption capabilities and hence can tackle high energy surges like those caused by lightning. Due to their high non-ohmic (non-linearity) exponent coefficient, current drawn under normal operating condition is very minimal. Metal oxides have a high non-linearity exponent when compared to other materials and hence they draw a very low standby current and have a lower clamping voltage.

Heat Sinks

Whenever a current I flows through a device or component, it generates heat equal to V×I or I^2R watts (joules per second). This heats up the device or component and its temperature starts to rise. The temperature goes on rising till it reaches a state where heat removed by the cool surroundings per second equals the heat generated per second. If no arrangements are made to remove the heat generated, the temperature can go high enough to burn out the component or the device.

Electronic devices like diodes, transistors, integrated circuits etc., are small in size and get heated even by small amounts of heat. Although in many cases the dissipation is small and no extra efforts for removing are required, there are power handling devices in which a lot of heat is generated. It is necessary to arrange for efficient removal of heat to keep them cool and ensure that the rise in temperature is within safe limits.

A heavy metallic chassis is a good heat sink, but it is not possible to mount all heat generating components directly on chassis. A wide variety of heat sinks are available commercially that can be fitted on semiconductor devices to keep them cool. Some of these heat sinks are shown in Figure E1.

When the power dissipated by advice is large (greater than 1W/sq. cm), the device can be mounted on a heat sink to enable removal of heat from the device. In this appendix, we describe the design of a heat sink. It is assumed that the heat sink is made of 4mm thick aluminium sheet which has been painted black.

Figure E1: Heat sink designs.

Heat Sink Design

When an electronic device dissipates up to 0.05 watts per sq. cm the removal heat can be handled easily by air cooling. In fact, chips which dissipate less than 1W/sq. cm require no special cooling consideration. Still air can transfer 0.001 W/sq. cm whereas forced air can transfer 0.01 to 0.03 W/sq. cm.

Sunlight contains 0.1 W/sq. cm. It can heat up objects appreciably. A typical 60 W bulb has an area of 120 sq. cm. Hence its power dissipation is 0.5 W/sq. cm. We know that a glowing 60 W bulb is too hot to be touched.

When the power dissipated by a package is less than 1W/sq. cm the cost of a high-quality power supply is the dominant consideration.

When the power dissipated by package is greater than 1 W/sq. cm the cost of cooling becomes the dominant consideration. For today's packages, 8W/sq. cm is the limit.

In particular, it is difficult to remove heat from a VLSI circuit, wherein the power dissipation is of the order of 20 W/sq. cm. Removal of heat from a VLSI device circuit can be done by cutting grooves in the back of silicon and forcing liquid such as freon through it. Freon can transfer 20 W/sq. cm.

The maximum bearable temperature for a chip in plastic package is 100°C and for a chip in ceramic package it is 120°C.

Thermal resistance θ is the difference in temperature in °C between a chip and ambient still air when 1 W is dissipated on the chip. The thermal resistance of commercially available IC packages varies from 30°C/W to 70°C/W.

A typical 40-pin DIP has θ of 62°C and a typical 64-pin ceramic chip has a θ of 36°C. Obviously, θ should be as small as possible. The ideal value of θ is zero in which case the device does not get heated up at all.

Let P_D be the average power in watts dissipated in a device; T_A be the ambient temperature in degrees centigrade; T_D be the temperature of device in °C; and θ_{DA} be the thermal resistance between the device and the ambient in °C/W. Then we can write:

$$T_D = P_D \theta_{DA} + T_A$$

θ_{DA} consists of three thermal resistors θ_{DC}, θ_{CH} and θ_{HA} in series where θ_{DC} is the thermal resistance between the device and its case; θ_{CH} is the thermal resistance between the case and the heat sink; and θ_{HA} is the thermal resistance between the heat sink and the ambient.

Hence, $\theta_{DA} = \theta_{DC} + \theta_{CH} + \theta_{HA}$

To make θ_{DA} small, we should make θ_{DC}, θ_{CH} and θ_{HA} as small as possible.

θ_{DC} can be controlled only during the manufacture of a device.

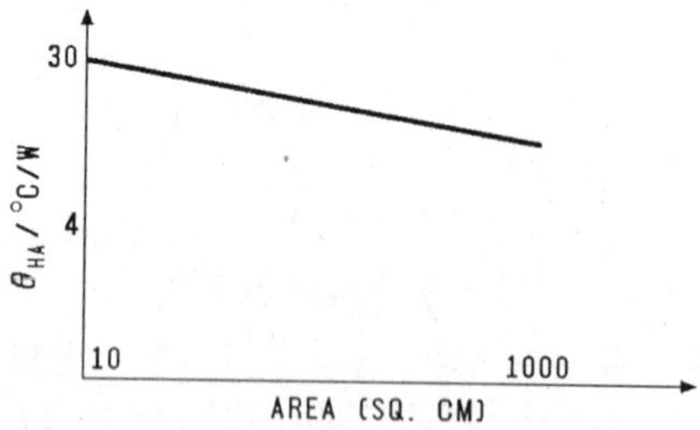

Figure E2: Thermal resistance and surface area.

From a critical examination of graph shown in Figure E2, we can easily see that θ_{DC} is made inversely proportional to power dissipation P_D. In fact, high-power transistors have a small θ_{DC}.

The heat sink is at ground potential. The case and the heat sink are separated by means of an electrical insulation. A thin mica spacer is used for this purpose. The thermal resistance of a mica spacer is directly proportional to its thickness and inversely proportional to the area of the casing in contact with the heat sink.

This thermal resistance can be reduced by using a silicon heat conducting grease between the heat sink and the mica spacer, and between the mica spacer and the casing as shown in Figure E3.

If this is done θ_{HC} can be calculated from the relation:

$$\theta_{HC} = \frac{80\,T}{A} \ \text{°C/W}$$

where T is the thickness of the mica spacer in mm and A is the area of casing in contact with the heat sink in sq. cm.

Thus, the thermal resistance of a mica spacer which is 0.05 mm thick with an area of 2 sq. cm is 2°C/W.

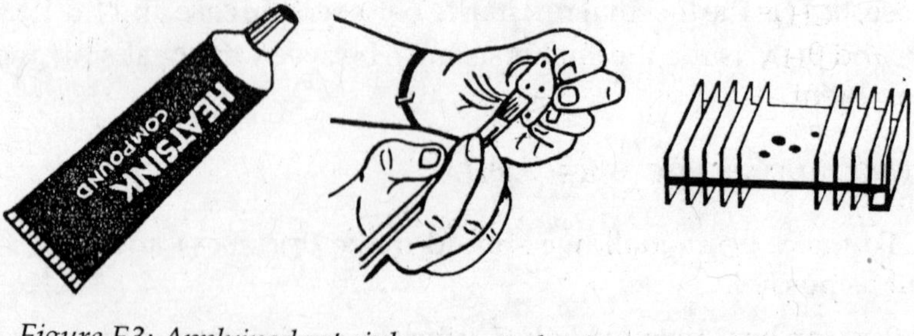

Figure E3: Applying heat sink compound reduces thermal resistance.

Appendix F

Semiconductors Data

Power supplies use a variety of semiconductor devices. These include ordinary rectifier diodes, high power rectifiers, Fast recovery diodes, Schottky diodes, low-power and high-power transistors, FETs, and thyristors. Before we decide to use any device, it is better to know its specific characteristics.

In this appendix we give some examples of the type of data furnished by manufacturers to make you familiar with the way the data is presented on the type of devices, applications, short-forms and symbols used to specify certain characteristics. Detailed information on particular types of devices can either be obtained from the manufacturer, or from the published data books such as those published by BPB Publications.

Rectifier Diodes (mains frequency diodes)

Type	V_{RRM} (V)	I_{FAV} (A)	I_{FSM} (A)	V_F (V) Max.	at	I_F (A)	t_{rr} (μs) Max.	Case JEDEC
1N 4001	50	1.1	50	1.2		1.0	5.0	DO-15
1N 4002	100							
1N 4003	200							
1N 4004	400							
1N 4005	600							
1N 4006	800							
1N 4007	1000							
BY 251	200	3.0	100	1.1		3.0	5.0	DO-201
BY 252	400							
BY 253	600							
BY 254	800							
BY 255	1300							
BY 256A	1500							
1N 5400	50	3.0	150	1.2		3.0	5.0	DO-201
1N 5401	100							
1N 5402	200							
1N 5403	300							
1N 5404	400							
1N 5405	500							
1N 5406	600							
1N 5407	800							
1N 5408	1000							
BY 10-200	200	10	150	1.7		30	5.0	TO-220
BY 10-400	400							
BY 10-600	600							
BY 10-800	800							
BY 10-1000	1000							

Fast Recovery Diodes

Type	V_{RRM} (V)	IFAV (A)	I_{FSM} (A)	V_F (V) Max.	at	I_F (A)	t_{rr} (ms) Max.	Case JEDEC
BA 157	400	1.0	35	1.3		1.0	0.3	DO-15
BA 158	600							
BA 159	1000							
BYX 05-100	100	5.0	80	1.3		5.0	0.3	TO-220
BYX 05-200	200							
BYX 05-400	400							
BYX 05-600	600							
BYX 05-800	800							
BYX 05-1000	1000							
BYT 08-100	100	8.0	100	1.85		20	0.3	TO-220
BYT 08-200	200							
BYT 08-500	400							
BYT 08-600	600							
BYT 08-800	800							
BYT 08-1000	1000							
BYT 12-100	100	12	150	1.9		30	0.3	TO-220
BYT 12-200	200							
BYT 12-400	400							
BYT 12-600	600							
BYT 12-800	800							
BYT 12-1000	1000							
BYT 25-200	200	25	400	1.9		50	0.5	DO-218
BYT 25-400	400							
BYT 25-600	600							
BYT 25-800	800							
BYT 25-1000	1000							

Schottky Diodes (T_{jmax} = 125°C)

Type	V_{RRM} (V)	I_{FAV} (A)	I_{FSM} (A)	V_F (V) Max.	at	I_F (A)	I_R (mA) Max.	Case JEDEC
BYS 01-40	40	1.0	40	0.70		1.0	0.1	DO-15
BYS 01-50	50	1.0	40	0.70		1.0	0.1	
BYS 01-60	60	1.0	40	0.70		1.0	0.1	
BYS 01-70	70	1.0	40	0.80		1.0	0.1	
BYS 01-80	80	1.0	40	0.80		1.0	0.1	
BYS 01-90	90	1.0	40	0.80		1.0	0.1	
BYS 01-100	100	1.0	40	0.80		1.0	0.1	
BYR 07-25	25	7.5	150	0.60		7.5	1.0	TO-220
BYR 07-35	35	7.5	150	0.60		7.5	1.0	
BYR 07-45	45	7.5	150	0.60		7.5	1.0	
BYR 07-40	40	7.5	150	0.75		7.5	0.1	
BYR 07-50	50	7.5	150	0.75		7.5	0.1	
BYR 07-60	60	7.5	150	0.75		7.5	0.1	
BYR 07-70	70	7.5	150	0.85		7.5	0.1	
BYR 07-80	80	7.5	150	0.85		7.5	0.1	
BYR 07-90	90	7.5	150	0.85		7.5	0.1	
BYR 07-100	100	7.5	150	0.85		7.5	0.1	
BYR 25-25	25	25.0	360	0.55		25.0	3	DO-218
BYR 25-35	35	25.0	360	0.55		25.0	3	
BYR 25-45	45	25.0	360	0.55		25.0	3	
BYS 50-40	40	50.0	600	0.74		50.0	2	
BYS 50-50	50	50.0	600	0.74		50.0	2	
BYS 50-60	60	50.0	600	0.74		50.0	2	
BYS 50-70	70	50.0	600	0.84		50.0	2	
BYS 50-80	80	50.0	600	0.84		50.0	2	
BYS 50-90	90	50.0	600	0.84		50.0	2	
BYS 50-100	100	50.0	600	0.84		50.0	2	

Zener Diodes
Glass, Package DO-35

Zener ½ W Diode Normal Value	Differential Resistance Rz, Iz = 5mA Ohm, Not more	Temperature Coefficient of regulating voltage Iz = 5mA, T_{amb} -60°C to 125°C, not more	Maximum reverse current I_r/Vr, μ_r/V	Maximum regulating current, $I_{Z\,max}$ at Vz = Vzn T_{amb} from -60° to35°C, mA
4V7	100	-0.01	2.0/1	85
5V1	75	+0.015	0.5/1	74
5V6	50	+0.03	0.5/1	70
6V2	35	+0.06	0.5/2	64
6V8	30	+0.06	0.5/3	58
7V5	20	+0.07	0.5/5	53
8V2	20	+0.08	0.5/6	47
9V1	30	+0.09	0.5/7.0	43
10V	30	+0.09	0.5/7.5	40
11V	30	+0.092	0.5/8.5	36
12V	30	+0.095	0.5/9.0	32
13V	30	+0.095	0.5/9.0	29
15V	40	+0.095	0.5/9.0	27
16V	40	+0.095	0.5/9.5	24
18V	55	+0.095	0.5/9.5	21
20V	55	+0.095	0.5/9.5	20

General Purpose Power Transistors

Type		V_{CBO} (V)	V_{CEO} (V)	I_C (A)	P_{tot} (W)	V_{CEsat} (V) max.	at I_C/I_B A/A	h_{FE} min	h_{FE} max	at V_{CE}/I_C V/A	Case
N	2N 3055	100	60	15	115	1.1	4/4	20	70	4/4	TO-3
N	2N 3055H	100	100	15	115	1.1	4/4	20	70	4/4	
N	2N 3442	160	140	10	117	1.0	10/2.0	20	70	4/3	
N	2N 3771	50	40	30	150	2.0	15/.15	15	60	4/15	
N	2N 3772	100	60	20	150	1.4	10/.1	15	60	4/10	
N	2N 3773	160	140	16	150	1.4	8.8	15	60	4/8	
N	2N 5038	150	90	20	140	2.5	20/5	20	100	5/12	
N	2N 6249	300	200	15	175	1.5	10/1	10	50	3/10	
N	2N 6250	375	275	15	175	1.5	10/1.25	8	50	3/10	
N	2N 6251	450	350	15	175	1.5	10/1.67	6	50	3/10	
N	URN O55	60	50	5	50	1.0	2/.1	25	100	2/2	
P	URP O55	60	50	5	50	1.0	2/.1	25	100	2/2	
P	2N 6107	80	70	7	40	1.0	2/.2	30	150	4/2	TO-220
P	2N 6109	60	50	7	40	3.5	7/3	30	150	4/2.5	
P	2N 6111	40	30	7	40	3.5	7/3	30	150	4/3.0	

Type	V_{CBO} (V)	V_{CEO} (V)	I_C (A)	P_{tot} (W)	V_{CEsat} (V) max.	at I_C/I_B A/A	h_{FE} min	h_{FE} max	at V_{CE}/I_C V/A	Case
N 2N 5294	80	70	4	36	1.0	5/.5	30	120	2/1.5	TO-220
N 2N 5296	60	40	4	36	1.0	1/.1	30	120	4/1.0	
N 2N 5298	60	50	4	36	1.0	.5/1.5	30	80	4/1.5	
N 2N 6292	80	70	4	40	3.5	7/3	30	150	4/2.0	
N TIP 31	40	40	3	40	1.2	3/.375	10	50	4/3	
P TIP 32	40	40	3	40	1.2	3/.375	10	50	4/3	
P TIP 32A	60	60	3	40	1.2	3/.375	10	50	4/3	
P 2SA 940	150	150	1.5	35	-	-	40	140	10/.5	
N 2SC 2073	150	150	1.5	25	-	-	40	140	10/.5	
N 2SC 2233	200	80	4	40	1.0	4/.4	30	150	5/1	
N MJE 3055T	70	60	10	75	1.1	4/.4	20	100	4/4	
P MJE 2955T	70	60	10	75	1.1	4/.4	20	100	4/4	
N MJE 13005	-	400	4	75	0.5	1/.2	10	60	5/1	
N MJE 13006	-	400	9	80	1.0	2/.4	8	60	5/2	
N MJE 13007	-	400	9	80	1.0	2/.4	8	60	5/2	

N = n-p-n P = p-n-p

Switching Power Transistors

	V_{CBO} (V)	V_{CEO} (V)	I_C (A)	P_{TOT} (W)	V_{CEsat} (V) max	t_f at (μs) max	I_C (A)	Case JEDEC
BUV 20	160	125	50	250	1.2	0.3	50	TO-3
BUV 21	250	200	40	250	1.5	0.4	25	
BUV 22	300	250	40	250	1.5	0.5	20	
BUV 23	400	325	30	250	1.0	1.2	16	
BUV 24	450	400	20	250	1.0	1.4	12	
BUV 25	500	500	15	250	1.0	1.6	8.0	
BUV 26	180	90	14	65	1.5	0.3	12	TO-220
BUV 26A	200	100	14	65	1.0	0.3	10	
BUV 27	240	120	12	65	1.5	0.4	8.0	
BUV 27A	300	150	12	65	1.0	0.4	7.0	
BUV 28	400	200	10	65	1.5	0.3	6.0	
BUV 28A	450	225	10	65	1.5	0.3	4.0	
BUV 39	160	90	25	120	1.2	0.25	20	TO-3
BUV 40	250	125	20	120	1.2	0.3	15	
BUV 41	300	200	15	120	1.2	0.3	8.0	
BUV 42	350	250	12	120	1.2	0.3	6.0	
BUV 42A	400	300	12	120	0.9	0.4	4.0	
BUV 50	250	125	25	150	0.9	0.3	20	
BUV 51	300	200	20	150	1.2	0.3	14	
BUV 52	350	250	20	150	1.2	0.3	12	
BUV 52A	400	300	20	150	0.9	0.4	7.0	
BUW 38	120	60	30	150	1.4	0.25	40	
BUW 39	160	80	30	150	1.2	0.25	30	
BUX 39	120	90	30	120	1.2	0.4		TO-3
BUX 40	160	125	20	120	1.6	0.4		
BUX 41	250	200	15	120	1.6	0.8		
BUX 42	300	250	12	120	1.6	1.2		
BUX 43	400	325	10	120	1.6	1.2		
BUX 44	450	400	8.0	120	1.2	2.0		

High Voltage Switching Power Transistors

Type	V_{CBO} (V)	V_{CEO} (V)	I_C (A)	P_{tot} (W)	V_{CEsat} (V) max	t_f (µs) max	at I_C (A)	Case JEDEC
BUL 46	700	400	5.0	70	0.4	0.8	2.0	TO-220
BUL 46A	800	450	5.0	70	0.4	0.8	2.0	
BUT 46	850	400	5.0	100	1.5	0.8	3.0	
BUT 46A	1000	450	5.0	100	1.5	0.8	2.5	
BUT 56	800	400	8.0	100	5.0	1.0	4.0	
BUT 56A	1000	450	8.0	100	5.0	1.0	3.3	
BUT 134	600	300	4.0	75	0.6	0.9	2.0	
BUT 135	700	400	4.0	75	0.6	0.9	2.0	
BUT 135A	850	400	4.0	75	0.6	0.9	2.0	
BUT 136	600	300	8.0	80	1.5	0.7	5.0	
BUT 137	700	400	8.0	80	1.5	0.7	5.0	
BUT 137A	850	450	8.0	80	1.5	0.7	5.0	
BUS 11	850	400	5.0	100	1.5	0.8	3.0	TO-3
BUS 11A	1000	450	5.0	100	1.5	0.8	2.5	
BUW 34	500	400	10	125	1.5	0.8	5.0	
BUW 35	800	400	10	125	1.5	0.8	5.0	
BUW 36	900	450	10	125	1.5	0.8	5.0	
BUW 44	500	400	15	175	3.0	0.8	10	
BUW 45	800	400	15	175	1.5	0.8	10	
BUW 46	900	450	15	175	1.5	0.8	10	
BUX 46	850	400	5.0	100	1.5	0.8	2.5	
BUX 46A	1000	450	5.0	100	1.5	0.8	2.0	
BUX47	850	400	9.0	125	1.5	0.8	6.0	
BUX 47A	1000	450	9.0	125	1.5	0.8	5.0	
BUX 48	850	400	15	175	1.5	0.8	10	
BUX48A	1000	450	15	175	1.5	0.8	8.0	
BUX 48B	1200	600	15	175	1.5	0.7	6.0	
BUX 48C	1200	700	15	175	1.5	0.7	6.0	
BUX 80	800	400	10	100	1.5	0.5	5.0	
BUY 89	1200	800	6.0	80	1.0	1.0	4.5	

Television SMPS Power Transistors

Type	V_{CBO} (V)	V_{CEO} (V)	I_C (A)	P_{tot} (W)	V_{CEsat} (V) max	at I_C/I_B A/A	t_f (μs) max	at I_C (A)	Case JEDEC
BU 126	750	300	3.0	40	5.0	2/1	0.7	4.0	TO-3
BU 204	1300	600	2.5	10	5.0	2/1	1.0	2.0	
BU 205	1500	700	2.5	10	5.0	2/1	1.0	2.0	
BU 206	1700	800	2.5	10	5.0	2/1.1	1.0	2.0	
BU 207	1300	600	5.0	12.5	3.0	4.5/2	1.0	4.5	
BU 208	1500	700	5.0	12.5	5.0	4.5/2	1.0	4.5	
BU 208A	1500	700	5.0	12.5	1.0	4.5/2	1.0	4.5	
BU 208D	1500	700	5.0	12.5	1.0	4.5/2	1.0	4.5	
BU 326	800	325	6.0	75	3.0	4/1.25	1.0	4.0	
BU 326A	900	400	6.0	75	3.0	4/1.25	1.0	4.0	
BU 526	900	400	8.0	86	5.0	8/3	1.0	8.0	
BU 536	1100	480	8.0	62	-	-	1.0	4.0	
BU 406	400	200	7.0	60	1.0	5/.5	0.5	5.0	TO-220
BU 407	330	150	7.0	60	1.0	5/.5	0.5	5.0	
BU 407D	330	150	7.0	60	1.0	5/.5	0.5	5.0	
BU 505	1500	700	2.5	75	5.0	2/1	1.0	2.0	
BU 506	1500	700	4.0	78	5.0	3/1.33	1.0	3.0	
BU 426	800	375	6.0	70	1.5	2.5/.5	0.75	2.5	TO-218
BU 426A	900	400	6.0	70	1.5	2.5/.5	0.75		
BU 508	1500	700	8.0	125	3.0	4.5/2.0	1.0	4.5	
BU 508A	1500	700	8.0	125	1.0	4.5/2.0	1.0	4.5	
BU 508D	1500	700	8.0	125	3.0	4.5/2.0	1.0	4.5	
BU 705	1500	700	2.5	75	5.0	2/.9	1.0	2.0	
BU 903	1350	550	8.0	125	2.0	3.2/5.3	1.0	3.2	
BU 908	1500	700	8.0	125	-	-	1.0	3.2	
BUV 70	1300	600	10	140	-	-	0.6	5.0	
BUV 71	1500	800	10	140	-	-	0.7	5.0	

Darlington Transistors

Type	V_{CBO} (V)	V_{CEO} (A)	I_C (A)	P_{tot} (W)	V_{CEsat} at (A)	I_C/I_B (A/A)	Case JEDEC
N BU 920	400	350	10	120	1.8	7/0.14	TO-3
N BU 921	450	400	10	120	1.8	7/0.14	
N BU 920P	400	350	10	105	1.8	7/0.14	TO-218
N BU 921P	450	400	10	105	1.8	7/0.14	
N BU 921HP	450	400	15	105	2.0	7/0.15	
N TIP 122	100	100	5	65	2.0	3/0.012	TO-220
P TIP 127	100	100	5	65	2.0	3/0.012	

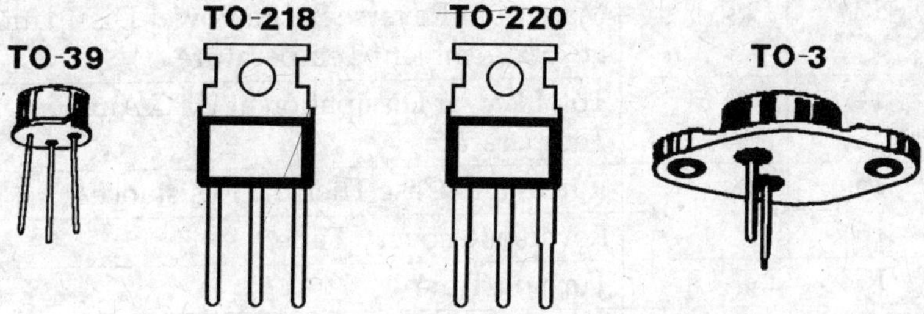

Figure F1: Outlines of semiconductor devices.

Terminology and Definition

Sl. No.	Symbol	Definition
1.	I_C/I_B	Current Gain
2.	h_{fe}	Current Gain
3.	h_{21E}	Current Gain
4.	I_C	Collector Current
5.	I_{CM}	Maximum collector Current
6.	I_B	Base Current
7.	I_F	Forward Current
8.	I_R	Reverse Current
9.	I_{FSM}	Peak Surge Forward Current
10.	I_{FAV}	Average Forward Current
11.	μs	Micro second
12.	ns	Nano second
13.	P_{RSM}	Maximum Reverse Surge Power Dissipation at 25°C Ambient Temperature
14.	P_{tot}	Total Power Dissipation at 25°C Ambient Temperature
15.	R_{thjct}	Junction to Case Thermal Resistance
16.	t_{rr}	Reverse Recovery Time
17.	t_{on}	Turn on Time
18.	t_s	Storage Time
19.	t_f	Fall Time
20.	$T_{j\ max}$	Maximum Junction Temperature
21.	V_{CBO}	Collector to Base Breadown Voltage
22.	V_{CEO}	Collector to Emitter Breadown Voltage
23.	V_{EBO}	Emitter to Base Breakdown Voltage
24.	V_{RRM}	Peak Reverse Repetitive Voltage
25.	$V_{CE\ (sat)}$	Collector Emitter Saturation Voltage
26	V_{CE}	Collector Emitter Voltage
27.	V_F	Forward Voltage

Dealers and Manufacturers

Address	Items
ALTEK SYSTEMS 114, Shan Brahma Complex Near Ratan Cinema, 631/632 Budhwar Peth, PUNE 411 002	Metal Oxide Varistors (MOV)
CRESTA INSTRUMENTS 9/A, Sitladevi Industrial Estate 140, Sitladevi Temple Road, Mahim Mumbai 400 016	Rectifiers, Schottky and ultrafast diodes, heat sinks and heat sink compuunds, varistors
CTR Manufacturing Industries Ltd. Nagar Road, Pune	Spike suppressors
DOMINION RADIOS 15, New Queens Road, Opera House Numbai 400 004	Diodes, transistors, thyristors, capacitors selenium rectifiers,
DEECEE ELECTROSYSTEMS PVT. LTD. 162/D/690 Lake gardens, Calcutta 700 045	Power conditioners
ELCOMAC TRADE LINKS Ltd. 52/D Chhotani Building, 2nd Floor Jyoti Estate, Proctor Road, Grant Road (E) Mumbai 400 007	Slow-blow and fast blow fuses
KHANDELWAL HERMANN ELECTRONICS LTD No. 2, Rahem Mansion, 44 Colaba CausewayMmubai 400 039	Selenium rectifiers, photovoltaic cells, rectifiers SCRs, triacs, diodes, (up to 150 A), and power transistors

Address	Items
Kudamm Corporation 214, Mansarovar, 90 Nehru Place New Delhi 110 019	Metal oxide varistors
International Rectifiers (India) Pvt. :Ltd/ 15/C-40, Connaught Place New Delhi 110001	Rectifiers, high speed diodes, Selenium rectifiers, Surge Klippers
KIRLOSKAR BATTERIES LIMITED Atur House, 1st Floor, 87, Worli Naka Mumbai 400 018	Maintenance free lead-acid batteires
MONTU ELECTRONICS 204. IInd Floor, 32-33 Kucha Chowdhry Chandni Chowk, Delhi 110006	Alkaline, Lithium and Zinc-air cells
NAINA SEMICONDUCTOR LTD C-33, Okhla Industrial Area Phase-I New Delhi 110 020	Rectifiers, zener diodes, high efficiency diodes, power transistors
Pan International Palghar, Marashtra	Pan surge protector plugs
Prime Products C-107, Synthofine Industrial Estate Off Aarey Road, Goregaon (East) Mumbai	Ni-Cd, Ag-Zn Batteries
Sakthi Batteries 231, 3rd Cross, 1st Main, KSTRC layout, JP Nagar, II Phase, Bangalore 560078	UPS Batteries
SOLLETEK 2 Skydeck Oberoi Complex Link Road, Andheri (West) Mumbai 400 053	Devices for protection from surges and voltage brown outs.
USHA INDIA LTD. 3, Bhikaji Cama Place, Ansal Chambers New Delhi 110066	Rectifiers, Schottky diodes, power and switching transistors
Usha Rectifier Corporation 12/1, Mathura Road, Faridabad (Haryana)	Surge protectors (KlipSel) and contact protectors

Index

N

O

P

R